Medicinal and Aromatic Plants of the World

Volume 6

Series Editor
Ákos Máthé
Faculty of Agriculture and Food Science
Széchenyi István University
Mosonmagyaróvar, Hungary

Medicinal and Aromatic Plants (MAPs) have been utilized in various forms since the earliest days of mankind. They have maintained their traditional basic curative role even in our modern societies. Apart from their traditional culinary and food industry uses, MAPs are intensively consumed as food supplements (food additives) and in animal husbandry, where feed additives are used to replace synthetic chemicals and production-increasing hormones. Importantly medicinal plants and their chemical ingredients can serve as starting and/or model materials for pharmaceutical research and medicine production. Current areas of utilization constitute powerful drivers for the exploitation of these natural resources. Today's demands, coupled with the already rather limited availability and potential exhaustion of these natural resources, make it necessary to take stock both of them and enrich our knowledge regarding research and development, production, trade and utilization, and especially from the viewpoint of sustainability. The series Medicinal and Aromatic Plants of the World is aimed to look carefully at our present knowledge of this vast interdisciplinary domain, on a global scale. In the era of global climatic change, the series is expected to make an important contribution to the better knowledge and understanding of MAPs. The Editor of the series is indebted for all of the support and encouragement received in the course of international collaborations started with his ISHS involvement, in 1977. Special thanks are due to Professor D. Fritz, Germany for making it possible. The encouragement and assistance of Springer Editor, Mrs. Melanie van Overbeek, has been essential in realizing this challenging book project. Thanks are due to the publisher - Springer Science+Business Media, The Netherlands - for supporting this global collaboration in the domain of medicinal and aromatic plants. We sincerely hope this book series can contribute and give further impetus to the exploration and utilization of our mutual global, natural treasure of medicinal and aromatic plants. Budapest, Prof. Dr. Ákos Máthé.

More information about this series at http://www.springer.com/series/11192

Ákos Máthé

Editor

Medicinal and Aromatic Plants of North America

 Springer

Editor
Ákos Máthé
Faculty of Agriculture and Food Science
Széchenyi István University
Mosonmagyaróvar, Hungary

ISSN 2352-6831   ISSN 2352-684X   (electronic)
Medicinal and Aromatic Plants of the World
ISBN 978-3-030-44928-5   ISBN 978-3-030-44930-8   (eBook)
https://doi.org/10.1007/978-3-030-44930-8

This Springer imprint is published by the registered company Springer Nature Switzerland AG
The registered company address is: Gewerbestrasse 11, 6330 Cham, Switzerland

Preface and Dedication

The present volume is part of the series Medicinal and Aromatic Plants of the World. Similar to the previous five volumes, the contributing authors are well-known personalities from their relevant scientific fields in different domains of the medicinal plant profession.

The aim of the volume is to introduce the wealth and diversity of Medicinal and Aromatic Plants (MAPs) on the North American continent. The authors would like to offer the readers an insight into the broad spectrum of medicinal and aromatic plants on this huge, diverse, as well as rich continent: from traditional uses to the prospective production and utilization. This aim is mainly served by a general introductory chapter, but specifically also other chapters reveal interesting aspects of the relevant medicinal plant species.

The topics have been chosen, with preference, in such a way as to present true American success stories on medicinal plants, e.g., American cranberry, American ginseng (*Panax quinquefolius*), Taxol *(Taxus brevifolia)*. To date, these are considered as classical examples of successful medicinal plant R&D and utilization. Wherever possible, these reviews will encompass the past, present, and future aspects of these commodities.

As demonstrated by this book, the USA, at present, is an excellent example to illustrate the wide-ranging activities that are needed to produce and use MAPs in a sustainable way. In this activity, special attention is paid to the conservation of their genetic resources.

The North American continent, more specifically the USA, is home to several important initiatives aimed at educating the herbal and dietary supplement industry about such important issues as ingredient and product adulteration, botanical raw material sourcing and sustainability, regenerative and sustainable agriculture, botany, ethnobotany, ethnobiology, horticulture, pharmacognosy, wild harvesting, herbal medicine, etc.

In a historical perspective, a separate chapter in the book deals with the impact of legislation (Dietary Supplement Health and Education Act of 1994), as well as issues related to the sustainable supply/conservation of good-quality botanicals

(Good Agricultural and Collection Practice – GACP) to be used with safety and with efficacy.

This volume is dedicated to the memory of the late botanist **Professor Dr. James (Jim) Duke**, who regretfully could not contribute to this volume any more. His memory should, however, be commemorated not only by this volume but also by the dozens of other milestone comprehensive volumes and scientific volumes on MAPs.

On a more personal note, the editor of the volume still recalls the memory of the friendly, hilarious musician personality of Jim Duke, the "herbal cowboy," whom he had the opportunity to meet at USDA ARS, in Beltsville, during his Fulbright scholarship, in 1984. Jim has written some 400 herbs-related poems and composed a vinyl record-full of "blue grass songs" that were not only written but also performed by him personally. A really unforgettable personal touch for the visiting young Hungarian Fulbright scholar, at USDA ARS, in Beltsville, who was trying to get orientated in the matters of horticultural research in America of the 1980s.

As rendered possible by fate, the Editor of this volume has had the opportunity to spend another half-year in the USA, teaching a course on the "production and biology of MAPs," at UMASS, Amherst. Another excellent chance to get acquainted, to study the North American traditions and system of medicinal and aromatic plant production and utilization.

There seem to be nearly endless approaches for a book to deal with the Medicinal and Aromatic Plants of North America. I am convinced, however, that the knowledge and experience of the authors compiled here has resulted in a useful as well as interesting volume that, regarding its contents, is sufficiently comprehensive and varied to serve as a good resource of orientation on Medicinal and Aromatic Plants of North America.

Finally, editor expresses special thanks to Prof. Dr. Lyle Craker, UMASS, Amherst and Josef Brinckmann, PhD, research fellow at Traditional Medicinals, Inc., Sebastopol, CA, for their support and critically reading relevant parts of the manuscript.

Mosonmagyaróvar, Hungary   Ákos Máthé
January 6, 2020

Contents

Chapter 1
Introduction to Medicinal and Aromatic Plants in North America

Ákos Máthé

Abstract Native American peoples developed a sophisticated "plant-based medical system" in the course of millennia before the European conquest of America. Despite the significant differences between the systems developed by the various native groups, there were also many broad similarities. Out of the approximately 28,000 species of plants in North America, native Americans used about 2500 medicinally.

Out of the thirty-six areas that qualify as biodiversity hotspots, for North America and Central America, only three regions qualify: the California Floristic Province, Madrean pine-oak woodlands, and Mesoamerica.

Recently, also the North American Coastal Plain (NACP) has been identified as a global hotspot: this finding they have based on its similarity with the classic definition of a region.

The Flora of North America North of Mexico (FNA), is conceived as a multivolume work describing the native plants of North America, in 30 volumes. 17 volumes of the Flora are already available online.

Germplasm conservation in the United States is not new. Realizing the importance of genetic diversity, the USDA has been collecting and preserving germplasm since the late 1800s. To date, the National Genetic Resources Program (NGRP) of USDA acquires, characterizes, conserves, documents, and distributes to scientists germplasm of all life forms important for food and agricultural production.

Á. Máthé (✉)
Faculty of Agriculture and Food Science, Széchenyi István University,
Mosonmagyaróvár, Hungary
e-mail: akos.mathe@upcmail.hu

© Springer Nature Switzerland AG 2020   1
Á. Máthé (ed.), *Medicinal and Aromatic Plants of North America*, Medicinal and Aromatic Plants of the World 6,
https://doi.org/10.1007/978-3-030-44930-8_1

The U.S. Department of Agriculture's Foreign Agricultural Service maintains statistics on imported herbs and spices but there is no systematic process for measuring wild-harvested (a.k.a. wildcrafted) or commercially cultivated herbs used in the medicinal plant market. In view of the increasingly strict quality requirements, the American Herbal Products Association (AHPA) published its Good Agricultural and Collection Practices (GACPs) and Good Manufacturing Practices (GMPs) for botanical materials. In the US, GACP implementation is mandatory only for botanical drug active ingredients: voluntary for botanical dietary supplement ingredients. As the U.S. Food and Drug Administration (FDA) has not yet established its own GACPs for botanical drugs, the agency specifically accepts GACP compliance according to guidelines published by the European Medicines Agency (EMA) and the World Health Organization (WHO).

In the US and Canada, pharmaceutical drugs are subject to strict regulations, like in most countries. Their availability is similarly restricted as either prescription drug products (Rx) or nonprescription over-the-counter (OTC) drug products. The regulation of medicinal and aromatic plant (MAP) production, trade and use in North America is the result of the influence of a multitude of contributing "actors" (governmental agencies, intergovernmental organizations, public institutions, standard-setting organizations, and other bodies). Much progress has been made since the 1800s, when the patent medicines (proprietary medicines made and marketed under a patent and available without prescription) first became popular. In the lack of Federal regulations for these products prior to the passage of the Pure Food and Drug Act of 1906 and subsequent United States Federal Food, Drug, and Cosmetic Act of 1938, consumers had no means to distinguish between false and valid claims. Many of these were of plant origin and often had "secret formulations". There is good reason to believe in the truth of Professor Varro Tyler's forecast saying: "Looking back, it is obvious that a great deal of progress has already been made. We now look to the future with eager anticipation and great expectations."

Keywords North America · Canada · Medicinal and Aromatic Plants · Biodiversity hotspots · Flora · Natural conservation · Official and public participants of legal regulation

1.1 Introduction

The present volume will deal with Medicinal and Aromatic Plants of Northern America. To allow for the fine tuning of the geographic scopes, in the present volume we set the boarder lines of North America to comprise the territory of the non-tropical area of the United States of America and Canada. As such, the volume will deal with MAPs in the conterminous 48 states of USA on the Canada, bordered on the east by the Atlantic Ocean, on the south by the Gulf of Mexico and Mexico, and on the west by the Pacific Ocean. The total area is ca. 9,809,630 square kilometers.

In the North, Canada is composed of ten provinces and three territories that extend from the Atlantic to the Pacific and northward into the Arctic Ocean, covering 9.98 million square kilometers (3.85 million square miles), making Canada the world's second-largest country by total area.

The US is home to many natural resources including MAPs.

In this context we would like to refer to the fact that the notion of Medicinal and Aromatic Plants (MAPs) seems to be fundamentally vaguely defined to denote either a group of plants that due to their biologically active substances are used to cure, ease or prevent diseases (Hoppe 2009) and/or plants that are distinguished due to the intensive, plant-specific fragrance (scent) of their leaves and flowers. This category is used only seldom in relevant surveys on natural resources and remarkably, Medicinal and Aromatic Plants generally do not qualify among the first ten most valuable resources, although since times immemorial, herbs and their use have thrived on this continent. Frequently used synonyms for MAPs are: herbs, botanicals, crude drugs, etc. Also, Essential Oil and Aromatic Plants is a term frequently used as a synonym to denote plants that contain volatile oils: aroma plants, however, distinguish themselves by their combined, mostly pleasant, scent and taste.

Similarly to the versatility of their natural occurrence and ecological roles (defense against insects, fungi, diseases and herbivorous animals) and due to their contents of phytochemicals with potential or established biological activity, the use of MAPs by man is also most versatile: herbal medicine, food-, feed additives, cosmetics, etc. (see: Máthé 2015).

In view of above-mentioned, the aim of this chapter is not to offer introduction to the vast scenery of MAPs utilization on the equally huge North American continent. It is meant to serve *quasy* as an "apetizer" demonstrating the nearly unlimited resources and values hidden in this group of economic plants on this large continent. All this will be presented, in a *quasi* historical perspective, so that in view of the past, present and the nearly unlimited prospective opportunities standing in front of MAPs can be projected It is our hope that the believed and/or legitimate shortcomings can remediated in the prospective volumes of the MAPW series.

1.2   Biodiversity Hotspots in North America

Biodiversity hotspots are biogeographic regions that are both biologically rich and highly threatened with destruction from urbanization, development, pollution, and diseases. For a region to be classified as a biodiversity hotspot, it must have at least 1500 vascular plants strictly endemic to the habitat and must have at most 30% of its original natural vegetation (Myers et al. 2000).

Biodiversity hotspots are not uniformly distributed on Earth. Out of the thirty-six areas that qualify as biodiversity hotspots, for North America and Central America, only three regions qualify: the **California Floristic Province**, **Madrean pine-oak woodlands**, and **Mesoamerica** (Molly 2018). These regions support nearly 60% of world's mammal, plant, amphibian, reptile, and bird species, a majority of which are endemic to the specific hotspots (Fig. 1.1) a great deal of conservation is required to preserve the remaining land of these species. Regarding the fact that Mesoamerica's span is mostly Central America, this hot-spot shall not be discussed here among the North American hot-spots.

Recently, however, Noss et al. (2015) have identified also the **North American Coastal Plain** (NACP) as a global hotspot: this finding they have based on its similarity with the classic definition of a region: i.e. >1500 endemic plant species

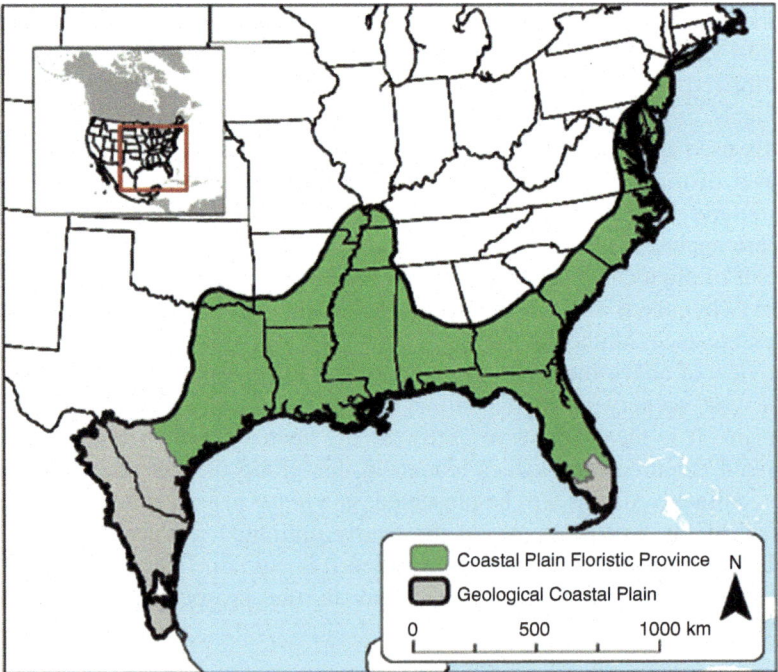

Fig. 1.1 The Geological Coastal Plain (GCP) and the Coastal Plain Floristic Province (CPFP). The CPFP is contained within the GCP. (Noss et al. 2015)

and > 70% habitat loss. In their proposal Noss et al. (2015) refer to their "suspicion" according to which there are systematic biases and misconceptions, in addition to missing information, that obscure the existence of similarly biologically significant regions world-wide.

1.2.1 California Floristic Province

The California Floristic Province is located along North America's Pacific coast. It is a zone of Mediterranean-type climate and is characterized by hot, dry summers and cool, wet winters (Molly 2018).

The California Floristic Province is home to 3488 native plant species, 2100 of which are endemic to the region. The region contains a wide variety of ecosystems, including sagebrush steppe, prickly pear shrubland, coastal sage scrub, chaparral, juniper-pine woodland, upper montane-subalpine forest, alpine forest, riparian forest, cypress forests, mixed evergreen forests, Douglas fir forests, sequoia forests, redwood forests, coastal dunes and salt marshes. Today, a quarter of the original vegetation has remained in a more or less pristine condition. The hotspot is home to the endangered giant sequoia *(Sequoiadendron giganteum)*, the planet's largest living organism and its taller but less massive relative, the endangered coastal redwood *(Sequoia sempervirens)*.

Located in the most populated and fastest growing regions of the US, the biodiversity of this region has been threatened by urbanization, pollution, agriculture, mining, logging, and even intentional fires. According to approximations, due to land reclamation, the salt and freshwater ecosystems have declined by 90%. As only 25% of the hotspot holds its pristine nature, great deals of conservation efforts are needed to protect the hotspot (Molly 2018).

1.2.2 Madrean Pine-oak Woodlands

This biodiversity hot-spot is located in the higher elevations of the mountain ranges of Mexico, as well as the US states of Arizona, New Mexico, and Texas. It is regarded as – probably - the world's richest and most endangered terrestrial ecosystem.

It comprises different habitats including tropical, subtropical, dry shrublands, and grasslands with 5300 flowering plants: 2000 species being unique to Madrean Sky Islands. The consistent pine, oak, Douglas fir, and fir species coupled with varying orientation of slopes, soil types, climate and geologic history give the hotspot its cohesive character.

Among the several threats to the Pine-Oak woodlands, illegal logging poses the greatest threats. Non-wood products including the vascular epiphyte are extracted and a variety of mushroom species found in the pine-oak forests are collected for culinary purposes. Intentional fires in favor of agricultural activity have largely altered habitats throughout the woodland forests .

1.2.3   North American Coastal Plain (NACP)

Recently, Noss et al. (2015) have identified also the North American Coastal Plain (NACP) as a new global hotspot. Their assumption / proposal they have based on similarities 6 compliances with the classic definition of a region with >1500 endemic plant species and > 70% habitat loss.

Based on their case study, they claim that this region has been bypassed in prior designations due to misconceptions and myths about its ecology and history. These fallacies include: (1) young age of the NACP, climatic instability over time and submergence during high sea-level stands; (2) climatic and environmental homogeneity; (3) closed forest as the climax vegetation; and (4) fire regimes that are mostly anthropogenic. They have demonstrated that the NACP is older and more climatically stable than usually assumed, spatially heterogeneous and extremely rich in species and endemics for its range of latitude, especially within pine savannas and other mostly herbaceous and fire- dependent communities.

Noss et al. (2015) do not only attempt to improve the understanding of this region and to promote its recognition as a hotspot. With this effort they also promote the appreciation and improved conservation of ecologically and evolutionarily similar regions globally.

1.3   Diversity of Plants in North America (Biomes)

The North American continent, makes up about 4.8% of the total surface area of the planet. This continent alone has 15 broad, level I ecological regions; 50 level II ecological regions intended to provide a more detailed description of the large ecological areas nested within the level I regions; and 182 Level III ecoregions, which are smaller ecological areas nested within level II regions (US Environmental Protection Agency Protection n.d.). A map of major terrestrial ecoregions is contained in Fig. 1.2.

Ecological hierarchy at this scale provides a context for revealing global and/or intercontinental patterns. Level I. **ecological regions** are: Arctic Cordillera, Tundra, Taiga, Hudson Plains, Northern Forests, Northwestern Forested Mountains, Marine West Coast Forests, Eastern Temperate Forests, Great Plains, North American Deserts, Mediterranean California, Southern Semi-Arid Highlands, Temperate Sierras, Tropical Dry Forests and Tropical Humid Forests.

In terms of biomes, from the overall 15 terrestrial and 12 aquatic biomes of the world, North America is frequently, broadly categorized into the following **six major biomes**: the Tundra biome, Coniferous forest biome, Prairie biome, Deciduous forest biome, Desert biome, and the Tropical rainforest biome.

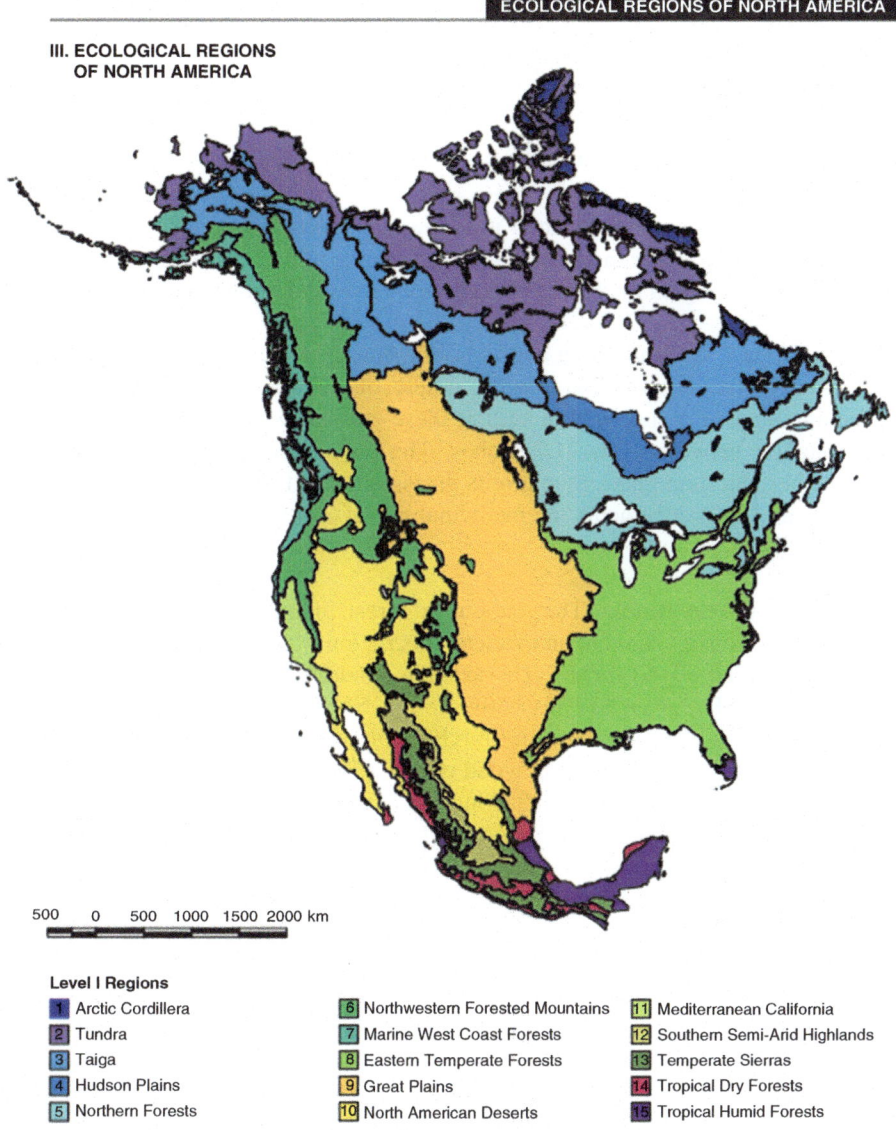

Fig. 1.2   III. Ecological regions of North America. (Source: CEC Secretariat 1997)

As a result of continental drift, the United States is located currently almost exclusively in the North temperate zone (Mauseth 2014) and receives much of its weather from the Pacific and Arctic Oceans in winter, and from the Pacific Ocean and the tropical Atlantic Ocean in summer.

Tundra Biome Typically, it is characterized by harsh climatic conditions, perma-frost (frozen soil) and as a consequence, the lack of evergreen vegetation. Despite the climatic conditions that are unsuitable for human habitation, several plants and animals have adapted themselves to the harsh climate prevailing in this region. The plant species found in the Tundra biome include various species of shrubs, mosses, grasses, and lichens.

Coniferous Forest Biome A North American biome that is typically characterized by a wide variety of coniferous trees, such as pine, fir, and spruce. The average temperature in this region in winter is as low as 14 °F, while the average rainfall in this region ranges between 14–29.5 inches. A range of herbivorous animals inhabit this region, feeding on the leaves of coniferous trees that grow here in abundance.

Prairie Biome Is also known as the North American prairie. This biome is pre-dominantly characterized by a wide variety of herbaceous plants and grasses. Spanning an area of 1.4 million sq. mi, these grasslands generally receive annual precipitation between 12.6–21.7 inches. The plant species found in this region include big bluestem grass, blue grama grass, buffalo grass, etc. The ethnobotanist Kindscher (1992) documents the medicinal use of 203 native prairie plants by the Plains Indians.

Deciduous Forest Biome The deciduous forest biome is found in the eastern region of North America. On an average, this region experiences an annual rainfall ranging between 30–60 inches. The average annual temperature in this region is around 50 °F. The plant species native to the North American deciduous biome include the American beech, white birch, white oak, etc. The medicinal species common to this biome include wood sorrel (*Oxalis montana*), but also many European to Eurasian species, that long ago escaped and naturalized in North America includingdandelion (*Taraxacum officinale*), stinging nettle (*Urtica dioica*), chickweed (*Stellaria media*), and broadleaf plantain (*Plantago major*),, etc. (Wisniewski 2017).

Desert Biome The desert biome, in North America, comprises deserts such as the Sonoran Desert of southern Arizona and the Mojave Desert of California. The region is dominated by drought-resistant species, such as the saguaro (*Carnegiea gigantean*) and Joshua tree (*Yucca brevifolia*) or *Agave* species, *Amaranthus* species, *Bowlesia* species, devil's claw (*Proboscidea* species, not to be confused with the native African devil's claw of *Harpagophytum* species), blue dlder (*Sambucus cerulea*) and Mexican elder (*Sambucus mexicana*), etc.

Tropical Rainforest Biome vs. Temperate Rain Forests Although in most dis-cussions, the Tropical rain forest biome is usually mentioned as a component of North American biomes, we also use this term for the sake of completeness, it must be emphasized that in our concept, North America is meant to denote and encom-pass mainly the territory of both Canada and the United States of America. (The territories north of the Darien Gap, will be discussed in a separate volume, in rela-tion to Central America.).

Temperate rainforests are coniferous or broadleaf forests that occur in the temperate zone and receive heavy rainfall. Temperate rainforests only occur in a few regions around the world, Most of these occur in oceanic moist climates: the Pacific temperate rain forests in Western North America (Southeastern Alaska to Central California). They also comprise the the Appalachian temperate rainforest of the Eastern U.S. The latter region is home to 125 plant species for medicinal use (Krochmal 1968), including black cohosh (*Actaea racemosa*), blue cohosh (*Caulophyllum thalictroides*), goldenseal (*Hydrastis canadensis*), North American wild yam (*Dioscorea villosa*), purple angelica (*Angelica atropurpurea*), and sassafras (*Sassafras albidum*), etc.

1.4 Flora of North America (FNA) vs. New Naturalized Species

The Flora of North America North of Mexico (usually referred to as FNA, is a multivolume work describing the native plants of North America. 17 volumes of the Flora are already available online (Flora of North America Editorial Committee 1993). It is expected to fill 30 volumes when completed and will be the first work to treat all of the known flora north of Mexico.

1.5 Natural Conservation of MAPS vs. Diversity Loss

According to Moerman (1996) native Americans used 2564 of 21,641 vascular species, or 11.8% of the available flora for medicinal purposes (Moerman 1996).

In an editorial under the title "A Census of Botanical Risk" of the New York Times. April 12, 1998), exactly 10 years after the Chiang Mai Declaration (1988), Johnston (1998) called attention to the serious losses in biodiversity (including MAPs). He stated that at least one out of every eight known plant species on Earth was either threatened with extinction or nearly extinct. In the United States, that probably has the planet's best-studied flora, about 29% of 16,000 species were at risk.

As a remarkable interpretation of these figures David G. I. Kingston (2011) explained "Plants have historically provided some of the most important drugs that we have," e.g.: morphine, aspirin, and quinine, as well as a number of less common drugs such as anti-cancer medications derived from the periwinkle, etc. "We've screened about 50,000 plant species so far, and gotten about 50 drugs," Kingston said, "so that's about one per thousand." If the same ratio of 1:1000 holds, the loss of 34,000 species could well doom development of 34 pharmaceuticals.

Nearly at the same time, in 1997, the International Union for Conservation of Nature (IUCN), in collaboration with the Smithsonian, the World Wildlife Fund (WWF) and 10 other government and independent research and conservation groups in a half-dozen countries compiled a 862-page report, titled "1997 IUCN Red List of Threatened Plants" (Walter et al. 1998).

To be classified as threatened, a species must have reached the point at which there are fewer than 10,000 individuals worldwide, or fewer than 100 locations in which it is found. Comparing the latest census against decades of field records and combined collections totaling 20 million specimens, experts found a pace of species decline far above the historic extinction rate. The study included only vascular plants, those with water and nutrient-conducting tissues. Algae, lichens, fungi and the like were not studied.

Concerning the Convention on International Trade in Endangered Species of Wild Fauna and Flora (CITES), several North American MAPs were listed in CITES Appendix II including wild American ginseng roots (*Panax quinquefolius* L., Araliaceae) in 1975, candelilla (*Euphorbia antisyphilitica* Zucc., Euphorbiaceae) in 1975, goldenseal (*Hydrastis canadensis* L., Ranunculaceae) in 1997, and guaicum (*Guaiacum sanctum* L., Zygophyllaceae) in 1975.

As an example, Baum et al. (2006) state that conservation of the natural *Echinacea* resources across North America has socio-political implications due to issues of private and public land tenure, especially on Aboriginal (i.e., Native American) land reserves. Governance of lands and natural resources tends to vary at the federal, state/provincial, or regional levels in both Canada and the United States. The majority (>90%) of natural *Echinacea* populations occur in the United States (Fig. 1.3), where there is a National Germplasm Conservation Program that addresses all *Echinacea* taxa among other resources and threatened species in collaboration with the Nature Conservancy, the State Departments of Natural Heritage/Conservation, and the US Department of Agriculture (USDA).

According to Brinker (2013), the popularity of *Echinacea* has resulted in regional overharvesting of wild *E. angustifolia*. Nonetheless, commercial cultivation of *E. purpurea* and conscientious wildcrafting can continue to provide a sustainable supply of these important botanical medicines. Cultivation of *Echinacea* species has increased rapidly because of the demand and its great value. Growth of the three major medicinal species, *E. angustifolia*, *E. pallida*, and *E. purpurea*, has been the most studied: among them *Echinacea purpurea* is easy to grow, as compared to the other two commercial species.

In an interesting paper by Foster (2017) in HerbalGram, the author expressively calls attention to the values of American medicinal trees, deeming them "Forest Gems". He states that rather than viewing trees as temples of life, humans – in the course of past civilizations – have tended to look at trees in terms of board feet for timber, cords for fuel, or even, as simple nuisances that prevent the soil from being pierced with a plow.

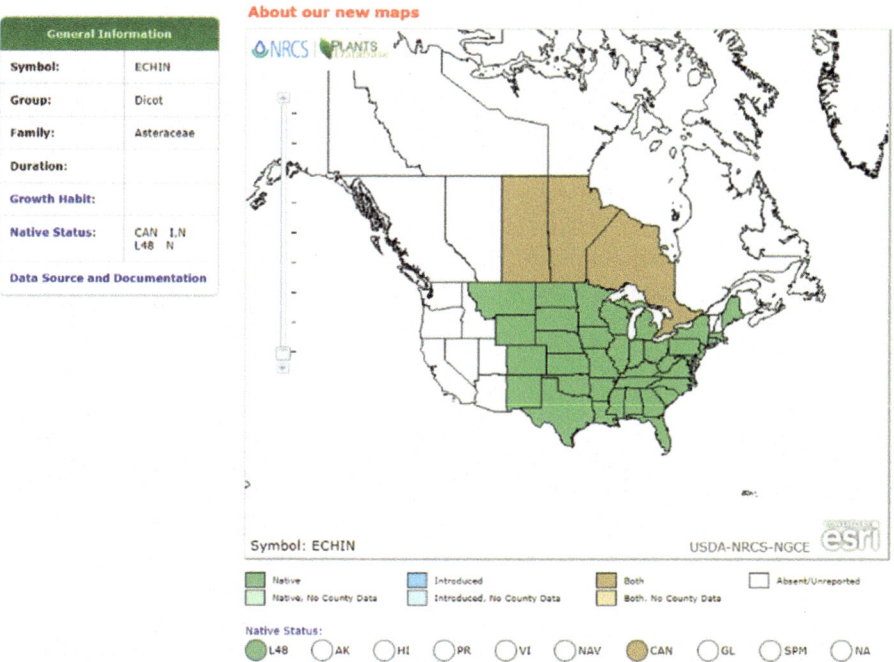

Fig. 1.3 North American distribution of Echinacea species. (USDA NRCS 2019)

1.6   Germplasm Conservation of MAPs in North America

Germplasm conservation in the United States is not   new. Seed saving, replanting and management of wild landscapes is part of the pre-Columbian era traditional ecological knowledge of indigenous peoples of North America (Anderson 2006). Realizing the importance of genetic diversity, the USDA has been collecting and preserving germplasm since the late 1800s (Byrne et al. 2007). To accomplish this huge task a network of more than 25 active germplasm conservation sites (germ-plasm repositories, so called genebanks) was established under the coordination of the US Department of Agriculture (USDA) within the Agricultural Research Service's (ARS) National Plant Germplasm System (NPGS).

To date, the National Genetic Resources Program (NGRP) of USDA acquires, characterizes, conserves, documents, and distributes to scientists germplasm of all life forms important for food and agricultural production. It is the task of the Germplasm Resources Information Network (GRIN) to document these (animal, microbial, and plant) collections through informational pages, searchable databases, and links to USDA-ARS projects that curate the collections. GRIN is operated by the National Germplasm Resources Laboratory in Beltsville, MD.

The U.S. National Plant Germplasm System (NPGS) is collaborative effort to safeguard the genetic diversity of agriculturally important plants, including medici-nal and aromatic species. NPGS's mission is to collect and conserve living plant material of plants for food, fiber, animal feeds, industrial and medicinal purposes, and for landscape and ornamental uses. It also aims to help solve immediate agricultural production problems as well as safeguard plant genetic diversity for future needs. This task is accomplished by:

- acquiring crop germplasm,
- conserving crop germplasm,
- evaluating and characterizing crop germplasm,
- documenting crop germplasm,
- distributing crop germplasm.

As the primary focus of germplasm collections has traditionally been the long-term conservation of major crop species and their wild relatives with the ultimate goal to develop new varieties, medicinal plants have been under-represented. Consequently, a comprehensive facility dedicated to their conservation and research does not currently exist. Even the North Carolina Arboretum Germplasm Repository (TNCAGR) that due to the extraordinary botanical and chemical diversity of the southern Appalachians could be an ideal location for a medicinal plant germplasm repository, leads diverse activities leading to conserve, study and utilize native plants, explore alternative and new crops (including MAPs) for economic development purposes. From the viewpoint of MAPs the viewpoint of MAPs the National Germplasm Repository, in Corvallis (Oregon) with its blueberry, cranberry, hop, mint collections could be notable example for the important contribution genebanks (repositories) provide for the conservation of valuable natural genetic resources.

1.7   Cultivation of Medicinal and Aromatic Plants in North America

Earlier it was reported by Blumenthal (2004) that obtaining statistics on the actual amount of bulk herbal material grown or wild-harvested within the United States was a challenge. The U.S. Department of Agriculture's Foreign Agricultural Service maintains statistics on imported herbs and spices but there is no systematic process for measuring wild-harvested (a.k.a. wildcrafted) or commercially cultivated herbs used in the medicinal plant market. In the past few years, in its attempt to begin to quantify the wild-harvested raw materials used in the manufacture of various herbal products, the American Herbal Products Association (AHPA) has published five Industry Tonnage Surveys. The sixth survey is presently in preparation.

As regards the main cultivated and wild-crafted North American botanical crops, in the lack of relevant official data, Brinckmann's following compilation can provide some guidance (Brinckmann 2007) (Table 1.1):

In view of the increasingly strict quality requirements, American farmers have been implementing Good Collection and Agricultural Practices (GCAP). It was

Table 1.1 Main North American cultivated and wild-collected crops

Name of the plant species	Cultivated	Wild collected
Aloe vera:	Cultivated in Texas, and Tamaulipas, Mexico	
American ginseng root:	Cultivated in Ontario, Wisconsin;	Wild collected in 19 States
Black cohosh rhizome:		Wild collected in Kentucky, Tennessee, Georgia, Ohio, North Carolina, Michigan, South Carolina, Virginia,
Capsicum fruit:	Cultivated in New Mexico, Texas, Arizona, California,	
Cranberry fruit:	Cultivated in Wisconsin, Massachusetts, Canada, New Jersey, Oregon, Washington	
Echinacea herb & root:	Cultivated in Oregon, Washington, British Columbia;	In many Midwestern and Mid-southern States
Flax seed:	Cultivated mainly in Canada; also North Dakota, South Dakota, Montana, Minnesota	
Garlic bulb:	Cultivated in California	
Ginger rhizome:	Cultivated in Hawaii	
Hop strobile:	Cultivated in Oregon, Washington, Idaho	
Jojoba seed:	Cultivated in Sonoran Desert in Arizona, California and Mexico	
Lavender flower:	Cultivated in Oregon and Washington	
Peppermint leaf:	Cultivated in Oregon, Washington, Idaho	
Saw Palmetto fruit:		Wild collected in Florida
Spearmint leaf:	Cultivated in Oregon, Washington	

After: Brinckmann (2007)

most recently the United Natural Products Alliance (UNPA) that has endorsed AHPA's Good Agricultural and Collection Practices and Good Manufacturing Practices for Botanical Material and will encourage member companies to adopt AHPA's guidance.

In harmony with above efforts, providing regulations and guidance for sustainable agriculture, the USDA National Organic Program (NOP) was established by Congress, as early as 2001. (see: later this chapter).

Search for Higher Value Crops

American farmers looking for higher-value new crops: one of the options is to introduce and grow plants used by American practitioners of Traditional Chinese Medicine (TCM). The practice of acupuncture and Chinese herbal medicine is authorized and regulated in 47 states and the District of Columbia. One of the heralds for this activity, Foster (Foster 2004), in a book co-authored with Professor Yue Chongxi,"Herbal Emissaries: Bringing Chinese Herbs to the West" documented the use of Chinese herbs in American gardens. It has been reported by Raterman (Raterman 2018) that the recently (in 2014) established Appalachian Herb Growers Consortium (AHGC) of small farmers, in southwestern Virginia, is engaged in using ecologically sustainable practices to grow Chinese medicinal herbs. Presently the consortium has some 50 small farmer members. Its mission is to bolster farmers' incomes and crop diversity; provide high-quality, effective herbs for practitioners of acupuncture and TCM; and grow and process herbs with respect for nature and the traditions of TCM.

Brinckman (Brinckmann 2015) also calls attention to the new trend to grow Chinese medicinal plants in North America. According to him, the areas mainly concerned are located in parts of rural eastern and southern United States where tobacco (*Nicotiana tabacum* L.; Solanaceae) leaf was once the main economic crop. Farmers in western North Carolina even claim that they share the same latitudes and elevations of the mountainous provinces in China so that are ideal for cultivation of certain Chinese medicinal plants.

Diversification of agricultural production has been an important issue for American farmers for decades. Presently, these activities are centered around university Extension and Outreach centers/programs. Probably the best known of these is the Center for New Crops & Plant Products, at Purdue University. Its web-site called NewCROP (https://www.hort.purdue.edu/newcrop/) provides windows to new and specialty crop profiles.

1.8 Medicinal plants in the North American Folk Medicine (Traditional Medicine)

Herbals/botanicals or herbal medicine (a plant or plant part or an extract or mixture of these) have been used to prevent, alleviate specific symptoms, or cure disease since ancient times of mankind.

According to Moerman (Moerman 2016), native American peoples developed a sophisticated "plant-based medical system" in the course of millennia before the European conquest of America. Despite the significant differences between the systems developed by the various native groups, there were also many broad similarities. Out of the approximately 28,000 species of plants in North America, native Americans used about 2500 medicinally (Moerman 1996, 1998); a comprehensive database of these plant uses is available in the database/book on **Native American Ethnobotany** (http://naeb.brit.org/).

The "Native American Ethnobotany" is perhaps best extensive on-line source of concise information about Native American uses of herbs, this site contains records of the historical uses of over 4000 species of plants by almost 300 North American aboriginal groups. Presently, database is maintained by the University of Michigan, as an expression of lifelong commitment of Dan Moerman to that school. In a practical way, the current version is linked to the USDA PLANTS database so that taxonomic and range information can also be readily found for most of the listed species.

A quick analysis of the database reveals that the richest sources of medicines are the sunflower family 12 (Asteraceae), the rose family (Rosaceae), and the mint family (Lamiaceae). As a remarkable contrast, the 13 grass family (Poaceae) and the rush family (Juncaceae) contain practically no medicinal species. This outstanding volume and extraordinary selectivity demonstrate without any doubt the falseness of degrading claims that according to which Native American medicines were chosen at random, quasi they "just used everything and stumbled on something useful once in a while."

1.8.1   Traditional Medicines Congress (TM)

The TM Congress is a forum for meetings and communications between various U.S.- based organizations with a common interest in preserving access to traditional medicines and improving the free flow of information about the traditional uses of these medicines.

The TM Congress was founded in 2004, when a diverse group of professional organizations met to initiate a coalition/cooperation aimed at the exchange of ideas about the future of traditional medicines in the United States. The result was the convening of the Traditional Medicines Congress, presently sponsored by the following nine national organizations: Acupuncture and Oriental Medicine Alliance (AOMA), American Association of Naturopathic Physicians (AANP), American Association of Oriental Medicine (AAOM), American Herbalist Guild (AHG), American Herbal Products Association (AHPA), Council of Colleges of Acupuncture and Oriental Medicine (CCAOM), Medicinal Herb Consortium (MHC), National Ayurvedic Medical Association (NAMA), National Certification Commission for Acupuncture and Oriental Medicine, (NCCAOM).

According to its declaration (Congress n.d.) the "goal of the Traditional Medicines Congress is to benefit public health by ensuring access to traditional medicines in a manner that provides a reasonable expectation of public safety." The members (sponsors) of the TM Congress have elaborated and published a Draft Proposed Regulatory Model for Traditional Medicines, in 2006. It is believed „to provide a rational framework for truthfully representing the value of traditional medicines and for assuring widespread consumer access to herbal products". Key components of the regulatory frameworks are expected „to clearly define traditional medicines and ensure access to these while addressing safety, in the retail marketplace and in clinical settings".

Table 1.2 On-line herb book resources of traditional medicinal plants

Culpeper, Nicholas: *The English Physician*. **1652**	http://doc.med.yale.edu/historical-old/culpeper/culpeper.htm
United States Pharmacopoeial Convention: *The Pharmacopoeia of the United States of America*. **1820**	https://collections.nlm.nih.gov/ext/dw/2567001R/PDF/2567001R.pdf
Wood GB, Bache F: The Dispensatory of the United States of America, 1st edition. **1833**	https://collections.nlm.nih.gov/catalog/nlm:nlmuid-10430300R-bk
Felter, Harvey Wickes and John Uri Lloyd: *King's American Dispensatory*, 18th edition, 3rd revision. **1898**	http://www.henriettesherbal.com/eclectic/kings/index.html
Lloyd, John Uri: History of the Vegetable Drugs of the Pharmacopeia of the United States. **1911**	http://www.swsbm.com/ManualsOther/USP_Drug_History_Lloyd.pdf
Hedrick, UP: Sturtevant's Notes on Edible Plants. **1919**	http://www.swsbm.com/Ephemera/Sturtevants_Edible_Plants.pdf
Grieve, Maude: *A Modern Herbal,* 2 vol. **1931**	http://www.botanical.com/botanical/mgmh/comindx.html
Remington, Joseph P., Horatio C. Wood, et al.: *The Dispensatory of the United States of America*, 20th edition. **1918**	http://www.swsbm.com/Dispensatory/USD-1918-complete.pdf
Woodville, William: *Medical Botany*, 4 vol. 1790–1793	http://www.illustratedgarden.org/mobot/rarebooks/title.asp?relation=QK91C7431790V1

1.8.2 Historical Herb Books on Line

Due to space limitations, in our introductory chapter we cannot go into details about the description and historical use of native American herbs. The following table (Table 1.2) offers a basic list of comprehensive **on-line resources** that provide sufficient information for getting oriented about the multiplicity of traditions and uses North American traditional medicinal plants (Table 1.2).

1.8.3 Patented Medicines vs. Integrative Medicine

In the United States, patent medicines (proprietary medicines made and marketed under a patent and available without prescription) first became popular in the 1800s. Many of these were of plant origin and often had "secret formulations". They were often sold directly to consumers. (As there was no Federal regulation of these products until passage of the Pure Food and Drug Act of 1906 and subsequent United States Federal Food, Drug, and Cosmetic Act of 1938, consumers had no means to distinguish between false and valid claims.) Nearly simultaneously, the pharmaceutical industry was developing, and marketing its products directly to health professionals. Remarkably, similarly to the patent medicines, there were no regulations in place to assure health professionals or consumers of a given product's effectiveness and safety.

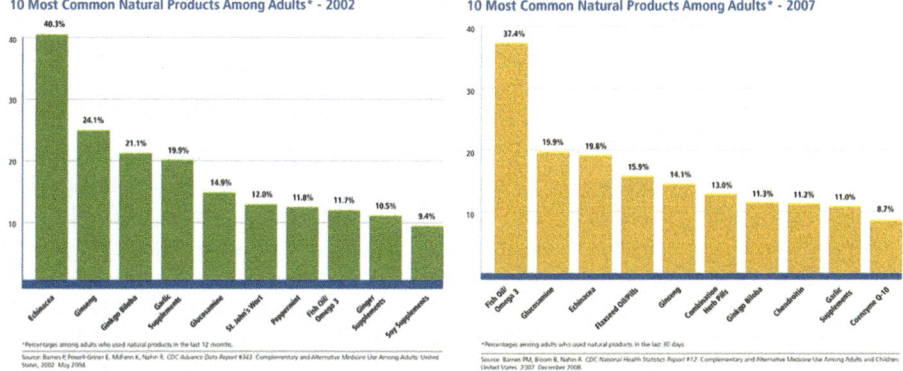

Fig. 1.4 10 most commonly used natural products among adults, in 2002 vs. 2007

By the early 1900s, the Federal Government enacted legislation to assure the characterization of effectiveness and safety of food and drugs. This step was largely driven by growing fraud within the respective industries (Barkan 1985).

In the subsequent decades, the interest of Americans population in self-care has been increasingly growing. In 2007, a spending of $33.9 billion out-of-pocket for CAM was reported which amounts to approximately 1.5% of total health care expenditures and more than 11% of total out-of-pocket health care expenditures. According to a survey by the National Center for Complementary and Integrative Health of National Institutes of Health (N.I.H.), in 2007, the most popular natural products were the following: fish oils/omega 3, glucosamine, echinacea, and flaxseed. In 2002, the most popular natural products were echinacea, ginseng, ginkgo, and garlic supplements (Barnes et al. 2008) (Fig. 1.4).

To-date, the National Center for Complementary and Integrative Health (NCCIH) is the Federal Government's lead agency for scientific research and on complementary and integrative health approaches: it is one of the 27 centers and institutes making up the National Institutes of Health (NIH) within the US Department of Health and Human Services.

NCCIH is striving to fulfil its mission „to define, through rigorous scientific investigation, the usefulness and safety of complementary and integrative health interventions and their roles in improving health and health care". As an "arbitrarily" chosen example from the multiplicity possible research topics, we refer to a recent study by Dettweiler et al. (Dettweiler et al. 2019) in which extracts from three plants (*Quercus alba*, *Liriodendon tulipifera* and *Aralia spinosa*) used as medicines during the American Civil War were tested and verified to show antimicrobial activity against multidrug-resistant bacteria associated with wound infections.

1.9   Botanical Drugs vs Herbal Dietary Supplements

In the U.S., botanical products, depending on the circumstances, may be regulated as botanical drugs (OTC, Rx, or Homoeopathic), non-drug cosmetics, herbal dietary supplements, or foods. American consumers' interest in health and self-care suggests that they are searching for alternatives to conventional foods for physical and mental well-being. There is a trend that with the increasing advent of food that is already worryingly synthetic and influenced by technology, a significant proportion of the US population is turning towards foods and medicines, including dietary supplements of plant origin. These they see as more natural (Mariott 2003). Dietary supplements of plant origin, roughly equate to botanical supplements or supplement ingredients.

As a result, to-date, there are more than 29,000 different dietary supplements available to consumers (Gibson and Taylor 2005), whereas the number of dietary supplement products made from plants is not exactly known.

The federal agency, Food and Drug Administration (FDA) is responsible for protecting the public health by ensuring the safety, efficacy, and security of human and veterinary drugs, biological products, including medicinal and aromatic plants. The scope of FDA's regulatory authority is very broad. In view of MAPs, the topic of this chapter, the following responsibilities of FDA's should be highlighted: dietary supplements, food additives, prescription (Rx) and non-prescription (over-the-counter (OTC)) drugs), non-drug cosmetics, veterinary drugs, etc. However, this is not an exhaustive list.

Today, the FDA regulates $1 trillion worth of products a year. It ensures the safety and effectiveness of all drugs, biological products and animal drugs and feed, etc.

1.10   Market of Botanicals

To-date, modern herb trade is seen as part of a multi-billion dollar herb industry. In 2014, for the 11th consecutive year, American consumers spent more on herbal dietary supplements than they did in the previous year. Since 2011, estimated herbal supplement sales in all channels have increased by more than $1 billion; in 2014, US consumers spent roughly $400 million more on plant-based supplements than in 2013. These figures suggest a clear trend: Americans are continuing to rely on botanical products for various aspects of their wellbeing and other personal needs (Smith et al. 2015).

The sales data presented by Smith et al. (2015) reflect consumer preferences. In 2014, retail sales of combination herbal supplements and total supplement sales in natural and health food stores seemed to increase faster than other herbal supplement products and channels, respectively.

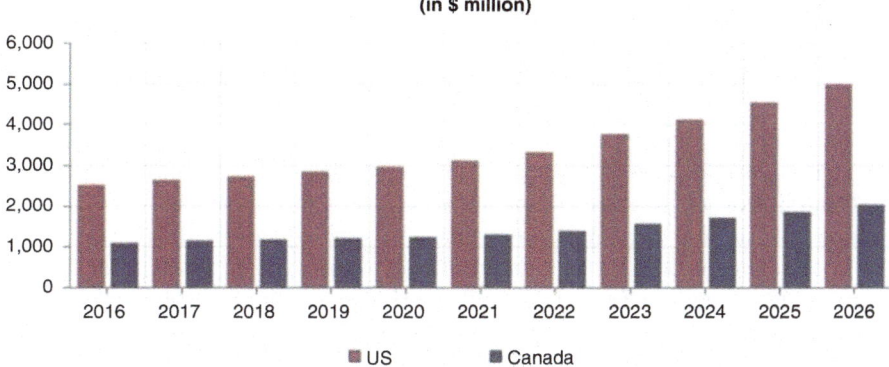

North America Botanical & Plant Derivative Drug Market, By Geography
(in $ million)

Fig. 1.5 North Amarican Botanical and Plant Derivative Drug Market. (Source: https://www.ink-woodresearch.com/reports/northamerica-botanical-plant-derivative-drug-market/)

According to estimates by Inkwood Research North America's Botanical and plant derivate drug market is deemed to see a farther steady growth (Fig. 1.5).

Total **retail sales of herbal dietary supplements** in the United States increased by an estimated 6.8% in 2014 marking the 11th consecutive year of growth (Smith et al. 2015). This increase is slightly less than 2013's 7.9% increase, which was the greatest percentage sales jump since the late 1990s.

Sales of herbal supplements have been increasing steadily since 2002–2003 when the market experienced a brief, two-year dip in total sales. American consumers spent roughly $6.4 billion on herbal supplements —over $1 billion more than was spent in 2011 and $400 million more than 2013 — according to aggregated market statistics provided by the *Nutrition Business Journal* (NBJ).

NOTE: A general comment regarding the content and structure of the following Sect. 1.11 on actors. You might consider organizing similar actors together. For example,

- **three compendia** (quality standards monographs) are discussed, the United States Pharmacopoeia (USP), the National Formulary (NF), and the American Herbal Pharmacopoeia (AHP). These could be listed one-after-the-other for coherence. There is another relevant official compendium not mentioned. The United States Pharmacopoeial Convention (USPC) actually produces three official compendia, the USP, the NF, and the Food Chemicals Codex (FCC): https://www.foodchemicalscodex.org/
- **two trade associations** are discussed, American Herbal Products Association (AHPA) and American Spice Trade Association (ASTA). These could be grouped together for coherence.
- **two regulatory agencies** are discussed, the U.S. Food and Drug Administration (FDA) and the U.S. Department of Agriculture (USDA). There are other governmental agencies that impact the production, trade and use of botanicals in the

United States, e.g. the U.S. Fish and Wildlife Service (USFWS) as the CITES authority and the U.S. Environmental Protection Agency (EPA) regarding the establishment of tolerances for pesticides that can be used on botanical crops. The U.S. Drug Enforcement Administration (DEA) is also involved for drug companies that import controlled botanical drugs like coca leaf from Peru or opium exudate (or registered churches that import ayahuasca tea from Brazil). You might also consider discussing the government-funded National Center for Natural Products Research: https://pharmacy.olemiss.edu/ncnpr/

- **one educational organization** is discussed, the American Botanical Council (ABC). Another educational non-profit organization you might consider discussing is the American Herbalists Guild (AHG): https://www.americanherbalists-guild.com/

1.11 Actors of North American Medicinal and Aromatic Plants Industry

Herbal products are most versatile. They can be sold in various forms, e.g.: as conventional foods or food additives (e.g. flavoring or coloring agents), as dietary supplements, as cosmetic ingredients, or as botanical drugs, etc. Ultimately, they may also be used as self-collected plants that are not marketed products IARC (2016).

In the US and Canada, pharmaceutical drugs are subject to strict regulations, like in most countries. Their availability is similarly restricted. Remarkably, however, this may not be the case with materials used in the preparation of herbal medicines, although the therapeutic benefit of such herbal products may have been recognized in certain communities for centuries. Moreover, herbal products are available in several regulatory models, ranging from foods and dietary supplements to cosmetics and over-the-counter (non-prescription) and prescription drugs. As a result, product quality and composition may vary from country to country and even within countries, (e.g.: different products bear the same name) (International Agency for Research on Cancer 2016).

In view of the abovementioned, the following compilation has the aim of providing an insight into the multitude of actors that are influential in shaping, regulating the MAP scenery in North America.

1.11.1 Food and Drug Administration (FDA)

The Office of Food and Drug Administration of U.S. Department of Health and Human Services (FDA) regulates both finished dietary supplement products and dietary ingredients. For dietary supplements a different set of regulations are in force than for those covering "conventional" foods and drug products. Under the Dietary Supplement Health and Education Act of 1994 (DSHEA):

No different than medicines, manufacturers and distributors of dietary supplements and dietary ingredients are prohibited from marketing products that are adulterated or misbranded. That means that these firms are responsible for evaluating the safety and labeling of their products before marketing to ensure that they meet all the requirements of DSHEA and FDA regulations.

FDA is responsible for taking action against any adulterated or misbranded dietary supplement, drug, or food product after it reaches the market.

The Dietary Supplement GMPs Final Rule and Interim Final Rule were effective August 24, 2007, introduced to help ensure the quality of dietary supplements so that consumers can be confident that the products they purchase contain what is on the label" (von Eschenbach 2007)" Farthermore, as a result of recent amendments to the Federal Food, Drug, and Cosmetic Act, by the end of that year, industry became required to report all serious dietary supplement related adverse events to FDA."

Remarkably, the final rule included flexible requirements that later on, can evolve with improvements in scientific methods used for verifying identity, purity strength, and composition of dietary supplements.

As a companion document, in an interim final rule, outlines for a petition process has also been issued for manufacturers to request an exemption to the cGMP requirement for 100% identity testing of specific dietary ingredients used in the processing of dietary supplements.

1.11.2   Pharmacopoeia of the United States of America (USP)

The United States Pharmacopoeial Convention (USPC) is responsible for several official compendia with monographs that are incorporated into regulations. These include the United States Pharmacopoeia (USP), the National Formulary (NF), and the Food Chemicals Codex (FCC). The Federal Food, Drug, and Cosmetic Act (FD&C Act) defines the term "official compendium" as the official USP, the official NF, and the official Homeopathic Pharmacopeia of the United States (HPUS). The USP-NF produces monographs for botanical ingredients that are used as active ingredients of botanical drugs (Rx and OTC), components of herbal dietary supplement products, and as excipients (food additives and pharmaceutical aids).

According to http://www.usp.org/about USP has the mission to improve global health through public standards and related programs that help ensure the quality, safety, and benefit of medicines and foods. USP approaches this vision with a sense of urgency and purpose, strengthened by its cadre of dedicated volunteers, members, and staff, and by working collaboratively with key stakeholders across the globe.

The USP enjoys a rich history with plants used as food, fiber and herbal medicine. The first publication of the Pharmacopoeia of the United States of America (USP), in 1820, included 254 botanical substances, as well as monographs for 246 herbal drug preparations, together accounting for over 80% of USP entries (Wahl and Brinckmann 2019). Subsequently, in 1833, the USP served as the basis for the first edition of the Dispensatory of the United States of America (USD).

After hardly more than 74 years, in February of 2016, the USP published a 'stimuli to the revision process' article for public comment on the topic of advisability and feasibility of developing new USP quality standards monographs (Giancaspro et al. 2016) for medical cannabis.

As a result of legalization of the medical use of cannabis in several U.S. states and internationally, at the end of the 2010's, and in the absence of modernized quality standards monographs, the USP has been requested to develop such standards.

USP monographs are not only official compendia of the U.S. but are used and incorporated by reference into drug regulations in many countries of the world, especially those lacking their own national pharmacopoeia (e.g. Australia and Canada), for specifications and quality control of medicines.

In 2009, USP published the first edition of the **USP Dietary Supplements Compendium** (the "USP Compendium"), a collection of public standards and technical information for dietary supplement manufacturers. The USP-Compendium includes information developed by the major trade associations and other organizations (Soller et al. 2012). The latest, 2019 issue of Dietary Supplements Compendium (DSC) is meant to be a source for monographs, regulatory guidance, and reference tools for the dietary supplement supply chain. It will contain 24 additional General Chapters, 72 new monographs, 59 updated HPTLC plates, etc. and will be available online only (https://www.usp.org/dietary-supplements-herbal-medicines). It also contains two important materials published by the American Herb Producers Association (AHPA) concerning the use of marker compounds for botanical identification and detection of adulterants (a), and a joint AHPA-American Herbal Pharmacopoeia (AHP) draft document concerning Good Agricultural and Collection Practices for herbal raw materials (b).

In Canada it is the *Compendium of Pharmaceuticals and Specialties* (CPS) that is the Canadian standard for conventional drug monographs. It offers an unrivalled compilation of product monographs that are developed by manufacturers, approved by Health Canada and optimized by CPhA editors. The information contained by CPC is current, as more than 60% of its content has changed since the last edition. Moreover, the digital formats are updated weekly. However, more relevant to OTC botanical drugs, which in Canada are regulated as Natural Health Products (NHPs) requiring pre-marketing authorization and issuance of a product license by the Natural and Non-prescription Health Products Directorate (NNHPD), are the NNHPD Compendium of Monographs (https://www.canada.ca/en/health-canada/services/drugs-health-products/natural-non-prescription/applications-submissions/product-licensing/compendium-monographs.html). Canada does not have its own national pharmacopoeia for the establishment of ingredient quality standards, The NNHPD accepts the following pharmacopoeias and international standards for use by product license applicants' for quality control of botanical ingredients contained in licensed NHPs: United States Pharmacopeia (USP), British Pharmacopoeia (BP), European Pharmacopoeia (Ph. Eur.), Pharmacopée française (Ph.f.), Pharmacopoeia Internationalis (Ph.I.), Japanese Pharmacopoeia (JP), and Food Chemicals Codex (FCC) (Natural and Non-prescription Health Products Directorate 2015).

1.11.3 National Formulary of the United States of America (N.F)

Although published annually in conjunction with USP (USP-NF), the National Formulary is a book of standards for certain pharmaceuticals and **preparations that are not included in the USP**. While some NF monographs provide specifications for active substances subject to FDA drug monographs, such as Cocoa Butter NF (fat obtained from the seed of *Theobroma cacao*, Sterculiaceae), a skin protectant active ingredient, most NF botanical monographs provide specifications for excipients (e.g., antioxidants, binders, bulking agents, coloring agents, flavoring agents, perfumes, pharmaceutical bases, sweetening agents). Examples of NF botanical excipient monographs include Candelilla Wax NF (purified wax from the leaves of *Euphorbia antisyphilitica*, Euphorbiaceae), Cardamom Seed NF (*Elettaria cardamomum*, Zingiberaceae), Carnauba Wax NF (wax from the leaves of *Copernicia cerifera*, Arecaceae), Stronger Rose Water NF (solution of odoriferous principles of the flowers of *Rosa centifolia*, Rosaceae), and Vanilla NF (cured, full-grown, unripe fruits of *Vanilla planifolia* or *V. tahitensis*, Orchidaceae), among others (Brinckmann 2011).

1.11.4 Dispensatory of the United States of America (U.S.D.)

It is a collection of monographs on unofficial drugs and drugs recognized by the United States Pharmacopeia, the British Pharmacopoeia, and the National Formulary. It describes indications for use and posology, general tests, processes, reagents, and solutions of the U.S.P. and N.F., as well as drugs used in veterinary medicine (Miller and Keane 2005).

1.11.5 American Herbal Products Association (AHPA)

Founded in 1982, the American Herbal Products Association (AHPA) is the national trade association and voice of the herbal products industry. AHPA's mission is to promote the responsible commerce of herbal products to ensure that consumers continue to enjoy informed access to a wide variety of herbal goods.

AHPA is comprised of more than 350 member companies. These are domestic and foreign companies doing business as growers, processors, manufacturers and marketers of herbs and herbal products as foods, dietary supplements, cosmetics, and non-prescription drugs. Some member companies provide expert services to the herbal trade.

1.11.6   American Herbal Pharmacopoeia (AHP)

AHP is a non-profit educational organization dedicated to promoting the responsible use of herbal products and herbal medicines. Its work relies on a worldwide network of experts, including botanists, chemists, herbalists, medical doctors, pharmacists, and pharmacologists. AHP develops comprehensive qualitative and therapeutic monographs on botanicals, including many of the Ayurvedic, Chinese, and Western herbs most frequently used in the United States.

1.11.7   American Botanical Council (ABC)

Founded in 1988, the American Botanical Council (abbreviated as ABC, also known as the Herbal Medicine Institute), is an independent, nonprofit research and education organization dedicated to providing accurate and reliable information for consumers, healthcare practitioners, researchers, educators, industry and the media.

ABC is a rich source of MAP related information, like a subscription to *HerbalGram* (the quarterly journal of the American Botanical Council), or Herbal News & Events (a weekly update on events and media). Of special significance is the Botanical Adulterants Monitor (a newsletter addressing Botanical Adulterants issues).

ABC has the declared mission to "Provide education using science-based and traditional information to promote responsible use of herbal medicine — serving the public, researchers, educators, healthcare professionals, industry and media".

A brief list of activities from the ABC Web-site: http://www.abc.org suffices to demonstrate the significant contributions that ABC has made in important areas of MAP issues: ABC Clinical Guide, Adulterants Program, Commission E, Expanded E, Healthy Ingredients, HerbalEGram, HerbalGram, HerbClip, Herbal MediaWatch, HerbMedPro, Medicinal Plant ID, Monographs, Virtual Garden. As a most recent and probably also one of the most important of these services the **Botanical Adulteration Prevention Program,** was initiated in 2014, by three non-profit organizations (ABC – AHPA – NCNPR). It aims to educate members of the herbal and dietary supplement industry about ingredient and product adulteration. Adulteration has been accompanying botanical/herbal product trade and consumption, since its very early beginnings.

1.11.8   American Spice Trade Association (ASTA)

American Spice Trade Association (ASTA) was founded in 1907 to represent the interests of approximately 175 members including companies that grow, dehydrate, and process spices. ASTA's members include U.S.-based agents, brokers and importers, and companies based outside of the U.S. that grow spices and ship them to the U.S. and other companies associated with the U.S. spice industry. ASTA members manufacture and market the majority of spices sold in the U.S. for industrial, food service and consumer use.

1.11.9   USDA National Organic Program (NOP)

USDA National Organic Program (NOP) was established by Congress, in 2001. It is Federal Regulatory Program that develops and enforces uniform national standards for organically-produced agricultural products, including herbs and spices, sold in the United States.

NOP is operating as a public-private partnership. It accredits private companies and helps train their inspectors to certify that farms and businesses meet the national organic standards.

USDA organic products have strict production and labeling requirements and must meet the following requirements:

- Produced without excluded methods, (e.g., genetic engineering, ionizing radiation, or sewage sludge).
- Produced using allowed substances.
- Overseen by a USDA National Organic Program-authorized certifying agent, following all USDA organic regulations.

The USDA seal may be used only for raw or processed agricultural products described in specific paragraphs of National Organic Program Regulations. Of particular relevance to sustainable production, trade and use of MAPs, is the NOP Wild-crop Harvesting Practice Standard and corresponding guidance for biodiversity conservation requirements in certified organic wild-collection areas.

1.12   Conclusions/Future

Out of the thirty-six areas that qualify as biodiversity hotspots in the World, for North America (including Central America), only three regions qualify: the **California Floristic Province**, **Madrean pine-oak woodlands**, and **Mesoamerica** (Molly 2018). Recently, Noss et al. (Noss et al. 2015) have identified also the North American Coastal Plain (NACP) as a new global hotspot.

The North American continent, with its area spanning 9,540,000 sq. miles, makes up about 4.8% of the total surface area of the planet. This continent alone has more than 150 identified eco-regions.

In the United States, that probably has the planet's best-studied flora, in which the richest sources of medicines are the sunflower family 12 (Asteraceae), the rose family (Rosaceae), and the mint family (Lamiaceae).

Native Americans used 2564 of 21,641 vascular species, or 11.8% of the available flora for medicinal purposes (Moerman 1996). Native American peoples developed a sophisticated "plant-based medical system" in the course of millennia before the European conquest of America. Although the exploitation of natural resources of MAPs has been steadily increasing, obtaining statistics on the actual amount of bulk herbal material grown or wild-harvested within the United States was a challenge (2004).

Since the early 1900s, the Federal Government has been making effort to assure the characterization of effectiveness and safety of food and drugs. Enacted steps of legislation, especially at the beginnings, have been largely driven by growing fraud within the respective industries (Barkan 1985).

American consumers' interest in health and self-care suggests that they are searching for alternatives to conventional foods for physical and mental well-being. There is a trend that with the increasing advent of food that is already worryingly synthetic and influenced by technology, a significant proportion of the US population is turning towards foods and medicines, including dietary supplements of plant origin.

As a result, to-date, there are more than 29,000 different dietary supplements available to consumers (Gibson and Taylor 2005), still the number of dietary supplement products made from plants is not exactly known.

This huge demand has resulted in the over-exploitation of natural resources of MAPs, so that there are already estimates declaring that about 29% of 16,000 species were at risk.

In the US and Canada, pharmaceutical drugs are subject to strict regulations. These, similarly to many other countries of the world. The existing state-led regulated landscape for medical Cannabis may be an opportunity for the U.S. market to explore a viable pathway to market for other botanical-based products at therapeutic levels (Wahl and Brinckmann 2019).

As exemplified by the recently regulated cannabis industry that allows to grow in the marketplace, and foster opportunities to discover therapeutic applications and innovative technologies – the state-regulated markets could be a viable model to bring also other plant-based products at therapeutic levels to market. Consumer interest, demand and need may warrant a due expansion of the state frameworks to include other botanicals.

There is good reason to believe in the truth of Professor Varro Tyler's forecast (Tyler 2000) saying: "There is no question in my mind that herbal medicine will continue to flourish and eventually become integrated into our materia medica. Looking back, it is obvious that a great deal of progress has already been made. We now look to the future with eager anticipation and great expectations."

References

American Herbal Products Association (2017) Good Agricultural and Collection Practices and Good Manufacturing Practices for Botanical Materials

Anderson MK (2006) Traditional Ecological Knowledge: An Important Facet of Natural Resources Conservation. Technical Note No.1. Series: Traditional Ecological Knowledge. Accessed 2 June 2020 https://www.nrcs.usda.gov/Internet/FSE_DOCUMENTS/stelprdb1045244.pdf

Barnes PM, Bloom B, Nahin R (2008) CDC National Health Statistics Report #12. Complementary and alternative medicine use among adults and children: United States, 2007

Barkan ID (1985) Industry invites regulation: the passage of the Pure Food and Drug Act of 1906. American Journal of Public Health, 75(1), 18–26

Baum BR, Binns SE, Amason JT (2006) Integrating recent knowledge about the genus Echinacea: morphology, molecular systematics, phytochemistry. HerbalGram 72:32–46. http://cms.herbalgram.org/herbalgram/issue72/article3058.html

Blumenthal M (2004) AHPA issues third tonnage survey of wild-harvested plants. HerbalGram 61:65–66. http://cms.herbalgram.org/herbalgram/issue61/article2649.html

Brinckmann J (2015) Geographical indications for medicinal plants: globalization, climate change, quality and market implications for geo-authentic botanicals. World J Tradit Chin Med 1(1):16–23. https://doi.org/10.15806/j.issn.2311-8571.2014.0020

Brinckmann J (2007) Market for U.S.-Grown Botanicals. North American Botanical Crops with Economic Potential. 2007. http://sfp.ucdavis.edu/files/137212.pdf

Brinckmann J (2011) Reproducible efficacy and safety depend on reproducible quality: matching the various quality standards that have been established for botanical ingredients with their intended uses in cosmetics, dietary supplements, foods, and medicines. HerbalGram 91:40–55. http://cms.herbalgram.org/herbalgram/issue91/FEAT_reproducible.html

Brinker F (2013) Echinacea differences matter: traditional uses of Echinacea angustifolia root extracts vs. modern clinical trials with. HerbalGram 97:46–57. http:cms.herbalgram.org/herbalgram/issue97/hg97.feat.echinacea.html

Byrne M, Haidet M, McCoy J-A (2007) Safeguarding the seeds of native plants. HerbalGram 75:30–37. http://cms.herbalgram.org/herbalgram/issue75/article3129.html

Congress, Traditional Medicines (n.d.) Frequently asked questions about the work of the traditional medicines congress. Accessed 28 Aug 2019. http://www.ahpa.org/Portals/0/Documents/06_0515_TMC.FAQ.pdf?ver=2015-12-18-165210-103

CEC (Commission for Environmental Cooperation) (1997) Ecological Regions of North America. Toward a Common Perspective. Report. pp.60. (Last Internet Access: 2020.06.02. at http://www3.cec.org/islandora/en/item/1701-ecological-regions-north-america-toward-common-perspective-en.pdf)

Dettweiler M, Lyles JT, Nelson K, Dale B, Reddinger RM, Zurawski DV, Quave CL (2019) American civil war plant medicines inhibit growth, biofilm formation, and quorum sensing by multidrug-resistant bacteria. Sci Rep 9(1):7692. https://doi.org/10.1038/s41598-019-44242-y

Eschenbach AC von (2007) FDA issues dietary supplements final rule. Office of the Commissioner 2007. http://wayback.archive-it.org/7993/20170111070400/http://www.fda.gov/NewsEvents/Newsroom/PressAnnouncements/2007/ucm108938.htm

Flora of North America Editorial Committee (ed) (1993) Flora of North America Editorial Committee, vol 1–17. New York/Oxford, Flora of North America Association. http://beta.floranorthamerica.org/Main_Page

Foster S (2004) He secret garden: important Chinese herbs in American horticulture: a photo essay. HerbalGram 64:44–51. http://cms.herbalgram.org/herbalgram/issue64/article2754.html

Foster S (2017) Forest gems: exploring medicinal trees in American forests. HerbalGram 116:50–71. http://cms.herbalgram.org/herbalgram/issue116/hg116-feat-medtrees.html

Giancaspro GI, Kim NC, Venema J, de Mars S, Devine J, Celestino C, ... & Jones Jr E (2016) The Advisability and Feasibility of Developing USP Standards for Medical Cannabis. Stimuli

Article, pp. 10. Accessed 2 June 2020 https://www.uspnf.com/notices/stimuli-article-advisability-and-feasibility-developing-usp-standards-medical-cannabis-posted-comment

Gibson JE, Taylor DA (2005) Can claims, misleading information, and manufacturing issues regarding dietary supplements be improved in the United States? J Pharmacol Exp Ther 314(3):939–944. https://doi.org/10.1124/jpet.105.085712

Hoppe B (2009) Handbuch Des Arznei- Und Gewürzpflanzenbaus. Band 1. Edited by B Hoppe. Saluplanta e.V. Bernburg, Bernburg

International Agency for Research on Cancer (2016) IARC monographs on the evaluation of carcinogenic risks to humans. No. 108. International Agency for Research on Cancer, Lyon, France. https://www.ncbi.nlm.nih.gov/books/NBK350452/

Johnston BA (1998) Major diversity loss: 1 in 8 plants in global study threatened. HerbalGram 43:54. http://cms.herbalgram.org/herbalgram/issue43/article1057.html

Kindscher K (1992) Medicinal wild plants of the prairie: an ethnobotanical guide. University Press of Kansas. https://kansaspress.ku.edu/978-0-7006-0527-9.html

Kingston DGI (2011) Modern natural products drug discovery and its relevance to biodiversity conservation. J Nat Prod 74(3):496. https://doi.org/10.1021/NP100550T

Krochmal A (1968) Medicinal plants and Appalachia. Econ Bot 22(4):332–337. https://doi.org/10.1007/BF02908128

Mariott BM (2003) Dietary supplements of plant origin. In: Maffei M (ed) Dietary supplements of plant origin. Taylor & Francis Inc, New York, pp 1–17. https://doi.org/10.4324/9780203351352

Máthé Á (2015) Botanical aspects of medicinal and aromatic plants. In: Máthé Á (ed) Medicinal and aromatic plants of the world, vol 1. Springer, Dordrecht, pp 13–33. https://doi.org/10.1007/978-94-017-9810-5_2

Mauseth JD (2014) Botany: an introduction to plant biology. Jones and Bartlett Publishers, Boston/Totonto/London/Singapore. https://books.google.hu/books?hl=hu&lr=&id=0BGEs95p5EsC&oi=fnd&pg=PR3&dq=mauseth+botany&ots=54P24AENwX&sig=meyf3TpMNKUDSZL618QxghESAlQ&redir_esc=y#v=onepage&q=mausethbotany&f=false

Miller CB, Keane BF (2005) Encyclopedia and dictionary of medicine, nursing and allied health, 7th edn. Saunders

Moerman DE (1996) An analysis of the food plants and drug plants of native North America. J Ethnopharmacol 52(1):1–22. https://doi.org/10.1016/0378-8741(96)01393-1

Moerman D (1998) Native North American Food and Medicinal Plants: Epistemological Considerations. Plants for food and medicine. In: Prendergast HDV, Etkin, NL, Harris DR and Houghton PJ. (eds.): Plants for Food and Medicine. (pp. 69–74). Kew: Royal Botanic Gardens

Moerman DE (2016) Ethnobotany in Native North America. In: Selin H (ed) Encyclopaedia of the history of science, technology, and medicine in non-western cultures. Springer Netherlands, Dordrecht, pp 1729–1736. https://doi.org/10.1007/978-94-007-7747-7_8580

Molly J (2018) The biodiversity hotspots found in North and Central America – WorldAtlas.Com. WorldAtlas, Aug. 10, 2018. https://www.worldatlas.com/articles/the-biodiversity-hotspots-found-in-north-and-central-america.html

Myers N, Mittermeler RA, Mittermeler CG, da Fonseca GAB, & Kent J (2000). Biodiversity hotspots for conservation priorities. Nature, 403(6772), 853–858.

Noss RF, Platt WJ, Sorrie BA, Weakley AS, Bruce Means D, Costanza J, Peet RK (2015) How global biodiversity hotspots may go unrecognized: lessons from the North American Coastal Plain. Edited by David Richardson. Divers Distrib 21(2):236–244. https://doi.org/10.1111/ddi.12278

Raterman K (2018) Appalachian herb growers consortium: fostering Chinese herb cultivation in the United States. HerbalGram 118:30–32. http://cms.herbalgram.org/herbalgram/issue118/hg118-orgnews-appherbgrow.html

Smith T, Lynch ME, Johnson J, Kawa K, Bauman H, Blumenthal M (2015) Herbal dietary supplement sales in US Rise 6.8% in 2014. HerbalGram 107:52–59. http://cms.herbalgram.org/herbalgram/issue107/hg107-mktrpt-2014hmr.html

Soller RW, Bayne HJ, Shaheen C (2012) The regulated dietary supplement industry: myths of an unregulated industry dispelled. HerbalGram 93:42–57. http://cms.herbalgram.org/herbalgram/issue93/FEAT_myth.html

Tyler VE (2000) Herbal medicine: from the past to the future. Public Heath Nutr 3(4A):447–452. https://pdfs.semanticscholar.org/7c21/188826fb3dc1d43d3eb347acb6f29b976afd.pdf

US Environmental Protection Agency Protection (n.d.) Ecoregions of North America. Accessed 27 Sept 2019. https://www.epa.gov/eco-research/ecoregions-north-america

USDA NRCS (2019) The Plants Database. Plant's Profile for Echinacea (purple cone flower). Accessed 2 June 2002 https://plants.usda.gov/java/reference?symbol=ECHIN

Wahl T, Brinckmann J (2019) A modern state-federal framework for a regulated US cannabis industry. HerbalGram 122:60–69. http://cms.herbalgram.org/herbalgram/issue122/hg122-feat-cannabisframe.html

Walter, KS, Gillett HJ, World Conservation Monitoring Centre, and International Union for Conservation of Nature and Natural Resources. Species Survival Commission (1998) 1997 IUCN Red List of Threatened Plants. IUCN – The World Conservation Union. https://portals.iucn.org/library/node/7377

Wisniewski A (2017) Healing plants you can forage – American forests. 2017. https://www.americanforests.org/blog/healing-plants-can-forage/

Chapter 2
Diversity, Conservation, and Sustainability of North American Medicinal Plants

Danna J. Leaman

Abstract Approximately 2000 species of medicinal and aromatic plants are native to North America. The conservation status of these species is currently being assessed and updated by the IUCN Medicinal Plant Specialist Group and NatureServe, in partnership with the Albuquerque BioPark. Conservation status assessments will contribute to the IUCN Plants for People initiative's objective to support conservation through species and habitat protection, sustainable wild harvest, and cultivation where warranted.

Keywords Medicinal and aromatic plants · Flora of North America · Species diversity · Taxonomic diversity · Conservation · Sustainable use · Crop wild relatives · IUCN red list · NatureServe rank · Wild harvest · Cultivation

2.1 Introduction

Native species of medicinal and aromatic plants comprise ten percent of the North American flora, and 0.05 percent of the known flora of the world. However, the North American native flora may provide nearly half of the medicinal and aromatic plants in global trade. Conservation and sustainable use of this resource is therefore of great importance for the domestic and international supply of this resource.

D. J. Leaman (✉)
Co-chair, IUCN SSC Medicinal Plant Specialist Group, Research Associate, Canadian Museum of Nature, Maberly, Canada
e-mail: djl@green-world.org

© Springer Nature Switzerland AG 2020   31
Á. Máthé (ed.), *Medicinal and Aromatic Plants of North America*, Medicinal and Aromatic Plants of the World 6,
https://doi.org/10.1007/978-3-030-44930-8_2

2.2 Diversity of the North American Medicinal Flora

2.2.1 *Number of Medicinal Plant Species*

The flora native to North America – considered in the *Flora of North America* to include the continental United States, Canada, the French islands Saint Pierre and Miquelon, and Greenland – comprises 20,000 species of vascular plants and bryophytes (Flora of North America 2019), approximately 5% of the world's known 400,000 plant species (World Flora Online 2019). At least 2000 of these species have reported historical or current medicinal uses. This estimate is derived from two data sources: the Native American Ethnobotany database (Native American Ethnobotany 2019) and the Medicinal and Aromatic Plants of the World (MAPROW) database (Schippmann 2019).

The Native American Ethnobotany database, compiled by Daniel Moerman, is a comprehensive digest of uses (medicinal as well as food, dyes, fibres) of native plant species by Native American peoples (Native American Ethnobotany 2003). The database was first published in 1977 (Moerman 1977), updated and republished in 1986 (Moerman 1986), and most recently updated in 2003 for publication online with its current content and format (Native American Ethnobotany 2019). Of 4029 North American plant species with documented uses, approximately half have medicinal uses (Native American Ethnobotany 2003).

The MAPROW database, compiled by Uwe Schippmann, is an off-line resource that digests taxonomic, use, and other conservation-relevant information about medicinal and aromatic plants world-wide based on published sources (Schippmann 2019). In October 2018, MAPROW included 26,000 taxa, of which 2000 were included in the Flora of North America (Flora of North America 2019). This estimate includes North American species that appear in major global, regional and national pharmacopoeias and those important in international trade, but likely does not adequately reflect the number and diversity of species historically and currently used by native Americans and indigenous peoples in Canada.

2.2.2 *Taxonomic Distribution*

Using his Native American Ethnobotany database, Moerman (1979, 1991) compared species frequencies of taxonomic orders in the North American medicinal flora with those in the general flora and used regression residuals to identify over-represented and under-represented taxa in the medicinal flora. He proposed that over-represented taxa are more likely to be biologically active than are taxa whose frequencies can be predicted from their frequencies in the general North American flora and taxa that are under-represented with respect to the general flora. Moerman (1991) found that the medicinal flora used by native North Americans emphasized trees and shrubs, and under-represented grasses and grass-like plants. He proposed that more complex plants, with more parts to defend, would have a richer array of defensive chemicals with potential medicinal applications.

2.2.3 Geographic Distribution

Also using the Native American Ethnobotany database (Native American Ethnobotany 2019), Canadian researcher Christina Turi produced an unpublished, preliminary evaluation of the regional distribution of the North American medicinal plant flora. Her work indicates that species richness is unevenly distributed across continental North America, with notably high diversity in California, low diversity in the Boreal and Tundra areas of Alaska and northern Canada, and greatest regionally unique diversity (i.e. greatest number of medicinal plant species found in no other region) in the western mountain states from Montana south to New Mexico and Arizona (Turi 2017a).

2.3 Conservation

2.3.1 NatureServe Ranks and IUCN Red List Status

The conservation status of a majority of plant species native to North America has been evaluated at the global, national, and state or provincial level by NatureServe, a non-government organization established in 1994 (NatureServe 2019). In North America, NatureServe compiles and analyzes species distribution and other population data from a network of partners, including state-supported natural heritage programmes throughout the United States, and provincially supported natural heritage conservation bodies throughout Canada. The conservation status of a species is expressed by NatureServe as a rank, or a range of rank values, that indicates the severity of the threat of extinction to the wild population, ranging from "presumed extinct" to "secure" (Table 2.1).

Summary information (rankings and rationale) is available for individual species on a searchable database managed by NatureServe (NatureServe Explorer 2019). Using this information-rich resource to obtain aggregate data – for example, to create a profile of level of threat to medicinal plant species according to geographic region – requires painstaking effort, since the database is not searchable by type of species use or by combinations of queries, such as geographic area and conservation rank. Such an effort by Turi has produced a preliminary profile of conservation information gaps and greatest conservation need for medicinal plants in Canada (Turi 2017b).

The IUCN Red List is organized and published by the International Union for Conservation of Nature (IUCN) with assessments contributed by a large network of

Table 2.1 Comparison of NatureServe global ranks and IUCN Red List threat categories[*]

	IUCN Red List category	NatureServe global (G) rank	Definitions
EXTINCT	Extinct (EX)[a]	Presumed Extinct (GX)[b]	(a) No reasonable doubt last individual has died [after] exhaustive surveys (b) Not located despite intensive searches; virtually no likelihood of rediscovery
	Extinct in the Wild (EW)[c]	Presumed Extinct in the Wild (GXC)[d]	(c) Species known to survive only in cultivation or captivity, or as naturalized population(s) well outside past range (d) Presumed extinct in the wild across entire native range, but extant in cultivation, in captivity, as naturalized population(s) outside historical native range, or as a reintroduced population not yet established
THREATENED	Critically Endangered (CR) (Possibly Extinct)[e]	Possibly Extinct (GH)[f] Possibly Extinct in the Wild (GHC)[g]	(e) Likely already Extinct, but confirmation is required (f) Known only from historical occurrences but still some hope of rediscovery (g) Possibly extinct in the wild across entire native range, but extant in cultivation, in captivity, as naturalized population(s) outside historical native range, or as a reintroduced population not yet established
	Critically Endangered (CR)[h] Endangered (EN)[i]	Critically Imperiled (G1)[j]	(h) Best available evidence indicates that species faces extremely high risk of extinction in the wild (i) Best available evidence indicates that species faces very high risk of extinction in the wild (j) At very high risk of extinction due to extreme rarity, very steep declines, or other factors
	Vulnerable (VU)[k]	Imperiled (G2)[l]	(k) Best available evidence indicates that species faces high risk of extinction in the wild (l) At high risk of extinction or elimination due to a very restricted range, very few populations, steep declines, or other factors

(continued)

Table 2.1 (continued)

	IUCN Red List category	NatureServe global (G) rank	Definitions
NOT THREATENED	Near Threatened (NT)[m]	Vulnerable (G3)[n]	(m) Does not meet criteria but is close to qualifying, or likely to qualify, for a threatened category in the near future (n) At moderate risk of extinction or elimination due to restricted range, relatively few populations, recent and widespread declines, or other factors
	Least Concern (LC)[o]	Apparently Secure (G4)[p] Secure (G5)[q]	(o) Does not meet criteria for threatened (e.g., widespread and abundant taxa) (p) Uncommon but not rare; some cause for long-term concern due to declines or other factors (q) Common: widespread and abundant
UNKNOWN	Data Deficient (DD)[r]	Unrankable (GU)[s]	(r) Inadequate information to make direct or indirect assessment of a species' extinction risk based on its distribution and/or population status (s) Currently unrankable due to lack of data or substantially conflicting information about status or trends
	Not Evaluated (NE)[t]	Unranked (GNR)[u]	(t) Not yet evaluated against the criteria (u) Global rank not yet assessed
	(No IUCN Red List equivalent)	Not Applicable (GNA)[v]	(v) Conservation status rank not applicable because species is not a suitable target for conservation activities

[*]Adapted from Westwood et al. (2017) and Frances et al. (2019), compiled by Oliver and Leaman (2018)

expert specialist groups, Red List authorities, and Red List partners (IUCN 2019a). It is considered to be the global standard for conservation status assessments. Red List assessors assign a category of extinction risk using a rule-based system of criteria and thresholds that accommodate both data-poor and data-rich species. Assessment categories and criteria are applied on a global scale (IUCN 2012a), with adjustments for national- and regional-scale applications (IUCN 2012b).

NatureServe's ranks and IUCN's Red List assessments consider similar factors related to each species' distribution, threats, and trends, but weigh and evaluate the data differently (Oliver and Leaman 2018) (Table 2.1). These differences produce complementary interpretations of extinction risk and provide users with nuanced perspectives to make informed conservation decisions. Recent publications have compared the two systems and provide a more detailed overview of the similarities and differences (Westwood et al. 2017; Frances et al. 2019). Both systems enable assessors to objectively apply rules for conservation status assessments, but also allow assessors to adjust the calculated rank or threat category.

A regional project is currently underway to produce global conservation status assessments for all North American medicinal plant species according to the IUCN

Red List categories and criteria. This project, led by the Medicinal Plant Specialist Group (MPSG) of the IUCN Species Survival Commission, is part of a global IUCN initiative – Plants for People – focused on conservation of plants important for health, food, and livelihoods (IUCN 2019b). The MPSG is a voluntary global network of experts working on conservation and sustainable use of medicinal plants.

The proportion of the world's medicinal flora for which the IUCN Red List status is known is small (IUCN 2019a), just 7% of the nearly 30,000 taxa included in the MAPROW database. In partnership with NatureServe and the New Mexico BioPark Society, where a North American "hub" for Red List assessment, assessment training, and conservation planning has recently been established, North American members of MPSG and other regional experts have begun to identify North American medicinal plant species with a high priority for new or updated Red List assessments and subsequent conservation planning and action. The IUCN Red List has search functions that will enable evaluation of the degree to which medicinal plants in North America are threatened with extinction compared with other regions of the world, which types of threat are most significant for medicinal plants, need for additional research and conservation action, and types of conservation action that will be most effective. The North American Plants for People initiative also supports updating NatureServe ranks for North American medicinal plants.

2.3.2 Health and Wellness Priorities

Before Europeans arrived, indigenous peoples in North America relied on medicines that could be found locally or in trade from other communities, derived mainly from plants (Moerman 2014). Moerman (2014) speculates that early health-care priorities were related to leading a "vigorous life": accidents resulting in "sprains, broken bones, cuts, and lacerations, and leading later to arthritic conditions; injuries during warfare, "problems associated with menstruation, pregnancy, childbirth, and lactation"; problems associated with "living in smoky houses", such as irritated eyes; and "the normal insults of daily life everywhere": colds, headaches, cold sores, and bruises. Since the late fifteenth century, indigenous systems of medicine have had to adapt to infectious diseases introduced by foreign explorers and settlers (plague, typhoid, smallpox cholera, etc.) to which there was no immunity, and to the health responses to changes in diet and environmental exposure that afflict all modern societies (carcinogens and other toxins, high fat-sugar-salt foods).

Searches for effective treatments for these modern health challenges have led to identification of components of North American medicinal plant species with new applications, such as Taxol from the bark of the Pacific Yew (*Taxus brevifolia*), effective in treating various cancers (National Cancer Institute 2019), and have also confirmed traditional knowledge, such as the use of Tamarack (*Larix laricina*) and other plant species amongst Canadian Cree communities to treat diabetes (Leduc et al. 2006; Fraser et al. 2007).

Between 1990 and 2009, the World Health Organization (WHO) published four volumes of *Monographs on Selected Medicinal Plants* in response to continuing popularity of traditional medicines in developing countries, "tremendous increase in the use of traditional medicine" throughout the world, and "urgent need" for information about safety and efficacy of globally important herbal medicines (WHO 1999, 2002, 2007, 2009). The 118 monographs included in these volumes refer to 25 plant species native to North America and 8 non-native species introduced to North America. These are indicated in Table 2.2, which lists some North American medicinal plant species of conservation interest.

2.3.3 Trade and Commerce Priorities

Approximately 3000 medicinal and aromatic plant species are found in international use and trade (Schippmann et al. 2006). The American Herbal Products Association includes nearly 2900 taxa in the second edition of "Herbs of Commerce" (McGuffin et al. 2000), of which approximately half are plant species native to North America.

Ten species of medicinal plants native to North America are included in Appendix II of the Convention on International Trade in Endangered Species of Fauna and Flora (CITES). These include American Ginseng (*Panax quinquefolius*), Goldenseal (*Hydrastis canadensis*), Venus' Flytrap (*Dionaea muscipula*) and all cactus and orchid species (Table 2.2). As signatories of this international convention, both the United States and Canada are required to determine whether international exports of bulk plant material or some processed products from these species are detrimental to survival of the species and to provide export permits only for shipments considered to be non-detrimental. Within the herbal products industry, inclusion of a plant species in CITES Appendix II is sometimes understood to be a determination that a species has been identified as threatened with extinction, and that international trade is considered the principal threat. This is not a correct interpretation of CITES Appendix II, however. Inclusion of any species is not a determination of conservation status or a ban on trade but reflects CITES members' evaluation that international trade of these species must be monitored and controlled to ensure that it does not contribute to the decline of wild populations (CITES 2019).

2.3.4 Species at Risk and to Watch

In 1997, the United Plant Savers (UpS), a not-for-profit organization dedicated to the conservation of native American medicinal plants (United Plant Savers 2019a), convened a group of "concerned herbalists and conservation-minded plant enthusiasts". This group proposed a list of "At-Risk" herbs in the United States to encourage practical and educational programs and to promote organic cultivation of these

Table 2.2 Some North American medicinal plant species of conservation interest

Species name	Common name	Family	Provenance	Conservation priority	NatureServe rank	IUCN Red List category
Actaea racemosa	Black Cohosh	Ranunculaceae	Native	1,3b	G4	NE
Althaea officinalis	Common Marsh-mallow	Malvaceae	Introduced	3b	GNR	NE
Arctostaphylos uva-ursi	Bearberry	Ericaceae	Native	3b	G5	NE
Centella asiatica	Asian Coinleaf	Apiaceae	Introduced	3a	G5	NE
Crataegus laevigata	Quick-set Hawthorn	Rosaceae	Introduced	3b	GNR	NE
Crataegus monogyna	English Hawthorn	Rosaceae	Introduced	3b	G5	NE
Cypripedium acaule	Pink Lady's-slipper	Orchidaceae	Native	1,2	G5	LC
Cypripedium arietinum	Ram's Head Lady's-slipper	Orchidaceae	Native	1,2	G3	NT
Cypripedium montanum	Mountain Lady's-slipper	Orchidaceae	Native	1,2	G4	VU
Cypripedium parviflorum	Small Yellow Lady's-slipper	Orchidaceae	Native	1,2	G5	LC
Cypripedium reginae	Showy Lady's-slipper	Orchidaceae	Native	1,2	G4G5	LC
Dionaea muscipula	Venus' Flytrap	Droseraceae	Native	1,2	G2	NE
Echinacea angustifolia	Narrowleaf Purple Coneflower	Asteraceae	Native	1,3a	G4	NE
Echinacea purpurea	Eastern Purple Coneflower	Asteraceae	Native	1,3a	G4	NE
Epipactis gigantea	Giant Helleborine	Orchidaceae	Native	1,2	G4	NE
Frangula purshiana	Cascara False Buckthorn	Rhamnaceae	Native	1,3b	G5	NE
Hamamelis virginiana	American Witch-hazel	Hamamelidaceae	Native	3b	G5	LC
Humulus lupulus	Common Hop	Cannabaceae	Native	3c	G5	NE
Hydrastis canadensis	Goldenseal	Ranunculaceae	Native	1,2,3c	G3G4	VU
Hypericum perforatum	Common St. John's-wort	Hypericaceae	Exotic	3b	GNR	NE
Lophophora williamsii	Peyote	Cactaceae	Native	1,2	G4	VU
Matricaria chamomilla	German Mayweed	Asteraceae	Exotic	3a	GNR	NE
Oenothera biennis	Common Evening-primrose	Onagraceae	Native	3b	G5	NE

Species name	Common name	Family	Provenance	Conservation priority	NatureServe rank	IUCN Red List category
Panax quinquefolius	American Ginseng	Araliaceae	Native	1,2,3b	G3G4	NE
Passiflora incarnata	Purple Passion-flower	Passifloraceae	Native/Exotic	3c	G5	NE
Polygala senega	Seneca Snakeroot	Polygalaceae	Native	3b	G4G5	NE
Sambucus nigra	European Black Elderberry	Viburnaceae	Native/Exotic	3b	G5T5	NE
Serenoa repens	Saw Palmetto	Arecaceae	Native	3b	G4G5	NE
Tanacetum parthenium	Common Feverfew	Asteraceae	Introduced	3b	GNR	NE
Taraxacum officinale	Common Dandelion	Asteraceae	Native/Exotic	3c	G5T5	NE
Urtica dioica	Stinging Nettle	Urticaceae	Native/Exotic	3b	G5	NE
Vaccinium macrocarpon	Large Cranberry	Ericaceae	Native/Exotic	3d	G5	NE
Valeriana officinalis	Common Valerian	Caprifoliaceae	Exotic	3a	GNR	NE

Key to conservation priority
1. United Plant Savers (2019c)
2. CITES (2019), Species+ (2019)
3. WHO (1999, 2002, 2007, 2009)

species (Cech 1997). Since then, UPS has added a "To-Watch" list of species that may be of conservation concern (United Plant Savers, 2019c). These lists currently include approximately 100 species of North American medicinal plants. Several of these species are also included in the WHO monographed species and in CITES Appendix II, as indicated in Table 2.2.

2.4 Conservation Action

2.4.1 Better Information

Determining or updating the conservation status of a species and taking appropriate action to ensure its survival require reliable and thorough information about the current and past distribution of the population, the size of subpopulations wherever it occurs, and an understanding of population trends. It also requires information about the species biology – its system and rate of reproduction, characteristics that make the species vulnerable or resistant to particular external events, such as drought, fire, extreme weather and temperature, or changes in quality and quantity of habitat. Information this comprehensive is rarely available for medicinal plant species, no matter how common or well-studied their use and efficacy. Although the IUCN Red List criteria (IUCN 2012b) are designed to provide robust assessments with limited available information, many species require additional and more current information for adequate assessments and conservation planning appropriate to the severity and type of threat to species survival.

2.4.2 Reducing Threats

Globally, although the greater *volume* of medicinal and aromatic plant material in trade may be sourced from cultivation (60–90%), the proportion of wild-harvested medicinal and aromatic plant species in trade is estimated to be 60–90% (Schippmann et al. 2006; Jenkins et al. 2018). This is likely the case for the North American medicinal flora. The American Herbal Products Association (AHPA) reported that, for 26 major herbal commodities derived from 22 species included in its survey of North American herbs in trade for the period 2006–2010, material from just four species was exclusively wild harvested, but cultivated sources were insignificant for the other 18 species (Dentali and Zimmermann 2012).

Cultivation is widely assumed to be the principal and straight-forward conservation action to reduce the pressure of wild harvest on medicinal plant species that are of conservation concern (Foster 1991). Bringing a wild plant species into cultivation can be a costly endeavour. The research required to cultivate Goldenseal (for example) was stimulated by the inclusion of this species in CITES Appendix II in 1997

(CITES 1997), and the associated costs to industry of applying for export permits. While exported Goldenseal has shifted from wild-harvest to cultivation (United States Fish and Wildlife Service 2012), much of the Goldenseal in domestic trade in the United States is from wild-harvested sources with uncertain sustainability (Oliver and Leaman 2018).

For some medicinal plant species, wild harvest may be sustainable. Sustainability of wild harvest is dependent not only on the ratio of harvest rate to regeneration rate of the harvested material, but also on the impact of harvest intensity and practices on the stability of the harvested population and the species itself (Peters 1994; Cunningham 2001; Peters 2018). For some commercially important North American medicinal plant species, wild harvest levels in the past are believed to have contributed to significant population declines and may be continuing to do so (Foster 1991).

Efforts of the organic sector to address sustainable wild harvest are insufficient to address the challenges of wild-harvested species (Brinckmann and Hughes 2010). The herbal industry in North America represented by the AHPA is responding to this challenge by promoting a set of voluntary principles for sustainable wild harvest of herbs (AHPA 2017). In October 2018, AHPA established a sustainability sub-committee of its botanical raw materials committee to "increase awareness of sustainability issues and best practices, provide a forum to discuss and develop industry-wide standards, and promote existing sustainability initiatives throughout the industry (AHPA 2018). Certification schemes that are designed particularly for wild-harvested (FairWild Foundation 2010) or forest-grown (United Plant Savers 2019b) plants in commercial trade are attracting the interest of companies and consumers concerned about the sustainability of wild-harvested medicinal and aromatic plants in North America (Brinckmann and Hughes 2010).

Degradation and loss of habitat is considered the principal historical threat to the Goldenseal population in the United States and Canada and may be the largest current threat to this species (Oliver and Leaman 2018), and to the majority of North American medicinal plant species included in ongoing assessments. As is the case in many countries, protection of habitat is distributed amongst national, state, and local authorities as well as private landowners in the United States and Canada. Land and other resource management objectives, such as timber extraction and recreation, often have priority over species and habitat protection. The likelihood that habitat management objectives incorporate species protection is greatly increased by knowledge of the conservation status of and threats to the species found there.

The impacts of climate change on survival of plant species, including medicinal plants, are increasingly apparent and understood to affect plant distributions, phenology, and community assemblages, as well as concentrations of the constituents that provide medicinal value, particularly for arctic and alpine species (Bauman et al. 2019).

2.5 Conclusion

While the North American native flora and its medicinal components are neither the most numerous nor the most diverse globally, the global and domestic importance of these species to health and economies must not be under-valued. The IUCN Plants for People initiative recognizes the value of medicinal and other useful plant species in even the least species-rich parts of the world. The North American medicinal Plants for People initiative brings together a dedicated volunteer network of experts (MPSG) with not-for-profit organizations committed to providing the best possible information about conservation status of this important resource. A deeper understanding of the major threats to these species, and the actions needed to support sustainable use, in situ and ex situ conservation, and research targeted to knowledge gaps, will strengthen the likelihood that these species will continue to be available to future generations.

References

AHPA (2017) Good agricultural and collection practices (GACP). American Herbal Products Association, Silver Spring. Available via: http://ahpa.org/Resources/GoodAgriculturalandCollectionPractices(GACP).aspx. Accessed 20 June 2018

AHPA (2018) AHPA launches sustainability subcommittee, formalizing ongoing efforts. American Herbal Products Association. http://www.ahpa.org/News/LatestNews/TabId/96/ArtMID/1179/ArticleID/1034/AHPA-launches-sustainability-subcommittee-formalizing-ongoing-efforts.aspx. Accessed 21 Aug 2019

Bauman H, Smith T, Yearsley C (2019) Plants in peril: climate crisis threatens medicinal and aromatic plants. HerbalGram 124:44–59

Brinckmann J, Hughes K (2010) Ethical trading and fair trade certification: the growing market for botanicals with ecological and social certification. HerbalGram 88:46–57

Cech RA (1997) The herbalists' United Plant Savers. HerbalGram 41:50

CITES (1997) Hydrastis canadensis. Inclusion in Appendix II. United States of America. Consideration of proposals for amendment of appendices I and II. Prop 10.73. Tenth meeting of the Conference of the Parties, Conven- tion on International Trade in Endangered Species of Fauna and Flora. Harare, Zimbabwe, June 9–20, 1997. Accessed at: https:www.cites.org/sites/default/files/eng/cop/10/prop/E-CoP10-P-73.pdf. Accessed 20 Mar 2018

CITES (2019) How CITES works. Convention on International Trade in Endangered Species of Fauna and Flora (CITES). https://www.cites.org/eng/disc/how.php. Accessed 21 Aug 2019

Cunningham AB (2001) Applied ethnobotany: people, wild plant use and conservation. Earthscan, London

Dentali S, Zimmermann M (2012) Tonnage surveys of selected North American wild-harvested plants, 2006–2010. American Herbal Products Association (AHPA), Silver Spring

FairWild Foundation (2010) FairWild Standard version 2.0. Available via https://www.fairwild.org/fairwild-standard-overview. Accessed 21 Aug 2019

Flora of North America (2019) Introduction. http://beta.floranorthamerica.org/Introduction. Accessed 17 Aug 2019

Foster S (1991) Harvesting medicinals from the wild: the need for scientific data on sustainable yields. HerbalGram 24:10–16

Frances A, Khoury C, Smith A (2019) Conservation status and threat assessments of crop wild relatives. In: Greene S, Williams K, Khoury C, Kantar MB, Marek L (eds) Crop wild relatives of North America. Springer, Dordrecht. (in press)

Fraser M-H, Cuerrier A, Haddad PS, Arnason JT, Owen PL, Johns T (2007) Medicinal plants of Cree communities (Quebec, Canada): antioxidant activity of plants used to treat type 2 diabetes symptoms. Can J Physiol Pharmacol 85(11):1200–1214

IUCN (2012a) IUCN Red List categories and criteria: version 3.1, second edition. International Union for Conservation of Nature (IUCN), Gland, Switzerland and Cambridge, UK. http://cmsdocs.s3.amazonaws.com/keydocuments/Categories_and_Criteria_en_web%2Bcover%2Bbckcover.pdf. Accessed 23 Mar 2018

IUCN (2012b) Guidelines for application of IUCN Red List criteria at regional and national levels, version 4.0. International Union for Conservation of Nature (IUCN), Gland, Switzerland and Cambridge, UK. https://www.iucnredlist.org/resources/regionalguidelines. Accessed 21 Aug 2019

IUCN (2019a) The IUCN Red List of Threatened Species. Version 2019-2. https://www.iucnredlist.org/about/background-history. Accessed 25 Nov 2019

IUCN (2019b) Plants for People. https://www.iucn.org/theme/species/our-work/iucn-red-list-threatened-species/plants-people. Accessed 21 Aug 2019

Jenkins M, Timoshyna A, Cornthwaite M (2018) Wild at home: exploring the global harvest, trade and use of wild plant ingredients. TRAFFIC International, Cambridge, UK. https://www.traffic.org/site/assets/files/7339/wild-at-home.pdf. Accessed 21 Aug 2019

Leduc C, Coonishish J, Haddad P, Cuerrier A (2006) Plants used by the Cree Nation of Eeyou Istchee (Quebec, Canada) for the treatment of diabetes: a novel approach in quantitative ethnobotany. J Ethnopharmacol 105(1–2):55–63

McGuffin M, Kartesz JT, Leung AY, Tucker AO (eds) (2000) Herbs of commerce, second edition. American Herbal Products Association, Silver Spring

Moerman DE (1977) American medical ethnobotany: a reference dictionary. Garland Publishing Company, New York

Moerman DE (1979) Symbols and selectivity: a statistical analysis of Native American medical ethnobotany. J Ethnopharmacol 1:111–119

Moerman DE (1986) Medicinal plants of Native America. Museum of Anthropology of the University of Michigan, Ann Arbor

Moerman DE (1991) The medicinal flora of Native North America: an analysis. J Ethnopharmacol 31:1–42

Moerman DE (2014) Ethnobotany in North America. In: Encyclopaedia of the history of science, technology, and medicine in non-western cultures. Springer, Dordrecht. https://doi.org/10.1007/978-94-007-3934-5_8580-2

National Cancer Institute (2019) Success story: Taxol® (NSC 125973). Developmental Therapeutics Program. https://dtp.cancer.gov/timeline/flash/success_stories/S2_taxol.htm. Accessed 21 Aug 2019

Native American Ethnobotany DB (2019.) Available via http://naeb.brit.org. Accessed 21 Aug 2019

Native American Ethnobotany: a database of plants used as drugs, foods, dyes, fibers, and more, by native Peoples of North America (2003) History. http://naeb.brit.org/about. Accessed 21 Aug 2019

NatureServe (2019) Our history. https://www.natureserve.org/about-us/our-history. Accessed 21 Aug 2019

NatureServe Explorer (2019) NatureServe, Arlington, Virginia. http://explorer.natureserve.org. Accessed 21 Aug 2019

Oliver LE, Leaman DJ (2018) Protecting Goldenseal: how status assessments inform conservation. HerbalGram 119:40–55

Peters CM (1994) Sustainable harvest of non-timber plant resources in tropical moist forest: an ecological primer. USAID Biodiversity Support Program

Peters CM (2018) Sustainable harvest of wild plant populations. HerbalGram 118:44–49

Schippmann U, Leaman D, Cunningham AB (2006) A comparison of cultivation and wild collection of medicinal and aromatic plants under sustainability aspects. In: Bogers RJ, Craker LE, Lange D (eds) Medicinal and aromatic plants. Springer, Dordrecht, pp 75–95

Schippmann U (2019) Medicinal and aromatic plants of the world (MAPROW) database. Not available online

Species+ (2019) UNEP-WCMC, Cambridge, UK. https://www.speciesplus.net. Accessed 21 Aug 2019

The IUCN Red List of Threatened Species (2019) Version 2019-2, ISSN 2307–8235. https://www.iucnredlist.org. Accessed 21 Aug 2019

Turi C (2017a) Assessment of USA and Canada distributions. Unpublished research

Turi C (2017b) Preliminary data analysis on the status of Canadian medicinal plants. Unpublished research

United Plant Savers (2019a). https://unitedplantsavers.org. Accessed 21 Aug 2019

United Plant Savers (2019b) Pennsylvania Certified Organic (PCO) is excited to announce the transfer of the PCO Forest Grown Verification Program to United Plant Savers. https://unitedplantsavers.org/pennsylvania-certified-organic-pco-is-excited-to-announce-the-transfer-of-the-pco-forest-grown-verification-program-to-united-plant-savers. Accessed 21 Aug 2019

United Plant Savers (2019c) Species At-Risk list. https://unitedplantsavers.org/species-at-risk-list. Accessed 21 Aug 2019

United States Fish and Wildlife Service (2012) Conference of the Parties to the Convention on International Trade in Endangered Species of Wild Fauna and Flora (CITES), Sixteenth Regular Meeting, Taxa Being Considered for Amendments to the CITES Appendices. Federal Register, the Daily Journal of the United States Government 11 April 2012, 77(70):21798–21802. Available via https://www.fws.gov/international/pdf/federal-register-notice-77-fr-21798-extended-version.pdf. Accessed 20 June 2018

Westwood M, Frances A, Man G, Pivorunas D, Potter KM (2017) Coordinating the IUCN Red List of North American tree species: a special session at the USFS gene conservation of tree species workshop. In: Sniezko RA, Man G, Hipkins V, Woeste K, Gwaze D, Kliejunas JT, McTeague BA (tech cords). Gene conservation of tree species—banking on the future. Proceedings of a workshop, Gen Tech Rep PNW-GTR-963. Department of Agriculture, Forest Service, Pacific Northwest Research Station, Portland, pp 12–23

WHO (1999) WHO monographs on selected medicinal plants, volume 1. World Health Organization, Geneva

WHO (2002) WHO monographs on selected medicinal plants, volume 2. World Health Organization, Geneva

WHO (2007) WHO monographs on selected medicinal plants, volume 3. World Health Organization, Geneva

WHO (2009) WHO monographs on selected medicinal plants, volume 4. World Health Organization, Geneva

World Flora Online (2019) World Flora Online Consortium. http://about.worldfloraonline.org. Accessed 21 Aug 2019

Chapter 3
Pharmacopoeial Wild Medicinal Plants of North America

Josef A. Brinckmann

Abstract In the time since the first publication of the *Pharmacopoeia of the United States of America* in 1820 and of the first edition of *Farmacopea Mexicana* in 1846, most pharmacopoeial articles of botanical origin, that are native North American species, have been, and continue to be, wild collected, to some extent or entirely, for commercial trade and therapeutic use.

While a much larger encyclopedial work could, and should, be prepared on wild collected medicinal and aromatic plants of North America, this chapter aims to consolidate essential information on 15 culturally, economically, functionally, and therapeutically important wild collected North American species utilized in the 23 countries and European overseas territories of North America.

For the first time, concise monographs are presented that juxtapose information published in quality and therapeutic compendia of Canada, the United Mexican States and the United States of America, such as current pharmacopoeial monographs and labeling standards monographs and associated authoritative or regulatory text. In some cases, *Farmacopeia Brasileira* and *Pharmacopée Française* are also referenced. Furthermore, the monographs in this chapter provide historical information on the compendial history of the article, standard common names used in the three main languages spoken in North America, English, French and Spanish, potential adulterants, and a summary of the conservation status of the species and whether cultivation is also occurring.

Keywords Medicinal and aromatic plants · Articles of botanical origin · Conservation · Herbal drugs · Officinal · Pharmacopoeia · Wild collection

J. A. Brinckmann (✉)
Traditional Medicinals, Sebastopol, CA, USA
e-mail: jbrinckmann@tradmed.com

© Springer Nature Switzerland AG 2020
Á. Máthé (ed.), *Medicinal and Aromatic Plants of North America*, Medicinal and Aromatic Plants of the World 6,
https://doi.org/10.1007/978-3-030-44930-8_3

45

Abbreviations

AHP	American Herbal Pharmacopoeia
CFR	U.S. Code of Federal Regulations
COFEPRIS	Comisión Federal para la Protección contra Riesgos Sanitarios
DER	Drug-to-extract ratio
DSP	dietary supplement product
FB	Farmacopeia Brasileira
FCC	Food Chemicals Codex
FDA	U.S. Food and Drug Administration
FHEUM	Farmacopea Herbolaria de los Estados Unidos Mexicanos
FR	U.S. Federal Register
GRASE	Generally Recognized as Safe and Effective
HMC	Herbal Medicines Compendium
IUCN	International Union for the Conservation of Nature
NF	U.S. National Formulary
NHP	Natural health product
NNHPD	Natural and Non-prescription Health Products Directorate, Health Canada
OTC	Over-the-counter drug product
PhEur	European Pharmacopoeia
PhFr	Pharmacopée française
USD	U.S. Dispensatory
USP	U.S. Pharmacopeia

3.1 Introduction

One cannot easily discuss the current uses of North American medicinal plants without first acknowledging that much of what we know today stems from thousands of years of use in indigenous systems of medicine during the Pre-Columbian era. As the settlements of European colonists expanded throughout the sixteenth to eighteenth centuries, North American botanical species and corresponding indigenous preparations and uses of the plant parts were observed, documented, classified in a European context, and, initially monographed in scientific publications and pharmacopoeias of the colonial powers. After the American War of Independence (1775–1783), much of the traditional ecological knowledge (TEK) and traditional medical knowledge (TMK) of the indigenous peoples was adopted into American medical practice and, by the early nineteenth century, codified in the first dispensatories, formularies and pharmacopoeias of the new country, the United States of America.

As was the case when the very first Pharmacopoeia of the United States of America was published in 1820 (United States Pharmacopoeial Convention 1820), most North American pharmacopoeial articles of botanical original were, and still are, for the most part, obtained by harvesting from wild populations in forests, meadows, mountains and deserts.

This chapter provides current information on 15 selected culturally, economically, functionally, and therapeutically important wild collected North American species that are monographed in authoritative and official compendia as well as in governmental rules and regulations of selected North American countries. Although there are 23 countries in North America, and some European overseas territories, the botanical entries in this chapter are articles of commerce that are included in various official compendia of Canada, the United Mexican States (Mexico) and the United States of America (U.S.). Sources of data include, for example, quality standards monographs of the Farmacopea Herbolaria de los Estados Unidos Mexicanos (FHEUM) and United States Pharmacopeia and National Formulary (USP-NF) and labeling standards monographs of Health Canada's Natural and Non-prescription Health Products Directorate (NNHPD) Compendium of Monographs and U.S. Food and Drug Administration (FDA) final monographs published in the Code of Federal Regulations (CFR) and tentative final monographs published in the Federal Register (FR).

In this consolidation of compendial data for 15 selected North American botanicals, it is worth noting that the very same herbal drug may be regulated differently in each country. In Canada, most of the articles described in this chapter are classified as medicinal ingredients for use in Natural Health Products (NHP) under Schedule 1 of the NHP regulations, requiring pre-marketing authorization and issuance of product license for over-the-counter (OTC) human use. As Canada does not have its own national pharmacopoeia, acceptable standards for establishing specifications for the quality of herbal drugs used in licensed NHPs include monographs of the United States Pharmacopeia and National Formulary (USP-NF) and European Pharmacopoeia (PhEur), among others (Natural and Non-Prescription Health Products Directorate 2015b). In Mexico, there are two categories of herbal medicinal products, both requiring pre-marketing authorization and issuance of a product registration, namely '*Medicamento Herbolario*' (herbal medicinal product with safety and efficacy supported by clinical and scientific data) and '*Remedio Herbolario*' (traditional herbal medicinal product). Specifications for the herbal drugs used in registered herbal medicinal products also need to conform with the quality standards established in the *Farmacopea Herbolaria de los Estados Unidos Mexicanos* (Comisión Permanente de la Farmacopea de los Estados Unidos Mexicanos 2013). In the U.S., many of the articles described in this chapter are classified as a generally recognized as safe and effective (GRASE) active ingredient for use in over-the-counter (OTC) drug products or in prescription only (Rx) medicines. Other botanicals, while regulated as medicinal substances in neighboring Canada and Mexico, are regulated as components of dietary supplement products (DSPs), requiring FDA notification and substantiation to support permissible nutrient content and/or structure/function claim statements. While use of official pharmacopoeial monographs as the basis of quality specifications for active ingredients of botanical drugs, OTC and prescription, is mandatory, USP standards are voluntary for specifications of botanical DSPs in the United States (Ma et al. 2018).

The 15 monographs in this chapter consolidate, for the first time, compendial information from Canada, Mexico and the United States, including not only descriptions of the pharmacopoeial article (identification, composition, quality) and preparations made from it, but also actions and indications for use, posology, ecology (habitat, distribution, conservation status), harvesting, production and trade scenarios, regulatory status and references to the governmental agencies, non-governmental organizations, scientific authorities and other related sources used for this work.

3.2   Selected Pharmacopoeial Articles

3.2.1   *Balsamum peruvianum*

Plant Source: *Myroxylon balsamum* var. *pereirae* (Royle) Harms (syn. *Myroxylon pereirae* Klotzsch), family *Fabaceae* (The Plant List (2013) Version 1.1).

Common Names: Eng: Balsam of Peru, Peru balsam, Peruvian balsam; Fr: Résine de baume Péruvien, Résine de Myroxylon pereirae (Baume péruvien); Port: Bálsamo do Peru; Span: Arbol del bálsamo, Bálsamo de El Salvador, Bálsamo de Perú, Bálsamo negro.

Compendial Monographs: Bálsamo de Perú FHEUM; Bálsamo do Peru FB.

Definition(s): The pharmacopoeial articles Bálsamo de Perú FHEUM and Bálsamo do Peru FB are both defined as the balsam obtained by contusion, incision or superficial burn of the cortex of *Myroxylon balsamum* (L.) Harms var. *pereirae* (Royle) Harms, containing not less than 45.0 per percent (m/m) and not more than 70.0% (m/m) of esters, mainly benzyl benzoate and benzyl cinnamate (Comisión Permanente de la Farmacopea de los Estados Unidos Mexicanos 2013; Comissão da Farmacopeia Brasileira 2010). Balsam Peru Oil FCC is defined as a yellow to pale brown, slightly viscous liquid having a sweet, balsamic odor, obtained by extraction or distillation of Peruvian Balsam obtained from *Myroxylon pereirae* Royle Klotzsche (United States Pharmacopeial Convention 2018a).

Compendial History: In 1820, this balsam was included in the primary list of Materia Medica of the first publication of the United States Pharmacopoeia (USP), with the officinal names Myroxylon (Latin) and Balsam of Peru (English), and botanical name *Myroxylon peruiferum* (United States Pharmacopoeial Convention 1820). In 1833, a monograph titled Myroxylon, U.S., Balsam of Peru was included in the first edition of the United States Dispensatory (USD) with the officinal synonym Balsamum Peruvianum (Wood and Bache 1833). In 1846, Bálsamo del Perú was included in the first edition of *Farmacopea Mexicana* (Academia Farmacéutica de la Capital de la República 1846; Schifter 2010). In 1888, Balsam of Peru was included as a component of several unofficinal preparations monographed in the first issue of the American Pharmaceutical Association's National Formulary (N.F. I) (Committee on National Formulary 1888).

Geographic Origin and Distribution: Native to Central America, occurring from southern Mexico into South America; introduced to Cuba (Vázquez-Yanes et al. 1999).

Conservation Status: This species has not yet been evaluated against the IUCN Red List categories and criteria.

Main Harvesting Areas: Peru Balsam is mainly produced in an area of approximately 400 km^2 in El Salvador (Berrera 2008), located between the port of La Libertad to the port of Acajutla (Panameño et al. 2011).

Material of Commerce: El Salvador is the main producer and exporter (Wichtl 2004). About 50% of all El Salvadoran production is exported to 2 countries, United States and Germany (Panameño et al. 2011).

Potential Adulterants: The Mexican and Brazilian pharmacopoeial monographs, respectively, include tests for determination of adulterants including artificial balsams, fatty oils and turpentine (Comisión Permanente de la Farmacopea de los Estados Unidos Mexicanos 2013; Comissão da Farmacopeia Brasileira 2010).

Herbal Drug Preparations: Balsam Peru Oil FCC (United States Pharmacopeial Convention 2018b). In Mexico, Bálsamo del Perú is a component of several officinal preparations including Linimento de Almendras dulces (Sweet almond liniment), Ungüento de Azufre y bálsamo de Perú (Sulfur and Peruvian balsam ointment), and Ungüento de Pomada de Poligonato (Poligonato ointment) (Comisión Permanente de la Farmacopea de los Estados Unidos Mexicanos 2017).

Actions and Indications for Use: In Canada, the distilled oil of Peru Balsam is permitted for use in licensed natural health products (NHPs) for aromatherapy (topical) application to help relieve cough and cold symptoms (Natural and Nonprescription Health Products Directorate 2018a). Extracts of Peru Balsam are also permitted for use as non-medicinal components of licensed NHPs for functions including film former, flavor enhancer, and fragrance ingredient. Crude exudates are not permitted for topical use. Extracts and distillates of Peru Balsam are permitted at concentrations equal to or less than 0.4%, in products intended for use on intact skin, in adults. Also permitted for use as a flavor enhancer at concentrations equal to or less than 0.0015%, in products intended for oral use, in adults (Natural and Nonprescription Health Products Directorate 2018b). In the U.S., oil of Peru Balsam, obtained by distillation or extraction is also classified as a generally recognized as safe (GRAS) food flavoring agent (United States Pharmacopeial Convention 2018b; US Food and Drug Administration 2017).

3.2.2 Cimicifugae racemosae rhizoma

Plant Source: *Actaea racemosa* L. (syn. *Cimicifuga racemosa* (L.) Nutt.), family *Ranunculaceae* (The Plant List (2013) Version 1.1).

Common Names: Eng: Black cohosh root; Fr: Actée noire, Actée à grappes, Actée à grappes noires; Span: Raíz de Cimicífuga racemosa, Raíz de culebra negra.

Compendial Monographs: Actée noire/Black Cohosh NNHPD, Black Cohosh USP, Cimicífuga Racemosa (Raíz de) FHEUM.

Definitions: The pharmacopoeial articles Black Cohosh USP and Raíz de Cimicífuga Racemosa FHEUM are both described as consisting of the dried rhizome and roots of *Actaea racemosa* L., containing not less than 0.4% of triterpene glycosides, calculated as 23-epi-26-deoxyactein ($C_{37}H_{56}O_{10}$; M_r 660.83) on the dried basis (United States Pharmacopeial Convention 2018a; Comisión Permanente de la Farmacopea de los Estados Unidos Mexicanos 2013).

Compendial History: In 1820, Cimicifugae Radix appeared on the secondary list of substances in the first publication of the *United States Pharmacopoeia* (USP), with the botanical name *Cimicifuga serpentaria* and common name black snake root (United States Pharmacopoeial Convention 1820). In 1833, a monograph titled Cimicifuga, U.S. Secondary, Black Snakeroot was included in the first edition of the United States Dispensatory (USD) with the botanical names *C. racemosa* and *C. serpentaria* (Wood and Bache 1833). In 1888, Syrupus Actaeae Compositus (Compound Syrup of Black Cohosh) was monographed as an unofficinal preparation in the first issue of the American Pharmaceutical Association's National Formulary (N.F. I) (Committee on National Formulary 1888).

Geographic Origin and Distribution: Native to the eastern United States, *Actaea racemosa* is found growing in rich soils on wooded hillsides, particularly in Appalachia, occurring in 25 U.S. states and part of one Canadian province, from Massachusetts and southern Ontario, west to Illinois and Missouri, south to Arkansas, central Alabama, through Georgia and South Carolina (NatureServe 2017; Lonner 2007).

Conservation Status: The NatureServe global conservation status rank for black cohosh is G4; Apparently Secure – meaning it is uncommon but not rare with some cause for long-term concern due to declines or other factors. While NatureServe assigns a national conservation status of N4, also apparently secure in the United

States, the status for Canada is N2; Imperiled – at high risk of extinction or elimination due to very restricted range, very few populations, steep declines, or other factors (NatureServe 2017). On the state level, other organizations list the species as endangered in Illinois and Massachusetts (Pengelly and Bennett 2012). The non-profit organization United Plant Savers (UpS) places black cohosh on its "Species At-Risk" list (United Plant Savers 2018).

Main Harvesting Areas: Kentucky, West Virginia, North Carolina, Missouri.

Material of Commerce: Most of the commercial supply of black cohosh rhizome (>98%) is obtained from wild collection in the United States (American Herbal Products Association 2007, 2012). There is a small amount of certified organic cultivation on farms in Maine, Missouri, New York, Oregon, Virginia, among other states.

Potential Adulterants: Known adulterants of black cohosh include other species in the genus *Actaea* imported from Asia, such as *Actaea cimicifuga* L. (syn: *Cimicifuga foetida* L.), *A. dahurica* (Turcz. ex Fisch. & C.A.Mey.) Franch. (syn: *C. dahurica* (Turcz.) Maxim.), *A. heracleifolia* (Kom.) J.Compton (syn: *C. heracleifolia* Kom.), *A. simplex* (DC.) Wormsk. ex Prantl (syn: *C. simplex* (DC.) Wormsk. ex Turcz.), and *A. brachycarpa* (P.K.Hsiao) J.Compton (syn: *C. brachycarpa* P.K.Hsiao) (Gafner 2016).

Herbal Drug Preparations: Black Cohosh Powder USP consists of Black Cohosh USP reduced to a powder or a very fine powder; Black Cohosh Dry Extract USP is prepared from Black Cohosh USP by extraction with hydroalcoholic mixtures or other suitable solvents; Black Cohosh Fluidextract USP is prepared from Black Cohosh USP by extraction with hydroalcoholic mixtures or isopropanol–water mixtures. Each mL contains the extracted constituents of 1 g of the plant material; Black Cohosh Tablets USP contain Black Cohosh Dry Extract USP (United States Pharmacopeial Convention 2018a).

Actions and Indications for Use: In Canada, licensed natural health products (NHPs) prepared from USP-quality black cohosh may be labeled with the following indications for use: "Used in Herbal Medicine to help relieve muscle and joint pain associated with rheumatic conditions (such as rheumatoid arthritis, osteoarthritis, and/or fibrositis), and of pain associated with neuralgia (such as sciatica);" "Used in Herbal Medicine to help relieve the pain associated with menstruation;" "Used in Herbal Medicine to help relieve premenstrual symptoms;" and "To help relieve symptoms associated with menopause" (Natural and Non-prescription Health Products Directorate 2008b). In Mexico, registered '*Medicamentos Herbolarios*' prepared from FHEUM-quality black cohosh may be labeled with therapeutic indications to the effect of "for treatment of symptoms related to menopause, such as hot flashes, sweating, palpitations, sleep disorders, anxiety, mood changes and nervousness" (Comisión Federal para la Protección contra Riesgos Sanitarios 2018).

3.2.3 Damianae folium

Plant Source: *Turnera diffusa* Willd. ex Schult., family Passifloraceae (formerly Turneraceae) (The Plant List (2013) Version 1.1).

Common Names: Eng: Damiana leaf; Fr: Feuille de damiane; Span: Hoja de damiana; Hierba de la pastora.

Compendial Monographs: Damiana (Hoja de) FHEUM.

Definition: The pharmacopoeial article Hoja de Damiana FHEUM consists of the dried leaves and stems of *Turnera diffusa* Willd (Fam. Turneraceae) (Comisión Permanente de la Farmacopea de los Estados Unidos Mexicanos 2013).

Compendial History: In 1846, Damiana was included in the first edition of *Farmacopea Mexicana*, albeit with the now antiquated botanical name of *Cineraria mexicana* (Academia Farmacéutica de la Capital de la República 1846; Schifter 2010).

Geographic Origin and Distribution: Arid and semi-arid forest regions of north-western Mexico (Gámez et al. 2010).

Conservation Status: This species has not yet been evaluated against the IUCN Red List categories and criteria.

Main Harvesting Areas: Damiana is harvested in 11 Mexican states. Damiana from the northwestern Mexican state of Baja California Sur is considered to be the highest quality due to a generally higher essential oil content compared to material from other states (Martínez 2013).

Material of Commerce: Almost all commercially traded damiana is obtained from wild collection in northwestern Mexico (Gámez et al. 2010).

Potential Adulterants: In the past, *Isocoma veneta* (Kunth) E. Greene, formerly known as *Aplopappus* or *Haplopappus discoideus* DC., was reported as an adulterant (Applequist 2006).

Herbal Drug Preparations: Elixir Turnerae, Extractum Turnerae Fluidum (Committee on National Formulary 1888), Turnerae folii extractrum fluidum, Turnerae folii extractum siccum, Turnerae folium pulveratum (Gámez et al. 2010).

Actions and Indications for Use: In Canada, damiana leaf, as an active ingredient of licensed natural health products (NHPs), prepared in forms of powdered dried leaf, fluidextract or tincture may be labeled with the following indications for use: "Traditionally used in Herbal Medicine as a nervine to help relieve nervousness and restlessness" (Natural and Non-prescription Health Products Directorate 2017a). In Mexico, there are products labeled for aphrodisiac and central nervous system stimulant effects (Gámez et al. 2010; Martínez 2013). In the United States, extract of damiana may be used as a natural flavoring substance (US Food and Drug Administration 2017), but damiana may also be used a component of dietary supplement products.

3.2.4   *Echinaceae angustifoliae radix*

Plant Source: *Echinacea angustifolia* DC., family Asteraceae (The Plant List (2013) Version 1.1).

Common Names: Eng: Echinacea angustifolia root, Narrow-leaved coneflower root; Fr: Racine d'échinacée angustifolia; Span: Raíz de equinácea angustifolia.

Compendial Monographs: Echinacea angustifolia USP, Equinácea Angustifolia, Raíz FHEUM.

Definition: Echinacea angustifolia USP consists of the dried rhizome and roots of *Echinacea angustifolia* DC. (Fam. Asteraceae), containing not less than 0.5% of total phenols, calculated on the dried basis as the sum of echinacoside ($C_{35}H_{46}O_{20}$), dicaffeoylquinic acid ($C_{25}H_{24}O_{12}$), and chlorogenic acid ($C_{16}H_{18}O_9$), and not less than 0.075% of dodecatetraenoic acid isobutylamides ($C_{16}H_{25}NO$) on the dried basis (United States Pharmacopeial Convention 2018a). Raíz de Equinácea Angustifolia FHEUM consists of the dried underground parts, whole or cut from *Echinacea angustifolia* DC. (Fam. Asteraceae), containing not less than 0.5% of echinacoside ($C_{35}H_{46}O_{20}$; M_r 786.5) with reference to the dried drug (Comisión Permanente de la Farmacopea de los Estados Unidos Mexicanos 2013).

Compendial History: Initially in 1916, Echinacea [defined as the dried rhizome and roots of *Brauneria pallida* (Nuttall) Britton (*Echinacea angustifoli*a De Candolle)] and the herbal drug preparation Fluidextractum Echinaceae were monographed in the fourth edition of the American Pharmaceutical Association's National Formulary (NF IV) (Committee on National Formulary 1916). Subsequently, in 1918, Echinacea NF IV was monographed in the twentieth edition of the Dispensatory of the United States of America (USD 20) (Remington and Wood 1918). While these early twentieth century monographs suggest that *E. angustifolia* was a synonym of *B. pallida*, today *B. pallida* is considered a synonym of *E. pallida*. Thus, it would appear that the pharmacopoeial article Echinacea NF IV permitted the use of rhizome and roots of *E. angustifolia* and/or *E. pallida*. In 2004, separate monographs for roots of *E. angustifolia*, *E. pallida*, and *E. purpurea* became official in the 27th revision of the United States Pharmacopoeia (USP 27) (United States Pharmacopeial Convention 2004).

Geographic Origin and Distribution: In Canada, *Echinacea angustifolia* occurs in the central provinces of Saskatchewan and Manitoba. In the United States, *E. angustifolia* occurs in the states of Montana, North Dakota, South Dakota, Minnesota, Wyoming, Nebraska, Iowa, Colorado, Kansas, Missouri, New Mexico, Texas, Oklahoma and Louisiana (NatureServe 2017).

Conservation Status: In Canada (provinces of Saskatchewan and Manitoba) the species is classified as vulnerable as well as in Iowa and Wyoming of the United States of America. It is probably extirpated in Missouri but apparently secure in Montana. Conservation status in the remaining areas of distribution are under review at the time of this writing and therefore not yet ranked (NatureServe 2017).

Main Harvesting Areas: Wild collection occurs within its range including, for example, Kansas. Commercial cultivation takes place on farms in the states of Oregon, Washington, Nebraska and North Carolina among other states, and Canadian province of Alberta, among other provinces.

Material of Commerce: The commercial supply is obtained from both wild collection and cultivation in the United States of America (American Herbal Products Association 2007, 2012), and Canada.

Potential Adulterants: Other *Echinacea* species, in particular *E. pallida* (Nutt.) Nutt. and *E. atrorubens* (Nutt.) Nutt. Historically, but not currently known to occur, the commercial supply had also been adulterated with roots of *Parthenium integrifolium* L., *Helianthus* spp., *Lespedeza capitata* Michx., *Rudbeckia nitida* Nutt., and *Eryngium aquaticum* L. (Upton 2010a).

Herbal Drug Preparations: Echinacea Angustifolia Powder USP consists of the dried rhizome and roots of Echinacea angustifolia USP reduced to powder. Echinacea Angustifolia Dry Extract USP is prepared from Echinacea Angustifolia USP by extraction with hydroalcoholic mixtures or other suitable solvents, with a drug-to-extract ratio range of 2:1–8:1, containing not less than 4.0% and not more than 5.0% of total phenols, calculated on the dried basis as the sum of chlorogenic acid, dicaffeoylquinic acids, and echinacoside, as well as containing not less than 0.1% of dodecatetraenoic acid isobutylamides on the dried basis (United States Pharmacopeial Convention 2018a).

Actions and Indications for Use: In Canada, licensed natural health products (NHPs) containing *Echinacea angustifolia* root and extracts of USP-quality as the active ingredients may be labeled with the indications for use statements "Traditionally used in Herbal Medicine to help relieve the symptoms of upper respiratory tract infections" and/or "Traditionally used in Herbal Medicine to help relieve sore throats" (Natural and Non-prescription Health Products Directorate 2017b). In Mexico, *Echinacea angustifolia* root is permitted as an active ingredient of '*Remedios Herbolarios*' (traditional herbal medicinal products) (Comisión Federal para la Protección contra Riesgos Sanitarios 2018). In the United States, *Echinacea angustifolia* root may be used as a component of dietary supplement products (DSPs).

3.2.5 Echinaceae pallidae radix

Plant Source: *Echinacea pallida* (Nutt.) Nutt., family Asteraceae (The Plant List (2013) Version 1.1).

Common Names: Eng: Echinacea pallida root, Pale coneflower root, Pale purple coneflower root; Fr: Racine d'Échinacée pallida, Échinacée pâle; Span: Raíz de Equinácea Pálida.

Compendial Monographs: Echinacea pallida USP, Equinácea Pálida, Raíz FHEUM.

Definitions: Echinacea pallida USP consists of the dried rhizome and roots of *Echinacea pallida* (Nutt.) Nutt. (Fam. Asteraceae), harvested in the fall after three or more years of growth, containing not less than 0.5% of total phenols, calculated on the dried basis as the sum of caftaric acid ($C_{13}H_{12}O_9$), chicoric acid ($C_{22}H_{18}O_{12}$), chlorogenic acid ($C_{16}H_{18}O_9$), and echinacoside ($C_{35}H_{46}O_{20}$) (United States Pharmacopeial Convention 2018a). Raíz de Equinácea Pálida FHEUM consists of the dried roots and rhizome of *Echinacea pallida* (Nutt.) Nutt. (Fam. Asteraceae), containing not less than 0.2% of echinacoside ($C_{35}H_{46}O_{20}$; M_r 786.5) calculated with reference to the dried drug (Comisión Permanente de la Farmacopea de los Estados Unidos Mexicanos 2013).

Compendial History: Initially in 1916, Echinacea [defined as the dried rhizome and roots of *Brauneria pallida* (Nuttall) Britton (*Echinacea angustifolia* De Candolle)] and the herbal drug preparation Fluidextractum Echinaceae were monographed in the fourth edition of the American Pharmaceutical Association's National Formulary (NF IV) (Committee on National Formulary 1916). Subsequently, in 1918, Echinacea NF IV was monographed in the twentieth edition of the Dispensatory of the United States of America (USD 20) (Remington and Wood 1918). While these early twentieth century monographs suggest that *E. angustifolia* was a synonym of *B. pallida*, today *B. pallida* is considered a synonym of *E. pallida*. Thus, it would appear that the pharmacopoeial article Echinacea NF IV permitted the use of rhizome and roots of *E. angustifolia* and/or *E. pallida*. In 2004, separate monographs for roots of *E. angustifolia*, *E. pallida*, and *E. purpurea* became official in the 27th revision of the United States Pharmacopoeia (USP 27) (United States Pharmacopeial Convention 2004).

Geographic Origin and Distribution: In Canada, *Echinacea pallida* occurs in the province of Ontario. In the United States, *E. pallida* occurs in Wisconsin, Nebraska, Iowa, Illinois, Indiana, Kansas, Missouri, Oklahoma, Arkansas, Tennessee, North Carolina, Alabama, Georgia, Louisiana and Texas. It also occurs as an exotic species in Michigan, New York, Connecticut, Rhode Island, Massachusetts, Maine, and Virginia (NatureServe 2017).

Conservation Status: The NatureServe global conservation status rank for *E. pallida* is G4 – Apparently Secure. However, according to NatureServe "Although still fairly abundant in parts of its range in the Great Plains and southern states, *Echinacea pallida* exhibits a declining trend over the past 30 years. It is threatened by root digging and excessive seed collection, as well as by impacts from road maintenance activities and urbanization in general." Its conservation status in Ontario, Nebraska, Tennessee, North Carolina and Georgia is critically imperiled, while imperiled in Alabama and vulnerable in Wisconsin (NatureServe 2017).

Main Harvesting Areas: Wild collection occurs within its range in prairies of the mid-western United States. There is some commercial cultivation, for example in Nebraska, among other states.

Material of Commerce: The commercial supply is obtained from both wild collection and cultivation in the United States of America (American Herbal Products Association 2007, 2012), and Canada.

Potential Adulterants: Mix-ups with other *Echinacea* species, in particular *E. angustifolia* DC. or *E. atrorubens* (Nutt.) Nutt. are possible because they are morphologically similar and because of the overlap in their range where wild collection takes place. It is also possible that commercial supplies of *E. pallida* may include hybrids with other species where wild collection takes place in areas where their ranges overlap, such as with *E. paradoxa* (Norton) Britton and/or *E. simulata* McGregor. Historically, but not currently known to occur, the commercial supply may also have been adulterated with roots of *Parthenium integrifolium* L. (Upton 2010b).

Herbal Drug Preparations: Echinacea Pallida Powder USP consists of Echinacea Pallida USP reduced to powder; Echinacea Pallida Dry Extract USP is prepared from Echinacea Pallida USP by extraction with hydroalcoholic mixtures or other suitable solvents, with a drug-to-extract ratio range of 2:1–8:1, containing not less than 4.0% and not more than 5.0% of total phenols, calculated as the sum of caftaric acid, chicoric acid, chlorogenic acid, and echinacoside, on the dried basis (United States Pharmacopeial Convention 2018a).

Actions and Indications for Use: In Canada, licensed natural health products (NHPs) containing *Echinacea pallida* root or extracts of USP-quality as the active ingredients may be labeled with the indications for use statements "Traditionally used in Herbal Medicine to help relieve cold symptoms," "Supportive therapy in the treatment of upper respiratory tract infections (e.g., common colds)" and "Helps to relieve the symptoms and shorten the duration of upper respiratory tract infections (e.g., common cold)" (Natural and Non-prescription Health Products Directorate 2017c). In Mexico, *E. pallida* root is permitted as an active ingredient of '*Remedios Herbolarios*' (traditional herbal medicinal products) (Comisión Federal para la Protección contra Riesgos Sanitarios 2018). In the United States, *E. pallida* root may be used as a component of dietary supplement products (DSPs).

3.2.6   *Eschscholziae herba*

Plant Source: *Eschscholzia californica* Cham., family Papaveraceae. There is one accepted subspecies, *Eschscholzia californica* subsp. *mexicana* (Greene) C.Clark. The subspecies '*californica*' is considered to be a synonym of *E. californica* (The Plant List (2013) Version 1.1).

Common Names: Eng: California poppy aerial parts; Fr: Pavot de Californie (parties aériennes fleuries d'); Span: Amapola de California, partes aéreas de.

Compendial Monographs: Eschscholziae herba PhFr.

Definition: Eschscholziae herba PhFr is defined as the dried flowering aerial parts of *Eschscholzia californica* Cham., containing not less than 0.50% and not more than 1.20% of total alkaloids, expressed as californidine ($C_{20}H_{20}NO^+_4$; M_r 338.4) (dried herbal substance) (Commission Nationale de Pharmacopée 2012).

Compendial History: Remarkably, California poppy has never been an officinal herbal drug of the Pharmacopoeia of the United States of America nor of the National Formulary. It is only briefly mentioned in early twentieth century editions of the Dispensatory of the United States (Remington and Wood 1918). There are, however, labeling standards monographs in force in Canada and France and a quality standards monograph published in *Pharmacopée française* (Commission Nationale de Pharmacopée 2012), which is among the accepted pharmacopoeias in Canada for the establishment of quality specifications for active ingredients of licensed Natural Health Products (NHPs) (Natural and Non-Prescription Health Products Directorate 2015b).

Geographic Origin and Distribution: Native to parts of northern Mexico (Baja Norte, Chihuahua, Sonora) south-central United States (New Mexico, Texas), southwestern United States (Arizona, California, Nevada, Utah) and northwestern United States (Oregon, Washington). The subspecies '*mexicana*' occurs mainly in Baja California Norte (Mexico), California (United States), Oregon and Washington (USDA Agricultural Research Service National Plant Germplasm System 2018).

Eschscholzia californica also occurs as an exotic species to the east of its native range, throughout much of Canada and the United States, apparently as an escape from cultivation, or due to "wildflower," roadside, or reclamation plantings (NatureServe 2017).

Conservation Status: The NatureServe global conservation status rank for California poppy is G4 – Apparently Secure, for reasons including the fact that the species is native in a moderately large range in the western U.S. states and north-western Mexico, that it has also become a widely cultivated species, frequently escaping cultivation and persisting outside of its native range through much of the rest of the U.S. and southern Canada. Threats in its native range reportedly include habitat alterations and genetic contamination of wild populations with genes from cultivars from roadside wildflower plantings (NatureServe 2017).

Main Harvesting Areas: While local small scale wild collection may still take place, for the commercial supply wild collection has been supplanted by cultivation. California Penal Code Section 384a requires written landowner permission to remove and sell California poppy material from land that a person does not own, However, California law does not prevent the collection of California poppy on private land by the landowner.

Material of Commerce: Mainly from cultivation in the United States and France. There is certified organic cultivation in California, Colorado, New Mexico, New York, North Carolina, Oregon, Vermont and Wisconsin (United States Department of Agriculture 2018).

Potential Adulterants: Not known to occur.

Herbal Drug Preparations: Infusion of Eschscholziae herba, Tincture of Eschscholziae herba.

Actions and Indications for Use: In Canada, California poppy flower, as an active ingredient of licensed natural health products (NHPs), prepared in forms of pow-dered herb, tea infusion or tincture may be labeled with the following indications for use: "Traditionally used in Herbal Medicine as a mild sedative and/or sleep aid (hypnotic)," and "Traditionally used in Herbal Medicine as an analgesic" (Natural and Non-prescription Health Products Directorate 2008a). Furthermore, if used as a sedative active ingredient of a licensed cognitive function NHP it may be labeled with the indications "Traditionally used in Herbal Medicine to help relieve nervous-ness (sedative and/or calmative)" or "Traditionally used in Herbal Medicine as a sleep aid (in cases of restlessness or insomnia due to mental stress)" (Natural and Non-prescription Health Products Directorate 2017a). In the United States, California poppy may be used as a component of dietary supplement prod-ucts (DSPs).

3.2.7 *Euphorbiae cera*

Plant Source: *Euphorbia antisyphilitica* Zucc., family Euphorbiaceae (The Plant List (2013) Version 1.1). In the past, plant species of the genus *Pedilanthus* were used as source plants for candelilla wax (Schneider 2009). See: Compendial history.

Common Names: Eng: Candelilla Wax; Fr: Cire de Euphorbia Cerifera (Candelilla); Span: Cera de candelilla.

Compendial Monographs: Candelilla Wax FCC, Candelilla Wax NF.

Definitions: Candelilla Wax FCC and Candelilla Wax NF are defined as a purified wax obtained from the leaves of the candelilla plant, *Euphorbia antisyphilitica* Zucc. (Fam. Euphorbiaceae), occurring as a hard, yellow-brown, opaque to translucent wax (United States Pharmacopeial Convention 2018a, b).

Compendial History: In 1874, a monograph for Candelilla first appeared in the first edition of the *Nueva Farmacopea Mexicana*, albeit with the Latin binomial *Pedilanthus pavonis* Boissier (Sociedad Farmacéutica de México 1874), which, today, is considered to be a synonym of *Euphorbia bracteata* Jacq. (The Plant List (2013) Version 1.1). The *Nueva Farmacopea Mexicana* monograph stated that the stems and leaves possessed emmenagogue and antisyphilitic actions.

Geographic Origin and Distribution: *Euphorbia antisyphilitica* occurs almost exclusively in the Chihuahuan Desert, a semi-arid region of North America, comprising parts of the Mexican states of Coahuila, Zacatecas, San Luis Potosi, Durango and Chihuahua, and extending north into the southern states of Texas, New Mexico and Arizona in the United States (Schneider 2009).

Conservation Status: All plant parts and derivatives of *Euphorbia antisyphilitica*, with the exception of finished products packaged and ready for retail trade, are protected under Appendix II of the Convention on International Trade in Endangered Species of Wild Flora and Fauna (CITES). CITES Appendix II lists species that are not necessarily now threatened with extinction but that may become so unless trade is closely controlled (Convention on International Trade in Endangered Species of Wild Flora and Fauna 2017).

Main Harvesting Areas: Wild collection takes place in the Mexican states of Durango, Zacatecas, Chihuahua, Nuevo León, San Luis Potosí, Tamaulipas, and Coahuila (Schneider 2009). The state of Coahuila accounts for about 80% of the total volume of candelilla wax traded globally (Candelilla Institute 2013).

Material of Commerce: Obtained entirely from wild collection in Mexico.

Potential Adulterants: With less expensive waxes such as rice bran wax or paraffin wax.

Herbal Drug Preparations: In pharmacy, candelilla wax is used mainly as a lubricant in tablet manufacturing but also as an excipient in coated tablets (Schneider 2009).

Actions and Indications for Use: Candelilla Wax FCC is used as a masticatory substance in chewing gum base and as a surface-finishing agent (United States Pharmacopeial Convention 2018b). In Canada, candelilla wax may be used in licensed Natural Health Products (NHPs) for purposes including cosmetic astringent (to induce a tightening or tingling sensation on skin), emulsion stabilizer, film former (materials which, upon drying, produce a continuous film), fragrance ingredient, masticatory substance (a substance chewed to increase salivation), polishing agent (substances used to impart an attractive sheen), skin-conditioning agent, emollient (ingredients which help to maintain the soft, smooth, and pliable appearance of skin), occlusive (ingredients which retard the evaporation of water from the skin surface), and viscosity increasing agent – nonaqueous (Substances used to thicken the lipid portions of products) (Natural and Non-prescription Health Products Directorate 2018b).

3.2.8 *Hamamelidis folium et cortex*

Plant Source: *Hamamelis virginiana* L., family Hamamelidaceae (The Plant List (2013) Version 1.1).

Common Names: Eng: Witch hazel, Hamamelis (leaf or bark); Fr: Hamamélis (feuille ou écorce), Hamamélis de Virginie; Span: Hamamelis.

Compendial Monographs: Hamamelis, Hoja FHEUM, Hamamelidis folium PhEur, Hamamelidis cortex PhEur.

Definitions: Hoja de Hamamelis FHEUM consists of dry, whole or fragmented leaf, of *Hamamelis virginiana* L. (Fam. Hamamelidaceae), containing not less than 3.0% of tannins, expressed as pyrogallol ($C_6H_6O_3$; M_r 126.1) calculated with reference to the dried drug (Comisión Permanente de la Farmacopea de los Estados Unidos Mexicanos 2013). It should be noted however that the current revision to the United States Pharmacopeia (USP 41) does not include monographs for the starting raw materials but only for the herbal drug preparation Witch Hazel USP, which is defined as a clear, colorless distillate prepared from recently cut and partially dried dormant twigs of *H. virginiana* L. (United States Pharmacopeial Convention 2018a). Therefore, of particular relevance to drug manufacturers in Canada are the raw material monographs of the European Pharmacopoeia (PhEur 9.6), Hamamelis Bark PhEur, the cut, dried bark from the trunk and branches of *H. virginiana* L., containing minimum 5.0% of tannins, expressed as pyrogallol, and Hamamelis Leaf PhEur, the whole or cut, dried leaf of *H. virginiana* L., containing minimum 3.0% of tannins, expressed as pyrogallol (European Pharmacopoeia Commission 2018).

Compendial History: In 1880, a monograph for witch hazel leaf (Hamamelis) and the fluidextract preparation made from it (Extractum Hamamelidis Fluidum) first appeared in the sixth decennial revision to the United States Pharmacopoeia (USP VI) (United States Pharmacopoeial Convention 1880). In 1888, Hamamelis Water (Aqua Hamamelidis) was monographed in the first issue of the American Pharmaceutical Association's National Formulary (N.F. I). The NF monograph noted that "This preparation should be made only from the fresh young twigs of *Hamamelis*, which are collected for this purpose preferably, when the plant is in flower, in the late autumn of the year" (Committee on National Formulary 1888). In 1900, monographs for witch hazel bark and twigs (Hamamelidis Cortex) and for the distillate made from the bark and twigs (Aqua Hamamelidis) entered the eighth decennial revision (USP VIII) (United States Pharmacopoeial Convention 1900).

Geographic Origin and Distribution: The tree or treelike shrub *Hamamelis virginiana* occurs in deciduous forests of the eastern half of Canada and the United States, from Nova Scotia to Wisconsin in the north and northern Florida to eastern Texas in the south (Engels and Brinckmann 2017).

Conservation Status: This species has not yet been evaluated against the IUCN Red List categories and criteria.

Main Harvesting Areas: New England region of the United States, especially Connecticut.

Material of Commerce: Historically, witch hazel raw materials have been obtained from wild collection near distillation facilities in the New England region (e.g., in Connecticut, Massachusetts, and Rhode Island), but also in Appalachia. There are certified organic wild collection operations in Connecticut, Kentucky, and Missouri. One of the largest harvesting operations is situated in Connecticut with 20,000 acres of wilderness area managed for certified organic wild harvesting of witch hazel (Engels and Brinckmann 2017).

Potential Adulterants: The Hamamelidis cortex PhEur monograph requires a test to rule out adulteration with bark and twigs of European hazel (*Corylus avellana* L., family Betulaceae) (European Pharmacopoeia Commission 2018).

Herbal Drug Preparations: Witch Hazel USP is a clear, colorless distillate prepared from recently cut and partially dried dormant twigs of *Hamamelis virginiana* L. (United States Pharmacopoeial Convention 2018a). Hamamelidis tinctura FB is obtained from witch hazel leaves containing minimum 0.6% of total tannins, expressed as pyrogallol ($C_6H_6O_3$, M_r 126.1), (w/w) (Comissão da Farmacopeia Brasileira 2010).

Actions and Indications for Use: In the United States, Witch Hazel USP is a Generally Recognized as Safe and Effective (GRASE) astringent active ingredient of hemorrhoid drug products. Permitted indications for use include statements to the effect of "for the temporary relief of anorectal itching and discomfort associated with hemorrhoids," "aids in protecting irritated anorectal areas," and "temporary relief of irritation or burning." Witch Hazel USP may also be used in combination with up to four anorectal protectant active ingredients, for example with Cocoa Butter NF (fat obtained from the seed of *Theobroma cacao* L.; Malvaceae) and/or with Topical Starch USP (granules separated from the mature grain of *Zea mays* L.; Poaceae), among others. FDA also permits Witch Hazel USP to be used in combination with any single analgesic, anesthetic, and antipruritic active ingredient such as with Camphor USP (ketone of *Cinnamomum camphora* (L.) J.Presl, Lauraceae) or with Juniper Tar USP (empyreumatic volatile oil obtained from the woody portions of *Juniperus oxycedrus* L., Cupressaceae). FDA also classifies Witch Hazel USP as a GRASE astringent active ingredient of skin protectant drug products indicated for the "relief of minor skin irritations due to insect bites, minor cuts, or minor scrapes" (US Food and Drug Administration 2017). Preparations of witch hazel are regulated similarly as OTC medicines in Canada and Mexico.

3.2.9   *Hintoniae latiflorae cortex*

Plant Source: *Hintonia latiflora* (Sessé & Moc. ex DC.) Bullock (The Plant List (2013) Version 1.1); formerly known as *Coutarea latiflora* Sessé & Moc. ex DC., *Coutarea pterosperma* (S.Watson) Standl., and *Portlandia pterosperma* S.Watson (Comisión Permanente de la Farmacopea de los Estados Unidos Mexicanos 2013).

Common Names: Eng: Copalchi bark; Fr: Écorce de Copalchi; Span: Corteza de copalchi.

Compendial Monographs: Copalchi, Corteza FHEUM.

Definition: Corteza de Copalchi FHEUM consists of the dried bark of *Hintonia latiflora* (Sessé & Moc. ex DC.) Bullock (Fam. Rubiaceae), containing not less than 6.4% of phenylcoumarin derivatives, expressed as 5-*O*-[β-D-apiofuranosyl-(1 → 6)-β-D-glucopyranosyl]-7-methoxy-3′,4′-dihydroxy-4-phenylcoumarin ($C_{27}H_{28}O_{15}$; M_r 592.50) and 5-*O*-[β-D-xylopyranosyl-(1 → 6)-β-D-glucopyranosyl]-7-methoxy-3′,4′-dihydroxy-4-phenylcoumarin ($C_{27}H_{30}O_{15}$; M_r 594.52), calculated with reference to the dried drug (Comisión Permanente de la Farmacopea de los Estados Unidos Mexicanos 2013).

Compendial History: In 1874, a monograph for Copalchi first appeared in the first edition of the *Nueva Farmacopea Mexicana*, albeit with taxonomic uncertainty. Several Latin binomials were included in the monograph including *Coutarea latiflora* DC., *Croton niveus* Jacq., *Croton pseudochina* Schltdl., and *Hedwigia balsamifera* (Sociedad Farmacéutica de México 1874).

Geographic Origin and Distribution: Tropical deciduous forest areas extending from parts of coastal northwestern Mexico to Central America including El Salvador and Guatemala (Beltrán-Rodríguez et al. 2015).

Conservation Status: The main threat to wild populations of *H. latiflora* has been harvesting intensity (overharvesting) and careless or destructive harvesting practices. There are now regulations in place to manage harvest and trade (Beltrán-Rodríguez et al. 2015).

Main Harvesting Areas: Northern part of Guerrero state in southwestern Mexico (Beltrán-Rodríguez et al. 2015).

Material of Commerce: The commercial supply is obtained from wild collection in Mexico.

Potential Adulterants: Possible with barks of other Mexican plants that are traded under the same vernacular name of Copalchi, including *Hintonia standleyana* Bullock and *Exostema caribaeum* (Jacq.) Schult., as well as the leaves of *H. latiflora* (Cristians et al. 2018).

Herbal Drug Preparations: Hintoniae latiflorae cortex extractum siccum.

Actions and Indications for Use: According to the Diabetes Canada Clinical Practice Guidelines Expert Committee, dry extract of *Hintonia latiflora* bark, a natural health product (NHP) for treatment of co-morbidities and complications of diabetes, has been shown to lower glycated hemoglobin (A1C) by at least 0.5% in adults with type 2 diabetes (Grossman et al. 2018). The expert committee cited a clinical study which concluded that there was a positive effect of a dry extract of *H. latiflora* bark (extraction solvent ethanol 32%; drug-to-extract-ratio (DER): 2.4:1) on blood glucose values, suggesting a potential benefit in the management of glucose metabolism in cases of type 2 diabetes (Korecova and Hladikova 2014). In the United States, the clinically studied extract is marketed as a dietary supplement product (DSP) for blood sugar balance and insulin function.

3.2.10 Hydrastis rhizoma

Plant Source: *Hydrastis canadensis* L., family Ranunculaceae (The Plant List (2013) Version 1.1).

Common Names: Eng: Goldenseal root and rhizome; Fr: Hydraste, Hydraste du Canada, Racine d'Hydrastis de Canada, Sceau d'or; Span: Sello de oro.

Compendial Monographs: Goldenseal USP.

Definition: Goldenseal USP consists of the dried roots and rhizomes of *Hydrastis canadensis* L. (Fam. Ranunculaceae), containing not less than 2.0% of hydrastine ($C_{21}H_{21}NO_6$) and not less than 2.5% of berberine ($C_{20}H_{18}NO_4$), calculated on the dried basis (United States Pharmacopeial Convention 2018a).

Compendial History: In 1830, Hydrastis Radix was included in the list of Materia Medica of the first decennial revision of the United States Pharmacopoeia (USP I) (United States Pharmacopoeial Convention 1830). In 1870, Hydrastis Radix was elevated from the secondary list to the primary list of Materia Medica and a new monograph for the liquid preparation Extractum Hydrastis Fluidum was added in the fifth decennial revision (USP V) (United States Pharmacopoeial Convention 1876). In 1888, Hydrastis Radix was included as a component of an unofficinal preparation monographed in the first issue of the American Pharmaceutical Association's National Formulary (N.F. I), Glyceritum Hydrastis (composed of powdered Hydrastis Radix, glycerin, alcohol and water) (Committee on National Formulary 1888).

Geographic Origin and Distribution: Native to Canada (province of Ontario only) and much of eastern half of the United States, goldenseal ranges from southern Vermont northward to Ontario, west to Minnesota and south to Georgia, Alabama, and Arkansas; common in Arkansas, Indiana, Illinois, Kentucky, Missouri, Ohio, and West Virginia (Oliver 2017).

Conservation Status: Underground parts (i.e. roots, rhizomes): whole, parts and powdered *Hydrastis canadensis* are protected under Appendix II of the Convention on International Trade in Endangered Species of Wild Flora and Fauna (CITES). CITES Appendix II lists species that are not necessarily now threatened with extinction but that may become so unless trade is closely controlled (Convention on International Trade in Endangered Species of Wild Flora and Fauna 2017). According to IUCN Red List categories and criteria, the conservation status of *Hydrastis canadensis* has been assessed as Vulnerable. According to the 2017 assessment report, "The species is considered Vulnerable based on a past and ongoing population reduction of at least 30% over three generations (approximately 21–27 years) inferred from past and ongoing declines in area of occupancy and habitat quality, and from documented levels of exploitation." (Oliver 2017). In Canada, for the purpose of marketing authorization of goldenseal medicinal products, the Natural and Non-prescription Health Products Directorate (NNHPD) states the following "The manufacturer must have proof of sale or a permit of the cultivated *Hydrastis canadensis* (Goldenseal) because the wild population is threatened: (i) *Hydrastis canadensis* is listed in Schedule 1 of the Species at Risk Act (SARA) as a "threatened" species and is afforded protection under this Act. Under section 32(2) of SARA, no person shall possess, collect, buy, sell or trade an individual of a wildlife species that is listed as an extirpated species, an endangered species or a threatened species, or any part or derivative of such an individual. Proof of purchase of cultivated *Hydrastis canadensis* is required. (ii) The Convention on International Trade in Endangered Species of Wild Fauna and Flora (CITES) sets controls on the movement of animal and plant species that are, or may be, "threatened" due to excessive commercial exploitation. To import *Hydrastis canadensis* (including the whole, part of, or the powdered root and/or rhizome) into Canada requires an accompanying CITES export permit from exporting countries" (Natural and Non-prescription Health Products Directorate 2010a).

Main Harvesting Areas: Wild collection within its range, including in the states of Missouri, Illinois, Indiana, Ohio, Kentucky, North Carolina, West Virginia, and Virginia, among others.

Material of Commerce: Most of the commercial supply of goldenseal is obtained from wild collection in the United States (American Herbal Products Association 2007, 2012). There is however commercial cultivation of goldenseal on farms in some states including Oregon, Washington, California, Georgia, Maine, Maryland, Michigan, Missouri, Pennsylvania, Wisconsin, and Vermont (United States Department of Agriculture 2018).

Potential Adulterants: Other plant parts with a yellowish color due to presence of berberine such as *Berberis* or *Mahonia* species, *Coptis groenlandica* L., *C. trifolia* (L.) Salisb., or *Xanthorhiza simplicissima* Marshall (Applequist 2006). According to the American Herbal Pharmacopoeia, common adulterants of goldenseal root include goldenseal leaf, Chinese goldthread (*Coptis* sp.), Oregon grape root (*Mahonia nervosa* (Pursh) Nutt.; syn. *Berberis nervosa* Pursh), Yellow dock root (*Rumex* sp.), and Yellow root (*Xanthorhiza simplicissima*) with leaves (Upton 2001).

Herbal Drug Preparations: Goldenseal Powder USP consists of Goldenseal USP reduced to a fine or very fine powder. Goldenseal Dry Extract USP is prepared from Goldenseal USP using suitable solvents, containing not less than 5% of hydrastine ($C_{21}H_{21}NO_6$) and not less than 10% of the sum of berberine ($C_{20}H_{18}NO_4$) and hydrastine, calculated on the dried basis (United States Pharmacopeial Convention 2018a).

Actions and Indications for Use: In Canada, licensed natural health products (NHPs) containing USP-quality goldenseal may be labeled with indications for use statements to the effect of "Traditionally used in Herbal Medicine to help alleviate infectious and inflammatory conditions of the digestive tract such as gastritis," and/ or "Traditionally used in Herbal Medicine as a bitter to aid digestion," and/or "Traditionally used in Herbal Medicine as a bitter to increase appetite," and/or "Traditionally used in Herbal Medicine to help relieve digestive disturbances such as dyspepsia," and/or "Traditionally used in Herbal Medicine as a mild laxative" (Natural and Non-prescription Health Products Directorate 2010a). In the United States, goldenseal products are regulated and labeled as dietary supplement products (DSPs).

3.2.11   Panacis quinquefolii radix

Plant Source: *Panax quinquefolius* L., family Araliaceae (The Plant List (2013) Version 1.1).

Common Names: Eng: American ginseng; Fr: Ginseng d'Amérique, Ginseng à cinq folioles; Span: Ginseng Americano.

Compendial Monographs: American Ginseng Root USP.

Definition: American Ginseng Root USP consists of the dried roots of *Panax quinquefolius* L. (Fam. Araliaceae), containing not less than 4.0% of total ginsenosides, calculated on the dried basis (United States Pharmacopeial Convention 2018a).

Compendial History: Possibly because the original trade route in the eighteenth to nineteenth centuries for wild collected American ginseng root was Québec to France, then on to China, one of the earliest pharmacopoeial listings appeared in the 1827 edition of "*Pharmacopée française, ou Code des médicamens*," albeit as "Panax de la Chine: *Panax quinquefolium* L., *Aureliana Canadensis*, *Iroquæis Garent-Oguen*, *Sinensibus ginseng*. Lajiteau" (Ratier 1827). In 1833, a monograph titled Panax Quinquefolium, U.S. – Ginseng was included in the first edition of the United States Dispensatory (USD) where it stated "Ginseng, though not included in any of the Pharmacopoeias, is deserving of a brief notice on account of its commercial importance, and from the circumstance that it is usually kept in the shops." Regarding medical uses, the USD stated "The extraordinary medical virtues formerly ascribed to ginseng, had no other existence than in the imaginations of the Chinese. It is little more than a demulcent; and in this country is not employed as a medicine" (Wood and Bache 1833). In 1840, Panax (described as the root of *Panax quinquefolium*) was added to the Secondary List of the second decennial revision of the Pharmacopoeia of the United States of America (USP II) (United States Pharmacopoeial Convention 1842). Secondary List drugs were articles deemed to be of secondary importance, not for use in official preparations.

Geographic Origin and Distribution: *Panax quinquefolius* occurs from the northeastern United States (Maine) west to adjacent provinces of Canada (Québec, Ontario and possibly Manitoba) and in the United States west to South Dakota, and south to Oklahoma, Louisiana, Alabama, Georgia, and South Carolina (NatureServe 2017; Upton 2012).

Conservation Status: Whole and sliced roots and parts of roots of *Panax quinquefolius*, excluding manufactured parts or derivatives, such as powders, pills, extracts, tonics, teas and confectionery, are protected under Appendix II of the Convention on International Trade in Endangered Species of Wild Flora and Fauna (CITES). CITES Appendix II lists species that are not necessarily now threatened with extinction but that may become so unless trade is closely controlled (Convention on International Trade in Endangered Species of Wild Flora and Fauna 2017). For the purpose of obtaining marketing authorization for medicinal products in Canada that contain American ginseng, the Natural and Non-prescription Health Products Directorate (NNHPD) states the following "Proof of sale or permit must be available for possession of the cultivated *Panax quinquefolius* because the wildlife species is endangered: (i) *Panax quinquefolius* is protected under the federal Species at Risk Act (SARA). Under section 32(2) of this Act, no person shall possess, collect, buy, sell or trade an individual of a wildlife species that is listed as an extirpated species, an endangered species or a threatened species, or any part or derivative of such an individual, (ii) *Panax quinquefolius* is protected by the Québec Act Respecting Threatened or Vulnerable Species. Under this Act, it is prohibited to possess, trade, or harm this species, or to disturb its habitat" (Natural and Non-prescription Health Products Directorate 2015a).

Main Harvesting Areas: About one-fourth of wild American ginseng root is harvested in the state of Kentucky, with smaller quantities wild collected in Tennessee, North Carolina, West Virginia, Indiana, Virginia, and Ohio. Wild collection of this species is not permitted in Canada (Upton 2012).

Material of Commerce: Both cultivated and wild collected. The commercial supply is obtained mainly from cultivation in Ontario, Canada and Wisconsin, United States. There is also some cultivation in the states of Maryland, Michigan, Minnesota, New York, Ohio, Pennsylvania, Vermont, and Washington (United States Department of Agriculture 2018). According to the American Herbal Pharmacopoeia, wild collected American ginseng roots account for less than 10% of total American ginseng root exports from the United States (Upton 2012). For the past 300 years, most American ginseng root, cultivated and wild, has been exported to China. There is also commercial American ginseng cultivation in three Chinese ecoregions, the Northeastern Provinces (Heilongjiang, Jilin, and Liaoning), Beijing, and Shandong (Huang et al. 2013).

Potential Adulterants: Powders may be adulterated with post-extracted spent marc or with flow agents or fillers such as dextrose or lactose. Extracts of American ginseng root may be adulterated with extractive of other ginsenoside-containing plant parts such as the leaves (Upton 2012).

Herbal Drug Preparations: American Ginseng Root Powder USP consists of American Ginseng Root USP reduced to a fine or a very fine powder. American Ginseng Root Dry Extract USP is prepared from American Ginseng Root USP using suitable solvents, dried to a powder, containing not less than 10.0% of total ginsenosides, calculated on the anhydrous basis, with a drug-to-extract ratio range of 3:1–7:1. American Ginseng Capsules USP and American Ginseng Tablets USP contain American Ginseng Root Dry Extract USP (United States Pharmacopeial Convention 2018a).

Actions and Indications for Use: In Canada, licensed natural health products (NHPs) composed of USP-quality American ginseng may be labeled with indications for use statements to the effect of "Used in Traditional Chinese Medicine (TCM) for deficiency of *qi* and *yin*, internal heat, cough, bloody phlegm, fire in the deficiency syndrome, dysphoria and tiredness, dry and thirsty mouth and throat," or "Used in Herbal Medicine as supportive therapy for the promotion of healthy glucose levels," or "Traditionally used in Herbal Medicine to help relieve nervousness (as mild sedative)," or "Traditionally used in Herbal Medicine to help relieve nervous dyspepsia/to help digestion in cases of nervousness and/or stress," or "Used in Herbal Medicine as an adaptogen to help maintain a healthy immune system" (Natural and Non-prescription Health Products Directorate 2015a). Additionally, American ginseng root is listed in the 'Cognitive Function Products' monograph as a sedative active ingredient (Natural and Non-prescription Health Products Directorate 2017a). In the United States, American ginseng products are regulated and labeled as dietary supplement products (DSPs).

3.2.12   Passiflorae herba

Plant Source: *Passiflora incarnata* L., family Passifloraceae.

Common Names: Eng: Passionflower herb; Fr: Passiflore (parties aériennes), Fleur de la passion; Span: Parte aérea de pasiflora.

Compendial Monographs: Pasiflora, parte aérea FHEUM.

Definition: Parte aérea de Pasiflora FHEUM consists of the dry aerial parts, fragmented or cut, of *Passiflora incarnata* L. (Fam. Passifloraceae), including flowers and/or fruits, containing not less than 1.5% of total flavonoids, expressed as vitexin ($C_{21}H_{20}O_{10}$; M_r 432.4) calculated with reference to the dried drug (Comisión Permanente de la Farmacopea de los Estados Unidos Mexicanos 2013). It should be noted that because there are no currently valid quality standards monographs for passionflower published in either the USP or NF, monographs of the European Pharmacopoeia (PhEur) are relevant for drug manufacturers in Canada, such as, in particular Passiflorae herba PhEur and Passiflorae herbae extractum siccum PhEur (European Pharmacopoeia Commission 2018).

Compendial History: In 1916, monographs for Passiflora (defined as the dried herbage of *Passiflora incarnata* Linné collected after some of the berries have matured) and the liquid preparation made from it (Tinctura Passiflorae) entered the fourth edition of the *National Formulary* (N.F. IV) (Committee on National Formulary 1916). Subsequently, in 1918, Passiflora NF IV was monographed in the twentieth edition of the Dispensatory of the United States of America (USD 20) (Remington and Wood 1918). Passiflora and Tinctura Passiflorae remained officinal through the fifth edition of the National Formulary (N.F. V) until 1936.

Geographic Origin and Distribution: *Passiflora incarnata* is native to the southeastern United States and is currently distributed from Florida west to Texas, north to southeastern Kansas, and east to Virginia (Engels and Brinckmann 2016).

Conservation Status: This species has not yet been evaluated against the IUCN Red List categories and criteria.

Main Harvesting Areas: Wild collection in southeastern United States.

Material of Commerce: While some passionflower is obtained from cultivation in the United States (Kentucky, Missouri, New York, North Carolina, Pennsylvania) and parts of southern Europe (France and Italy), much of the commercial supply of passionflower is still wild-collected in the southeastern United States (Engels and Brinckmann 2016).

Potential Adulterants: *Passiflora incarnata* is occasionally confused with other cultivated species of *Passiflora* (Applequist 2006). Because passionflower is a climbing vine, it has become problematic that highly invasive Asian climbing vines, including kudzu (*Pueraria montana* var. *lobata* (Willd.) Sanjappa & Pradeep, Fabaceae) and precatory (*Abrus precatorius* L., Fabaceae), now share habitat with *P. incarnata* in the southeastern United States. Other climbing vines may get tangled up with passionflower vines, making it difficult for harvesters to separate out plant parts of non-target species. Precatory vine plant parts have been implicated in the past as a toxic contaminant of wild collected passionflower (Engels and Brinckmann 2016). Machine harvested passionflower herb, as a cultivated crop in southern Europe, is listed by the European Medicines Agency (EMA) as one of ten medicinal herb crops at highest risk of pyrrolizidine alkaloid (PA) contamination from presence of PA-containing weeds in the fields (Committee on Herbal Medicinal Products (HMPC) 2016).

Herbal Drug Preparations: Passiflorae herbae extractum siccum PhEur, Tinctura Passiflorae NF V.

Actions and Indications for Use: In Canada, labels of licensed natural health products (NHPs) prepared from PhEur-quality passionflower may carry the claim statement "Traditionally used in Herbal Medicine as a sleep aid (in cases of restlessness or insomnia due to mental stress)" (Natural and Non-prescription Health Products Directorate 2008d). Additionally, passionflower is listed in the 'Cognitive Function Products' monograph as a sedative active ingredient "Traditionally used in Herbal Medicine to help relieve nervousness (sedative and/or calmative)" (Natural and Non-prescription Health Products Directorate 2017a). In Mexico, depending on the preparation and levels of evidence to support claims, submitted by the applicant, passionflower products may be regulated as either '*Medicamento Herbolario*' (herbal medicinal product with safety and efficacy supported by clinical and scientific data) or as '*Remedio Herbolario*' (traditional herbal medicinal product) (Comisión Federal para la Protección contra Riesgos Sanitarios 2018). In the United States, passionflower may be used as a natural flavoring substance, so long as the minimum quantity to produce the intended effect is used (US Food and Drug Administration 2017), and is also permitted for use as a component of dietary supplement products (DSPs).

3.2.13   Rhamni purshianae cortex

Plant Source: *Rhamnus purshiana* DC.; according to FHEUM, formerly known as *Frangula purshiana* (D.C.) A. Gray ex J.C. Cooper (Comisión Permanente de la Farmacopea de los Estados Unidos Mexicanos 2013). According to The Plant List, *Frangula purshiana* Cooper is an accepted name (The Plant List (2013) Version 1.1).

Common Names: Eng: Cascara sagrada bark; Fr: Écorce de cascara sagrada; Span: Corteza de cáscara sagrada.

Compendial Monographs: Cascara Sagrada Bark USP, Cáscara sagrada, Corteza FHEUM.

Definition: Cascara Sagrada Bark USP is the dried aged bark [collected not less than 1 year before use] of *Frangula purshiana* (DC.) A.Gray [syn. *Rhamnus purshiana* DC.] (Fam. Rhamnaceae), that yields not less than 7.0% of total hydroxyanthracene derivatives, calculated as cascaroside A, calculated on the dried basis. Not less than 60% of the total hydroxyanthracene derivatives consists of cascarosides, calculated as cascaroside A (United States Pharmacopeial Convention 2018a). Corteza de Cáscara Sagrada FHEUM is defined similarly, other than it is required to contain not less than 8.0% of total hydroxyanthracene derivatives, calculated as cascaroside A (Comisión Permanente de la Farmacopea de los Estados Unidos Mexicanos 2013).

Compendial History: In 1880, a monograph for cascara sagrada (described as the dried bark of *Rhamnus Purshiana* De Candolle (Fam. Rhamnaceae), collected at least one year before being used) was adopted in the sixth decennial revision to the United States Pharmacopoeia (USP VI) (United States Pharmacopoeial Convention 1880). In 1888, cascara sagrada was included as a component of several unofficinal preparations monographed in the first issue of the American Pharmaceutical Association's National Formulary (N.F. I) including Extractum Rhamni Purshianae Fluidum, Elixir Rhamni Purshianae (composed of Fluidextract of Rhamni Purshianae with Elixir of Glycyrrhiza and Compound Elixir of Taraxacum), and Elixir Rhamni Purshianae Compositum (composed of Fluidextract of Rhamni Purshianae with Fluidextract of Senna, Fluidextract of Juglans, Fluidextract of Glycyrrhiza, Compound Tincture of Cardamom, Aromatic Spirit, Purified Talcum and Water) (Committee on National Formulary 1888). In 1890, the liquid preparation Extractum Rhamni Purshianae Fluidum (Fluidextract of Cascara Sagrada) was adopted in the seventh decennial revision of the pharmacopoeia (USP VII) (United States Pharmacopoeial Convention 1890). Added to the eighth decennial revision (USP VIII) in 1900 were Extractum Rhamni Purshianae (Extract of Cascara Sagrada powder) and Fluidextractum Rhamni Purshianae Aromaticum (Aromatic Fluidextract of Cascara Sagrada) (United States Pharmacopoeial Convention 1900).

Geographic Origin and Distribution: This species is native to British Columbia, Canada and parts of northwestern United States (California, Idaho, Montana, Oregon, Washington). Occurring from British Columbia south to northern California, it is less common in northern Idaho and Montana (Stritch 2018a).

Conservation Status: According to IUCN Red List categories and criteria, the conservation status of *Frangula purshiana* has been assessed as LC (Least Concern). A taxon is Least Concern when it has been evaluated against the criteria and does not qualify for Critically Endangered, Endangered, Vulnerable or Near Threatened. Widespread and abundant taxa are included in this category (Stritch 2018a).

Main Harvesting Areas: Oregon, Washington, British Columbia.

Material of Commerce: The commercial supply of cascara sagrada bark is obtained almost entirely from wild collection in the Pacific Northwest (American Herbal Products Association 2007, 2012).

Potential Adulterants: Barks of similar trees including California coffeeberry (*Frangula californica* (Eschsch.) A. Gray; syn. *Rhamnus californica* Eschsch.) bark as well as bark of the stems and branches of European frangula (*Rhamnus frangula* L.; syn. *Frangula alnus* Miller) (Applequist 2006).

Herbal Drug Preparations: Casanthranol USP is obtained from Cascara Sagrada Bark USP, each 100 g containing not less than 20.0 g of total hydroxyanthracene derivatives calculated on the dried basis, calculated as cascaroside A. Not less than

80% of the total hydroxyanthracene derivatives consists of cascarosides, calculated as cascaroside A. Cascara Sagrada Dry Extract USP contains, in each 100 g, not less than 10.0 g and not more than 12.0 g of total hydroxyanthracene derivatives, of which not less than 50% consists of cascarosides, both calculated as cascaroside A. Cascara Sagrada Fluidextract USP is an aqueous extract with Alcohol USP 18.0–20.0% added after extraction, drug-to-extract-ratio of 1:1. Aromatic Cascara Fluidextract USP is prepared from a mixture of coarsely powdered Cascara Sagrada Bark USP, magnesium oxide, suitable sweetening agent(s), suitable essential oil(s), suitable flavoring agent(s), alcohol, and purified water, according to the instructions provided in the monograph. Cascara Tablets USP are prepared from Cascara Sagrada Dry Extract USP, containing not less than 9.35% and not more than 12.65% of total hydroxyanthracene derivatives, calculated as cascaroside A, in the labeled amount of Cascara Sagrada Extract. Not less than 50% of the hydroxyanthracene derivatives are cascarosides, calculated as cascaroside A (United States Pharmacopeial Convention 2018a).

Actions and Indications for Use: In Canada, licensed natural health products (NHPs) composed of USP-quality cascara sagrada may be labeled with indications for use to the effect of "Traditionally used in Herbal Medicine as a stimulant laxative," and "Used in Herbal Medicine for the short-term relief of occasional constipation," and/or "Used in Herbal Medicine to promote bowel movement by direct action on the large intestine" (Natural and Non-prescription Health Products Directorate 2008c). In the United States, cascara sagrada products are regulated and labeled as dietary supplement products (DSPs).

3.2.14   Sabalis serrulatae fructus

Plant Source: *Serenoa repens* (W.Bartram) Small, family Arecaceae (The Plant List (2013) Version 1.1).

Common Names: Eng: Saw palmetto berry; Fr: Fruit du palmier de Floride, Palmier nain, Sabal; Span: Fruto de Serenoa, Fruto de sabal.

Compendial Monographs: Saw Palmetto Fruit USP, Serenoa, Fruto FHEUM.

Definition: Saw Palmetto Fruit USP consists of partially dried, ripe fruit of *Serenoa repens* (W. Bartram) Small (Fam. Arecaceae) [*Serenoa serrulatum* Schult.; *Sabal serrulata* (Michx.) Nutt. ex Schult. & Schult. f.], containing not less than 2% (v/w) of volatile oil, not less than 7% of lipophilic extract, and not less than 9.0% of total fatty acids, determined on the dried basis (United States Pharmacopeial Convention 2018a). Fruto Serenoa FHEUM consists of the dried, mature fruit, containing not less than 11.0% of total fatty acids (Comisión Permanente de la Farmacopea de los Estados Unidos Mexicanos 2013).

Compendial History: In 1900, a monograph for Sabal (defined as the partially dried ripe fruit of *Serenoa serrulata*) entered the eighth decennial revision of the United States Pharmacopoeia (USP VIII) (United States Pharmacopoeial Convention 1900). In 1910, the preparation Fluidextractum Sabal (Fluidextract of Sabal) entered the subsequent ninth decennial revision (USP IX) (United States Pharmacopoeial Convention 1910). In 1926, Sabal and preparations made from it entered the fifth edition of The National Formulary (N.F. V) (Committee on National Formulary 1926).

Geographic Origin and Distribution: Southeastern United States including Florida, Georgia, South Carolina, Alabama, Mississippi and Louisiana.

Conservation Status: The NatureServe global conservation status rank for saw palmetto is G4; Apparently Secure – meaning that the species is at fairly low risk of extinction or elimination due to an extensive range and/or many populations or occurrences, but with possible cause for some concern as a result of local recent declines, threats, or other factors. While saw palmetto is ranked as Apparently Secure in the state of Florida, it is ranked as Critically Imperiled in Louisiana. It's status is under review and has not yet been ranked in the other states where it occurs (NatureServe 2017).

Main Harvesting Areas: Wild collected in the southeastern states, mainly Florida, but also Georgia and South Carolina to a much lesser extent.

Material of Commerce: Most of the commercial supply of saw palmetto berries (>98%) is obtained from wild collection in the United States (American Herbal Products Association 2007, 2012).

Potential Adulterants: Fruits of closely-related palm species, or dilution with exhaustively extracted saw palmetto berry powder, or use of unripe saw palmetto berries, or admixing of vegetable oils to saw palmetto berry extracts, and/or full substitution of saw palmetto extract with other vegetable oils. One study found a dietary supplement product (DSP), labeled as containing saw palmetto berry, to actually contain everglades palm, also known as silver saw palm (*Acoelorrhaphe wrightii* (Griseb. & H.Wendl.) H.Wendl. ex Becc., family Arecaceae). Extracts of saw palmetto berry have also been adulterated with other plant oils, such as canola (*Brassica napus* ssp. napus, Brassicaceae) oil, coconut (*Cocos nucifera*, Arecaceae) oil, olive (*Olea europaea*, Oleaceae) oil, African palm (*Elaeis guineensis*, Arecaceae) oil, peanut (*Arachis hypogaea*, Fabaceae) oil, and sunflower (*Helianthus annuus*, Asteraceae) oil (Gafner and Baggett 2017).

Herbal Drug Preparations: Saw Palmetto Fruit Powder USP consists of Saw Palmetto Fruit USP reduced to a fine or a very fine powder; Saw Palmetto Fruit Extract USP is obtained from comminuted Saw Palmetto Fruit USP by extraction with hydroalcoholic mixtures or solvent hexane, or by supercritical extraction with carbon dioxide, with a drug-to-extract ratio range of 8.0: 1 to 14.3: 1. The extract contains not less than 80.0% of fatty acids, not less than 0.2% of sterols, and not less than 0.1% of β-sitosterol, on the anhydrous basis. The lipophilic extract contains not less than 0.15% and not more than 0.35% of long-chain alcohols. The hydroalcoholic extract contains not less than 0.01% and not more than 0.15% of long-chain alcohols. It contains no added substances (United States Pharmacopeial Convention 2018a).

Actions and Indications for Use: In Canada, licensed natural health products (NHPs) composed of USP quality saw palmetto may be labeled with indications for use to the effect of "Used in Herbal Medicine to help relieve the urologic symptoms (e.g. weak urine flow, incomplete voiding, frequent daytime and night time urination) associated with mild to moderate benign prostatic hyperplasia" (Natural and Non-prescription Health Products Directorate 2010b). In Mexico, depending on the preparation and levels of evidence to support claims submitted by the applicant, products composed of FHEUM-quality saw palmetto may be registered as '*Medicamento Herbolario*' (herbal medicinal product with safety and efficacy supported by clinical and scientific data) (Comisión Federal para la Protección contra Riesgos Sanitarios 2018). In the United States, saw palmetto products are regulated and labeled as dietary supplement products (DSPs).

3.2.15   Ulmi rubrae cortex

Plant Source: *Ulmus rubra* Muhl., family Ulmaceae (The Plant List (2013) Version 1.1).

Common Names: Eng: Slippery elm inner bark; Fr: Écorce interne de l'orme rouge; Span: Corteza interna de olmo resbaladizo.

Compendial Monographs: Elm USP.

Definition: Elm USP is the dried inner bark of *Ulmus rubra* Muhl. (*Ulmus fulva* Michx.) (Fam. Ulmaceae), containing not-more-than 2% of adhering outer bark (United States Pharmacopeial Convention 2018a).

Compendial History: In 1820, Ulmus (the inner bark of slippery elm) was included in the primary list of Materia Medica of the first publication of the United States Pharmacopoeia (USP), as well as a monograph for the preparation Infusion of Slippery Elm (Infusum Ulmi USP); "*Take of Slippery elm, sliced, one ounce. Boiling water, one pint. Infuse for twelve hours in a covered vessel, near the fire with frequent agitation, and strain*" (United States Pharmacopeial Convention 1820). In 1833, a monograph titled Ulmus, U.S., Slippery Elm Bark was included in the first edition of the United States Dispensatory (USD) indicated for use as a demulcent. The first USD also included a monograph for Infusum Ulmi U.S. (Infusion of

Slippery Elm Bark) stating *"This infusion may be used ad libitum, as a demulcent and nutritious drink in catarrhal and nephritic diseases, and in inflammatory affections of the intestinal mucous membrane"* (Wood and Bache 1833). In 1916, a monograph for the preparation Trochisci Ulmi (Troches of Elm), made with slippery elm inner bark powder, tragacanth (*Astragalus gummifer* Labill., family Fabaceae) sugar, methyl salicylate, and water, entered the fourth edition of the *National Formulary* (N.F. IV) (Committee on National Formulary 1916).

Geographic Origin and Distribution: Native to parts of Canada (Ontario, Québec) and almost all of the eastern half of the United States, *Ulmus rubra* is distributed from "southwestern Maine west to extreme southern Quebec, southern Ontario, New York, northern Michigan, central Minnesota, eastern North Dakota; south through South Dakota, central Nebraska, southwestern Oklahoma, and central Texas; then east to northwestern Florida and Georgia" (Stritch 2018b).

Conservation Status: According to IUCN Red List categories and criteria, the conservation status of *Ulmus rubra* has been assessed as LC (Least Concern). A taxon is Least Concern when it has been evaluated against the criteria and does not qualify for Critically Endangered, Endangered, Vulnerable or Near Threatened. Widespread and abundant taxa are included in this category (Stritch 2018b).

Main Harvesting Areas: Slippery elm bark is harvested from wild populations in eastern Canada and the United States (United States Pharmacopeial Convention 2018a).

Material of Commerce: Obtained from wild collection mainly in the states of Missouri, Kentucky and West Virginia.

Potential Adulterants: The most common adulterant of slippery elm inner bark powder is outer bark which does not provide mucilage. The Elm USP monograph requires that it contain not-more-than 2% adhering outer bark. Historically, but not known to occur presently, adulteration with various starches, flours and/or powders of other non-mucilaginous barks reportedly occurred (Upton 2011).

Herbal Drug Preparations: Elm lozenges CFR, Elm and Menthol lozenges CFR.

Actions and Indications for Use: In the United States, Elm USP is a Generally Recognized as Safe and Effective (GRASE) oral demulcent active ingredient of OTC Oral Discomfort Drug Products. Permitted indications for use include statements to the effect of "For temporary relief of minor discomfort and protection of irritated areas in sore mouth and sore throat." Permitted combinations include Elm USP with an anesthetic/analgesic active ingredient such as Menthol USP (US Food and Drug Administration 1991). Preparations of slippery elm inner bark are regulated similarly as licensed natural health products (NHPs) in Canada.

References

Academia Farmacéutica de la Capital de la República (1846) Farmacopea Mexicana, 1ª edn. Imprenta de Manuel N. de la Vega, México

American Herbal Products Association (2007) Tonnage survey of select North American wild-harvested plants, 2004–2005. American Herbal Products Association, Silver Spring

American Herbal Products Association (2012) Tonnage surveys of select North American wild-harvested plants, 2006–2010. American Herbal Products Association, Silver Spring

Applequist W (2006) The identification of medicinal plants: a handbook of the morphology of botanicals in commerce. Missouri Botanical Garden Press, St. Louis

Beltrán-Rodríguez L, Romero-Manzanares A, Luna-Cavazos M, Vibrans H, Manzo-Ramos F, Cuevas-Sánchez J, García-Moya E (2015) Historia natural y cosecha de corteza de quina amarilla *Hintonia latiflora* (Rubiaceae). Bot Sci 93(2):261–272

Berrera M (2008) Ficha de Producto de El Salvador hacia el Mercado de la Unión Europea. Deutsche Gesellschaft für Internationale Zusammenarbeit (GIZ) GmbH, Bonn

Candelilla Institute (2013) Mexican researchers working on reinforced candelilla plants to ensure successful reforestation. Candelilla Institute Science Briefs (Art. 3.1)

Comisión Federal para la Protección contra Riesgos Sanitarios (2018) Listados de Registros Sanitarios de Medicamentos. Ciudad de México

Comisión Permanente de la Farmacopea de los Estados Unidos Mexicanos (2013) Farmacopea Herbolaria de los Estados Unidos Mexicanos (FHEUM), Segundo edición. Secretaría de Salud, México

Comisión Permanente de la Farmacopea de los Estados Unidos Mexicanos (2017) Consulta a Usuarios de la FEUM 2017-1 Medicamentos magistrales y oficinales. Secretaría de Salud, Ciudad de México, México

Comissão da Farmacopeia Brasileira (2010) Farmacopeia Brasileira 5ª edição. Agência Nacional de Vigilância Sanitária, Brasília

Commission Nationale de Pharmacopée (2012) Pharmacopée Française, 11th edn. Wolters Kluwer France, Rueil-Malmaison Cedex

Committee on Herbal Medicinal Products (HMPC) (2016) Public statement on contamination of herbal medicinal products/traditional herbal medicinal products with pyrrolizidine alkaloids. Transitional recommendations for risk management and quality control. European Medicines Agency, London

Committee on National Formulary (1888) The National Formulary of Unofficinal Preparations. First Issue. American Pharmaceutical Association, Washington, DC

Committee on National Formulary (1916) The National Formulary Fourth Edition. American Pharmaceutical Association, Washington, DC

Committee on National Formulary (1926) The National Formulary Fifth Edition. The American Pharmaceutical Association, Washington, DC

Convention on International Trade in Endangered Species of Wild Flora and Fauna (2017) Appendices I, II and III. CITES Secretariat, Geneva

Cristians S, Bye R, Nieto-Sotelo J (2018) Molecular markers associated with chemical analysis: a powerful tool for quality control assessment of Copalchi Medicinal Plant Complex. Front Pharmacol 9(666). https://doi.org/10.3389/fphar.2018.00666

Engels G, Brinckmann JA (2016) Passionflower: *Passiflora incarnata* L. Family: Passifloraceae HerbalGram (112):8–17

Engels G, Brinckmann JA (2017) Witch Hazel: *Hamamelis virginiana* Family: Hamamelidaceae HerbalGram (116):8–19

European Pharmacopoeia Commission (2018) European Pharmacopoeia, Ninth Edition, Supplement 9.6 (PhEur 9.6). European Directorate for the Quality of Medicines, Strasbourg

Gafner S (2016) Botanical adulterants bulletin on adulteration of *Actaea racemosa*. Botanical Adulterants Bulletin

Gafner S, Baggett S (2017) Botanical adulterants bulletin on saw palmetto (*Serenoa repens*) adulteration. Botanical Adulterants Bulletin

Gámez AE, Ivanova A, Martínez JA (2010) La comercialización mundial de damiana y los pequeños productores de Baja California Sur. Revista de Comercio Exterior 60(3):209–220

Grossman LD, Roscoe R, Shack AR (2018) Complementary and alternative medicine for diabetes. Can J Diabetes 42:S154–S161. https://doi.org/10.1016/j.jcjd.2017.10.023

Huang LF, Suo FM, Song JY, Wen MJ, Jia GL, Xie CX, Chen SL (2013) Quality variation and ecotype division of *Panax quinquefolium* in China [Article in Chinese]. Yao Xue Xue Bao 48(4):580–589

Korecova M, Hladikova M (2014) Treatment of mild and moderate type-2 diabetes: open prospective trial with *Hintonia latiflora* extract. Eur J Med Res 19(1):16. https://doi.org/10.1186/2047-783x-19-16

Lonner J (2007) Medicinal plant fact sheet: Cimicifuga racemosa/Black Cohosh. A collaboration of the IUCN Medicinal Plant Specialist Group, PCA-Medicinal Plant Working Group, and North American Pollinator Protection Campaign. PCA-Medicinal Plant Working Group, Arlington

Ma C, Oketch-Rabah H, Kim N-C, Monagas M, Bzhelyansky A, Sarma N, Giancaspro G (2018) Quality specifications for articles of botanical origin from the United States Pharmacopeia. Phytomedicine 45:105–119. https://doi.org/10.1016/j.phymed.2018.04.014

Martínez JA (2013) Evaluación económico-financiera de un plan de negocios para la damiana seca. Revista mexicana de ciencias forestales 4(16):86–100

Natural and Non-prescription Health Products Directorate (2008a) Monograph – California poppy. Health Canada, Ottawa

Natural and Non-prescription Health Products Directorate (2008b) Monograph: Black Cohosh. Health Canada, Ottawa

Natural and Non-prescription Health Products Directorate (2008c) Monograph: Cascara Sagrada. Health Canada, Ottawa

Natural and Non-prescription Health Products Directorate (2008d) Passionflower. Health Canada, Ottawa

Natural and Non-prescription Health Products Directorate (2010a) Monograph: Goldenseal – Oral. Health Canada, Ottawa

Natural and Non-prescription Health Products Directorate (2010b) Monograph: Saw Palmetto. Health Canada, Ottawa

Natural and Non-prescription Health Products Directorate (2015a) Monograph: Ginseng, American. Health Canada, Ottawa

Natural and Non-Prescription Health Products Directorate (2015b) Quality of natural health products guide, version 3.1. Health Canada, Ottawa

Natural and Non-prescription Health Products Directorate (2017a) Natural health product – cognitive function products. Health Canada, Ottawa

Natural and Non-prescription Health Products Directorate (2017b) Natural health product – Echinacea Angustifolia. Health Canada, Ottawa

Natural and Non-prescription Health Products Directorate (2017c) Natural health product – Echinacea Pallida. Health Canada, Ottawa

Natural and Non-prescription Health Products Directorate (2018a) Licensed Natural Health Products Database (LNHPD). Health Canada, Ottawa

Natural and Non-prescription Health Products Directorate (2018b) Natural health products ingredients database. Health Canada, Ottawa

NatureServe (2017) NatureServe Explorer: an online encyclopedia of life [web application]. Version 7.1. NatureServe, Arlington

Oliver L (2017) *Hydrastis canadensis.* The IUCN red list of threatened species 2017: e.T44340011A44340071. https://doi.org/10.2305/IUCN.UK.2017-2.RLTS. T44340011A44340071.en. Downloaded on 08 July 2018

Panameño DJM, Roldán MMP, Abrego AS (2011) Seminario de Trabajo de Investigación Tesina "Plan de Negocios para Bálsamo del Mojón, Municipio de Tepecoyo" Universidad Dr. José Matías Delgado, El Salvador

Pengelly A, Bennett K (2012) Appalachian plant monographs. Black cohosh *Actaea racemosa* L. Published online at http://www.frostburg.edu/aces/appalachian-plants/. Frostburg, MD

Ratier F-S (1827) Pharmacopee francaise, ou, Code des medicamens: nouvelle traduction de Codex medicamentarius, sive pharmacopoea gallica. Chez J.-B, Bailliere

Remington JP, Wood HC (1918) The dispensatory of the United States of America, 20th edn. J.B. Lippincott Co., Philadelphia/London

Schifter L (2010) La farmacopea mexicana. Guardiana de un patrimonio nacional viviente. Casa del Tiempo III (29):63–67

Schneider E (2009) Trade survey study on succulent *Euphorbia* species protected by CITES and used as cosmetic, food and medicine, with special focus on Candelilla Wax. Convention on Internatinoal Trade in Endangered Species of Wild Flora and Fauna, Eighteenth meeting of the Plants Committee Buenos Aires, Argentina

Sociedad Farmacéutica de México (1874) Nueva Farmacopea Mexicana. 1ª edn. Imprenta de Ignacio Escalante, Bajos de San Agustin, México

Stritch L (2018a) *Frangula purshiana*. The IUCN red list of threatened species 2018: e.T61957071A61957074.. Downloaded on 08 July 2018

Stritch L (2018b) *Ulmus rubra*. The IUCN red list of threatened species 2018: e.T61967382A61967384. Downloaded on 08 July 2018

The Plant List (2013) Version 1.1. Published on the Internet: http://www.theplantlist.org/

United Plant Savers (2018) Species at-risk. https://www.unitedplantsavers.org/species-at-risk. Accessed 21 Apr 2018

United States Department of Agriculture (2018) Organic INTEGRITY database. USDA Agricultural Marketing Service, Washington, DC

United States Pharmacopeial Convention (2004) The pharmacopeia of the United States of America, twenty-seventh revision and the National Formulary, twenty-second edition (USP 27–NF 22). United States Pharmacopeial Convention, Rockville

United States Pharmacopeial Convention (2018a) The pharmacopeia of the United States of America, forty-first revision and the National Formulary, thirty-sixth edition (USP 41–NF 36). United States Pharmacopeial Convention, Rockville

United States Pharmacopeial Convention (2018b) Food chemicals codex, eleventh edition (FCC 11). United States Pharmacopeial Convention, Rockville

United States Pharmacopoeial Convention (1820) The pharmacopoeia of the United States of America 1820. Charles Ewer, Boston

United States Pharmacopoeial Convention (1830) The pharmacopoeia of the United States of America 1830. S. Converse, New York

United States Pharmacopoeial Convention (1842) The pharmacopoeia of the United States of America, second decennial revision. Grigg & Elliot, Philadelphia

United States Pharmacopoeial Convention (1876) The pharmacopoeia of the United States of America, fifth decennial revision. J. B. Lippincott & Co., Philadelphia

United States Pharmacopoeial Convention (1880) The pharmacopoeia of the United States of America, sixth decennial revision. William Wood & Company, New York

United States Pharmacopoeial Convention (1890) The pharmacopoeia of the United States of America, seventh decennial revision. P. Blakiston's Son & Company, Philadelphia

United States Pharmacopoeial Convention (1900) The pharmacopoeia of the United States of America, eighth decennial revision. P. Blakiston's Son & Company, Philadelphia

United States Pharmacopoeial Convention (1910) The pharmacopoeia of the United States of America ninth decennial revision. P. Blakiston's Son & Company, Philadelphia

Upton R (ed) (2001) Goldenseal root – *Hydrastis canadensis*. In: American Herbal Pharmacopoeia® and Therapeutic Compendium. American Herbal Pharmacopoeia, Scotts Valley

Upton R (ed) (2010a) Echinacea Angustifolia root – *Echinacea angustifolia* DC. In: American Herbal Pharmacopoeia® and Therapeutic Compendium. American Herbal Pharmacopoeia, Scotts Valley

Upton R (ed) (2010b) Echinacea Pallida root – *Echinacea pallida* (Nutt.) Nutt. In: American Herbal Pharmacopoeia® and Therapeutic Compendium. American Herbal Pharmacopoeia, Scotts Valley

Upton R (ed) (2011) Slippery elm inner bark – *Ulmus rubra* Muhl. In: American Herbal Pharmacopoeia® and Therapeutic Compendium. American Herbal Pharmacopoeia, Scotts Valley

Upton R (ed) (2012) American ginseng root – *Panax quinquefolius* L. In: American Herbal Pharmacopoeia® and Therapeutic Compendium. American Herbal Pharmacopoeia, Scotts Valley

US Food and Drug Administration (1991) Oral health care drug products for over-the-counter human use; amendment to tentative final monograph to include OTC relief of oral discomfort drug products. Fed Regist 56(185):48302–48347

US Food and Drug Administration (2017) Code of Federal Regulations. National Archives and Records Administration, Washington, DC

USDA Agricultural Research Service National Plant Germplasm System (2018) Germplasm Resources Information Network (GRIN-Taxonomy). National Germplasm Resources Laboratory, Beltsville

Vázquez-Yanes C, Batis Muñoz AI, Alcocer Silva MI, Gual Díaz M, Sánchez Dirzo C (1999) Árboles y arbustos potencialmente valiosos para la restauración ecológica y la reforestación. Reporte técnico del proyecto J084. CONABIO – Instituto de Ecología, UNAM

Wichtl M (ed) (2004) Herbal drugs and phytopharmaceuticals: a handbook for practice on a scientific basis third edition (trans: Brinckmann JA, Lindenmaier MP). Medpharm Scientific Publishers, Stuttgart

Wood GB, Bache F (1833) The dispensatory of the United States of America. Grigg & Elliot, Philadelphia

Chapter 4
American Cranberry (*Vaccinium macrocarpon* Ait.) and the Maintenance of Urinary Tract Health

Thomas Brendler and Amy Howell

Abstract

The juice of the fruit of the American cranberry (*Vaccinium macrocarpon* Ait.) is by far the most popular and widely used botanical preparation for the prevention and treatment of urinary tract infections. To date, most of the medical research on cranberry consists of observational clinical trials yielding mixed results or preclinical studies, the latter focusing on mechanisms of action, many of which are relevant to the putative benefits attributed to cranberry. Cranberry benefits are primarily linked to the presence of proanthocyanidin (PAC) oligomers, also referred to as condensed tannins or polyflavan-3-ols, and their capacity to prevent bacteria, particularly *E. coli*, from adhering to uroepithelial cells. In the following, we present a summary of the state of cranberry research in the context of urinary tract health with focus on data derived from clinical settings. The overall consensus from recent reviews and meta-analyses is that there are beneficial effects of cranberry consumption on reducing risk of UTI recurrence.

This text has been adapted and updated with kind permission from a recently published revision of the American Herbal Pharmacopoeia's cranberry monograph (Upton & Brendler 2016).

T. Brendler (✉)
Plantaphile, Collingswood, NJ, USA

Department of Botany and Plant Biotechnology, University of Johannesburg, Johannesburg, South Africa

Traditional Medicinals Inc, Rohnert Park, CA, USA
e-mail: txb@plantaphile.eu

A. Howell
PE Marucci Center for Blueberry and Cranberry Research, Rutgers University, Chatsworth, NJ, USA

© Springer Nature Switzerland AG 2020
Á. Máthé (ed.), *Medicinal and Aromatic Plants of North America*, Medicinal and Aromatic Plants of the World 6,
https://doi.org/10.1007/978-3-030-44930-8_4

81

Keywords Cranberry · Urinary tract infection · Proanthocyanidins · Anti-adhesion
Prevention · Antibiotics · Clinical trials · Vulnerable populations

4.1 Introduction

The fruit, or more specifically the juice of cranberry is by far the most popular and
widely used botanical preparation for the prevention and treatment of urinary tract
infections. Cranberry juice is popular among health practitioners, is given freely to
residents of nursing care facilities, and is one of the most commonly used home
remedies among consumers.

The genus name of cranberry *Vaccinium* is believed to derive from the Latin vac-
cinus, meaning dun-colored, while the species name is from the Greek *makro*, or
large, and *karpós*, meaning fruit. The common name cranberry is thought to be a
derivation of the early name craneberry, which may be due to the stamens resem-
bling the beak of a crane or to the reported fondness of marsh cranes for the berry.
Writings of the middle 1600s variably use the names cranberry, cramberry, and
craneberry.

4.2 Traditional Uses and Early Research

Indigenous to North America, various species of *Vaccinium*, including *Vaccinium
macrocarpon*, *V. oxycoccos*, and *V. vitis-idaea* were consumed widely by numerous Native American tribes living in northern regions for their tart flavor and
nutritive value. Cranberry's popular use as a food, especially as a sauce — considered integral to a uniquely North American holiday, Thanksgiving — is well
known. According to the oral history of the Wampanoag of Cape Cod,
Massachusetts, the originating location of the first colonial Thanksgiving, cranberries (sasumuneash or ibimi) were cooked in stews. In his detailed history of
cranberry, American Cranberry (1931), Eck reports that Captain John Smith
(1580–1631), a seminal figure in the founding of America's first colony,
Jamestown, Virginia, may have been the first European to write about the cranberry in 1614: "The Herbes and Fruits are of many sorts and kinds: as Alkermes,
currants, mulberries... Of certain red berries, called Kermes..." According to
Eck, early Colonists found cranberry growing profusely on peat bogs off Cape
Cod, and in marshes along the Sudbury, Concord, Charles, and Neponset rivers in
Massachusetts, in the marshes of Barnegat Bay and the swamps of Great Pine
Barrens (now known as the Pinelands) in New Jersey, northward into the Maritime
provinces of Canada, as far south into the Carolinas, occurring sporadically in the
Allegheny Mountains and from Southern Pennsylvania to peat swamps of Virginia.
As Europeans migrated west, they found wild cranberries growing in the wetlands
of Indiana, Michigan, Wisconsin, and Minnesota. Wild cranberries also grow into

the boreal forests of Canada and west to the Haida Gwaii islands and Alaska. In 1789, the fruit was so in demand that over-harvesting by early European immigrants forced the New Jersey legislature to restrict the harvesting of unripe berries (Eck 1931).

Cranberry does not appear in the American Materia Medicas of the 1700s. It is first referenced in the first herbal produced and printed in America, the American Herbal of Samuel Stearns (1801). Stearns credits this fruit as being anti-scorbutic, "good in the sarcy and similar complaints." He goes on to say; "they are much used at the table and when eaten freely prove laxative," and also records cranberry's use for fevers. John Monroe in his The American Botanist and Family Physician (1824) records cranberry as "a very wholesome and agreeable tart, which is good in fevers, and helps the appetite." The renowned naturalist Constantine Rafinesque mentions cranberry, as *Oxycoca macrocarpa*, in his Medical Flora (1830), citing its use as a food in tarts and preserves, as well as medicinally as a mild laxative, refrigerant, diuretic, anti-pyretic, and anti-scorbutic. Henry Hollembaek (1865) appears to be the first Eclectic practitioner to include cranberry, under the name of *Oxycoca macrocarpa*, in his writings. Felter and Lloyd (1905), in their King's American Dispensatory, note the use of the fruit in domestic practice as a poultice for erysipelas, inflammatory swellings, swollen glands, indolent and malignant ulcers, tonsillitis, and for boils on the tip of the nose. Despite its medicinal use by Native Americans, European immigrants, and Eclectics, cranberry did not immediately find its way into mainstream medical practice as it was not included in early editions of the United States Pharmacopeia, National Formulary, or United States Dispensatory.

Early research (e.g., Blatherwick 1914; Blatherwick and Long 1923; Fellers et al. 1933) reported on the effect of cranberries on increasing urinary acidity. In 1931, researchers isolated an anthocyanin from cranberry, naming it "oxycoccicyanin," after the then Latin name Oxycoccus macrocarpus. This provided one of the earliest references postulating the presence of anthocyanins in cranberry (Grove and Robinson 1931).

Numerous studies from 1959 to the 1980s continued to support the health benefits of cranberry juice, primarily for the urinary tract, as well as for other indications. One of the earliest formal investigations of the antibacterial activity occurred in 1959 (Bodel et al. 1959). These researchers suggested that hippuric acid acidifying the urine causing a bacteriostatic effect was the mechanism behind the ongoing folkloric use of cranberries in UTIs and reported on the successful prophylactic treatment of chronic pyelonephritis.

In subsequent studies, focus was placed on the antibacterial effects of cranberry in relationship to urinary tract health. A variety of mechanisms were reported, including the ability of cranberry to decrease urinary pH, which both increased the efficacy of other antibacterial agents (Brumfitt and Percival 1962) and was beneficial in preventing and treating some renal problems (Sternlieb 1963); inhibition of growth of *E. coli* (Kraemer 1964); antifungal activity (Swartz and Medrek 1968; Ujvary et al. 1961); reduction of urinary ionized calcium in patients with kidney stones (Light et al. 1973); and antiviral activity (Borukh et al. 1972; Konowalchuk and Speirs 1978; Ibragimov and Kazanskaia 1981). While many of these reports lacked the methodological strength of formal modern clinical studies, they clearly suggest a trend for benefit and clinical relevance.

4.3 Cranberry in Modern Medicine

The primary health-promoting benefit for which cranberry consumption has been used is to maintain urinary tract health. Several clinical trials, including meta-analyses, support its use to prevent urinary tract infections (UTIs). Modern herbal practitioners and consumers similarly use cranberry, predominantly as juice products or dietary supplements for UTI prevention and treatment, although data are lacking on its effectiveness for treatment. Additional work has investigated the use of cranberry as an antioxidant, antiviral, anticancer, anticariogenic, anti-ulcerogenic, cholesterol-lowering, and vasorelaxant agent. To date, most of medical research on cranberry consists of observational clinical trials yielding mixed results or preclinical studies, the latter focusing on mechanisms of action, many of which are relevant to the putative benefits attributed to cranberry.

Cranberry ("juice preparation") was included in the 19th edition of the United States Pharmacopeia-National Formulary (USP 24-NF 19, 1999) and a suite of new and revised cranberry and cranberry product monographs has recently been published in the Pharmaceutical Forum (USP-NF PF 2019) for public comment. The American Herbal Pharmacopoeia developed a Cranberry Fruit Monograph and Therapeutic Compendium in 2002, which received an update in 2016 (Upton and Brendler 2016).

Over the years there have been many systematic and critical reviews of the clinical trials, with the most recent ones concluding that there is benefit to consuming cranberry for prevention of UTIs in most populations (Chen et al. 2019; Mantzorou and Giaginis 2018; Luís et al. 2017; Fu et al. 2017). Very recently, a panel of urologists from the American Urological Association (AUA)/Canadian Urological Association (CUA)/Society of Urodynamics, and Female Pelvic Medicine & Urogenital Reconstruction (SUFU) reviewed the results of over 200 cranberry studies including clinical trials conducted prior to 2018. As a result of this review, the publication "Recurrent Uncomplicated Urinary Tract Infection in Women: AUA/CUA/SUFU Guideline (2019)" now includes cranberry as a prophylaxis for women for UTI prevention. The panel cited the positive clinical outcomes and need to reduce overuse of antibiotics for prevention and treatment of UTI (Anger et al. 2019). Cranberry's preventative effects, without subsequent bacterial resistance issues, offer an alternative strategy for maintenance of urinary tract health. Clearly, the overwhelming trend of the data and individual studies supports efficacy.

4.4   Cranberry Proanthocyanidins

Cranberry proanthocyanidin (PAC) oligomers, also referred to as condensed tannins or polyflavan-3-ols are largely made up of epicatechin extender units. The stereochemistry of the flavonol monomers is predominantly of the 2,3-cis type with a small proportion of 2,3-trans units (Foo et al. 2000a, 2000b). Some studies also report the presence of epigallocatechin and catechin units in PAC oligomers (Foo et al. 2000a, 2000b; Howell et al. 2005; Neto et al. 2006; Porter et al. 2001; Reed et al. 2005). There are 2 common series of PAC dimers. The B-type series are dimers linked either in the C4–C6 or C4–C8 position whereas the A-type series are dimers linked in the C4–C8 position with an additional C2–O–C7 ether linkage. Cranberry PAC oligomers with a degree of polymerization (DP) greater than 2 may incorporate both A-type and B-type interflavan linkages. By extension of this definition, and for purposes of discussing differences among oligomers, PAC that contain one or more A-type interflavan bonds in their structure are referred to as A-type PAC whereas PAC oligomers that contain only B-type interflavan bonds are referred to as B-type PAC (Krueger et al. 2013a). Feliciano et al. (2012) applied a method to deconvolute matrix-assisted laser desorption/ionization time-of-flight mass spectrometry (MALDI-TOF MS) isotope patterns and determined that more than 91% of cranberry PAC molecules had at least one A-type linkage. The ratios of A-type to B-type linkages in PAC are product-specific and therefore constitute information that can be used to authenticate cranberry content and help prove if adulteration has occurred. PACs are either soluble (extractable in polar solvents) or insoluble, which affects analysis of the PAC content with quantitation and bioactivity assays. Soluble PACs are present in the juice portion of the fruit and are easily analyzed by assays such as 4-Dimethylaminocinnamaldehyde (DMAC) (Krueger et al. 2013b). Insoluble PACs mainly occur in the cranberry skins or pumice and are often bound to complex cellulose-based carbohydrate cell wall components or proteins. The average degree of polymerization of PACs differs and is variably reported as 4.7 (Foo et al. 2000a), 8.5–15.3 (Gu et al. 2003), and up to 23 (Blumberg et al. 2013; Reed et al. 2005), in part due to variable findings using different analytical methods that employ thiolysis and pholoroglucinolysis (Karonen et al. 2007; Zhou et al. 2011). Overall PAC levels in cranberries can vary, due to variations among genetic cultivars, post-harvest handling and processing, and utilization of different analytical techniques to assess PAC content. Gu et al. (2004) reported PAC content of 418.8 ± 75.3 mg/100 mg of fresh fruit and 231 ± 2 mg/L in cranberry juice cocktail using a normal phase HPLC method that was traditionally used to analyze foods containing all B-type PACs (such as grape and chocolate), but is now known to be unable to resolve the A-type linkages in cranberry. In both matrices, the majority of the PAC had a > 10 degrees of polymerization. Some colorimetric methods such as Bate-Smith cannot differentiate anthocyanins from PACs, resulting in an overestimation of PAC levels up to 5 times higher than DMAC (with Procyanidin A2 reference standard), which is the industry-standard method. Use of methods other than DMAC for soluble PACs results in inaccurate reporting and significant overestimation of PAC content (Krueger et al. 2013b). This gives some product manufacturers an unfair market advantage and misleads consumers that rely on accurate reporting of product specifications to select foods with health functionality.

Howell et al. (2001) demonstrated for the first time how cranberry PAC ingestion resulted in urine with bacterial anti-adhesion activity *in vivo*. Urine collected from mice that were fed purified cranberry PACs had bacterial anti-adhesion activity against P-fimbriated *E. coli*. In another study, rats fed 108 mg PACs/animal/day produced urine that prevented adhesion of *E. coli* by 83 and 52%, respectively (Risco et al. 2010). In one study, very low levels of PAC A2 dimer (0.541 ± 0.10 ng/mL) were found in rat plasma samples 1 h after cranberry administration Rajbhandari et al. (2011) and in low levels in human urine (McKay et al. 2015). In another study, healthy rats were orally supplemented with 100 mg/kg of cranberry extract daily (11.3 mg/kg PAC-A; 4.3 mg/kg PAC-B) for 35 days and urines collected over time (Peron et al. 2017). PAC metabolites (valerolactone derivatives) were present in urine that exhibited bacterial anti-adhesion activity. PAC metabolites were also found in urines of 10 people following cranberry juice consumption in a double-blind randomized controlled trial (Feliciano et al. 2017).

While it has generally been assumed that cranberry exerts its effect on UTIs directly through the urine, an alternative and untested hypothesis is that it also works preventatively through affecting adhesive properties of bacteria in the large intestine and colon (Kontiokari et al. 2001; Ofek et al. 1996; Sobota 1984; Zafriri et al. 1989). Recent studies have suggested other possible mechanisms of action of cranberry components in the prevention of UTI (Krueger et al. 2013b; Shanmuganayugari et al. 2013) involving decreases of the relative levels of 8 proteins/peptides in the urine of human subjects post-supplementation with cranberry (Krueger et al. 2003). The functions of these proteins are not fully understood and the implications of these shifts in UTI are unclear. Recently, it has been shown that exposure of extraintestinal pathogenic *E. coli* to cranberry PACs inhibits their invasiveness into enterocytes, disrupts surface structures of the *E. coli*, and increases killing of *E. coli* by macrophages (Shanmuganayugari et al. 2013). The action of PACs on prevention of adhesion, invasion and immune function are pharmacologically more complex than previously thought and require further study to determine more about the metabolism and precisely how these various biological activities exert effects in the urinary tract and gut to maintain urinary tract health.

4.5   Cranberry and UTI Prevention

Consumption of cranberries and cranberry products has been widely recommended for the maintenance of urinary tract health in general (Bone and Morgan 1999; Henig and Leahy 2000; Kerr 1999; Patel and Daniels 2000; Reid 1999;

Wang et al. 2012), as well as for the prevention of UTIs (Bruyère et al. 2019; Ledda et al. 2017; Wan et al. 2016; Ledda et al. 2016; Singh et al. 2016; Bonetta and Di Pierro 2012; Haverkorn and Mandigers 1994; Hess et al. 2008; Rogers 1991; Stothers 2002; Walker et al. 1997). A variety of preparations have been used in the various studies including cranberry juice, cranberry juice cocktail (~27% juice), and varying cranberry extracts (see review of select studies below). Some studies do not fully characterize the preparations used, while other studies report low compliance and high dropout rates (see Jepson et al. 2012). Patients with recurrent UTIs appear to prefer "natural" therapies such as cranberry to avoid prophylactic antibiotic use (Mazokopakis et al. 2009; Nowack and Schmitt 2008) underscoring the importance for health care providers to understand the benefits, limitations, dosage, and product characterizations to maximize the efficacy of cranberry preparations. One dose shown to be efficacious in a clinical trial of nursing home patients (Avorn et al. 1994) used approximately 300 mL of cranberry juice cocktail (27% cranberry juice) that, when calculated, delivered approximately 36 mg of soluble PACs (analyzed according to DMAC with an A2 standard). This 36-mg dosage has been the target dose for clinical efficacy ever since and has proven to be effective for preventing urinary bacterial anti-adhesion following cranberry consumption in a multicenter trial (Howell et al. 2010) and clinically in children (Uberos et al. 2012).

A variety of mechanisms for cranberry's putative effects have been articulated, indicating the complex nature of the biological benefits for maintenance of urinary tract health. For many years, it was assumed that hippuric acid excreted in the urine following cranberry consumption was responsible for the effect on prevention of UTIs, as hippuric acid can be bacteriostatic against *E. coli* (at concentrations of 1–2 mg/L; Bodel et al. 1959; Hamilton-Miller and Brumfitt 1976). In human trials, urinary pH levels were somewhat reduced following consumption of cranberry juice, but to achieve a bacteriostatic effect, urinary pH must be reduced to at least 5.0 with a minimum hippuric acid concentration of 0.02 M (Bodel et al. 1959; Blatherwick 1914; Blatherwick and Long 1923; Jackson and Hicks 1997; Kinney and Blount 1979; Nickey 1975; Papas et al. 1966; Schultz 1984a, b). To attain these levels, humans would need to consume at least 1500 mL of cranberry juice per day (Kahn et al. 1967). To date, researchers are continuing to study the *in vivo* mechanisms of action, and there is substantial *in vitro* and *ex vivo* evidence indicating that cranberry and cranberry A-type PACs stimulate an activity that results in inhibiting bacteria, particularly *E. coli*, from adhering to uroepithelial cells (Gupta et al. 2007; Howell et al. 1998; Sobota 1984; Zafriri et al. 1989), prevent formation of bacterial biofilms (LaPlante et al. 2012), hinder motility, and downregulate the enzyme urease and the transcription of flagellin, which are important virulence factors (McCall et al. 2013). Live bacteria must attach and gain entry to uroepithelial cells in the urinary tract to grow and cause infection, and they do so by binding to certain cell receptors with filamentous appendages called pili or fimbriae. The anti-adhesion activity of cranberry was first recognized by Sobota (1984). A series of experiments using cranberry juice and uropathogenic strains of *E. coli* demonstrated that cranberry juice contains one or more compounds that inhibit *in vitro* bacterial adherence to uroepithelial cells (Howell et al. 1998; Zafriri et al. 1989). It appears that by preventing the *E. coli* from adhering to uroepithelial cells, the bacteria will not grow and cause infection, but be flushed out in the urine stream. Since this mechanism is not killing the bacteria, it is unlikely to result in bacterial resistance to cranberry. Anti-adhesion activity of cranberry has been demonstrated in humans (Di Martino et al. 2006; Howell et al. 2005, 2010; Lavigne et al. 2008; Tempera et al. 2010; Valentova et al. 2007). More recently, focus has been given to PACs as the primary compounds inducing anti-adherence activity. The A-type PACs in cranberry have demonstrated greater bacterial anti-adhesion activity than the B-type PACs (Foo et al. 2000a; Foo et al. 2000b; de Llano et al. 2015).

Other mechanisms for UTI prevention utilizing cranberry include anti-inflammatory effects through induction of gene expression (Hannon et al. 2016), modulation of innate and acquired immune responses in macrophages by PAC-protein complexes (Carballo et al. 2017) and reducing uropathogenic bacterial invasion in the GI tract (Alfaro-Viquez et al. 2019; Feliciano et al. 2014). Cranberries contain a rich and diverse mixture of polyphenolic compounds which potentially contribute to these activities to promote maintenance of urinary tract health. It is likely that a suite of compounds contributes to cranberry's biological activity.

4.6   Cranberry in Clinical Investigations

UTIs are among the most common bacterial infections in the ambulatory setting (Schappert and Rechtsteiner 2011). Although both males and females can develop a UTI, infections occur more frequently in women (Foxman and Brown 2003). It is estimated that more than 50% of women will experience at least one UTI in their lifetime (Griebling 2005), and 20–30% of women who experience a UTI will have 2 or more recurrent episodes (Foxman 1990). Other populations at risk for developing UTIs include children, pregnant women, the elderly, patients with spinal cord injuries, catheterized patients and those with chronic and/or immune-compromising diseases such as diabetes and HIV/AIDS (Foxman 2002). Cranberry consumption potentially offers these populations an alternative prophylaxis for prevention of UTIs, but there is no clinical data to support treatment effects.

The clinical trial data on prevention of UTI is mixed, with both positive and negative outcomes in various patient populations Table 4.1. Unlike pharmaceutical drug trials which can effectively be compared in meta-analyses, clinicals testing functional foods, such as cranberry, are difficult to compare because different product forms (juice, powder, dried fruit) and formulations with poorly defined dosages of actives are utilized (Howell 2013). Different quantification assays that produce highly variable, often inaccurate results are used to report levels of actives in these trials. Therefore, without proper standardization of cranberry products utilized in clinicals for UTI prevention, it is extremely difficult to draw broad

Table 4.1 Clinical trials of cranberry preparations in the prevention of urinary tract infections (UTIs)

Reference	Study design	Patient population	Product and daily dosage	Treatment duration	Outcome
Rogers (1991)	OBS	Children with neuropathic bladders; $n = 17$	CJ, 360–480 mL for 1 week and 540–660 mL for the 2nd week	2 weeks	Observed reduction in red and white cell counts; *E. coli* still present in samples at end of study
Avorn et al. (1994)	DBRPCT	Elderly women; $n = 153$	CJC (27%, saccharin-sweetened), 300 mL	6 months	Lower odds ratio for bacteriuria with pyuria in treatment group ($P = 0.004$)
Haverkorn and Mandigers (1994)	RX, control water	Elderly men and women; $n = 17$	CJ, 30 mL	1 month	Fewer incidences of bacteriuria ($P = 0.004$)
Foda et al. (1995)	PCTX	Children with neuropathic bladders; $n = 21$	CJC, 15 mL/kg	6 months	No reduction in incidence of UTIs
Walker et al. (1997)	DBRPCTX	Women (28–44 years) with history of recurrent UTI; $n = 10$	CE (Cranactin®), 1 capsule equivalent to 400 mg cranberry solids daily	3 months	Significantly fewer UTIs in treatment group ($P = 0.005$)
Dignam et al. (1998)	OBS, cross-sectional	Elderly men / women; $n = 538$	CJ, 120 mL or 6 Azo-cranberry capsules	8 months	Fewer UTIs during the treatment period ($P = 0.008$)
Dignam et al. (1998)	OBS, longitudinal	Elderly men / women; $n = 113$	CJ, 120 mL or 6 Azo-cranberry capsules	16 months	No reduction in incidence of UTIs
Schlager et al. (1999)	DBPCTX	Children with neuropathic bladders; $n = 15$	CC; 60 mL (= 300 mL CJC)	3 months	No reduction in bacteriuria
Kontiokari et al. (2001), (2005)	PRCT, 3 arm, Control Lactobacillus rhamnosus drink	Young women suffering from UTI at recruitment; $n = 150$	Cranberry-lingonberry concentrate, 50 mL in 200 mL water	6 months	56% fewer UTIs in cranberry group after 6 months ($P = 0.02$)

(continued)

Table 4.1 (continued)

Reference	Study design	Patient population	Product and daily dosage	Treatment duration	Outcome
McGuinness (2002)	PRCT, control beetroot powder	Multiple sclerosis patients; $n = 126$	Cranberry containing tablet product (NOW Natural Foods): 8000 mg tablet, one tablet/daily	6 months	No significant advantage of cranberry over control
Stothers (2002)	DBRPCT	Women (21–57 years) with history of UTIs; $n = 150$	CJ or CE, brands and dosage unspecified	1 year	Mean number of symptomatic UTIs reduced in both cranberry groups ($P \leq 0.05$)
Linsenmeyer et al. (2004)	PRCTX	Patients with neurogenic bladders secondary to spinal cord injury; $n = 21$	Cranberry tablets: 400 mg standardized tablets	9 weeks	No statistically significant treatment (favorable) effect for cranberry supplement beyond placebo
Waites et al. (2004)	PRCT	Men and women at least 1 year post spinal cord injury; $n = 48$	Concentrated cranberry extract: 2 g in capsule form	6 months	No reduction in bacteriuria and pyuria
McMurdo et al. (2005)	PRCT	Elderly men and women; $n = 376$	Cranberry juice: 300 mL	35 days	Between-group differences not significant
Lee et al. (2007)	PRCT, control methenamine hippurate	Men and women with spinal cord injury; $n = 305$	Treatment group 1, Methenamine hippurate: 2 g; cranberry: 1600 mg, Treatment group 2, Methenamine hippurate: 2 g; cranberry placebo, Treatment group 3, Cranberry: 1600 mg; methenamine hippurate placebo	6 months	No difference in UTI-free period with either treatment
Bailey et al. (2007)	OBS	Women with a history of recurrent infections of a minimum of 6 UTIs in the preceding year; $n = 12$	One capsule twice daily containing 200 mg of a concentrated cranberry extract standardized to 30% phenolics	3 months	No UTIs during study period

Hess et al. (2008)	RCTX, control rice flour	Men and women with spinal cord injury; $n = 47$	Cranberry tablet: 500 mg twice daily	6 months	Reduction in the likelihood of UTI and symptoms for any month while receiving the cranberry tablet ($P < 0.05$ for all)
Wing et al. (2008)	PRCT, 3 arms	Women < 16 weeks gestation; $n = 115$	Treatment group 1, Cranberry juice: 240 mL at breakfast, placebo juice at other meals, Treatment group 2, Cranberry drink: 240 mL, 3 times/daily, reducing to twice daily after 52 enrollments because not well tolerated, Control group, Placebo: 3 daily doses of matched juice product	~5 months	Non-significant trend for reduction in asymptomatic bacteriuria and symptomatic urinary tract infections in pregnancy
Ferrara et al. (2009)	PRCT, 3 arms	Girls 3–14 years; $n = 80$	Cranberry-lingonberry concentrate (97.5 g and 1.7 g respectively); Lactobacillus GC drink	6 months	Significant reduction in the risk of repeated UTIs in the cranberry group ($P < 0.05$) compared with the Lactobacillus group and the control group
Mazokopakis et al. (2009)	OT	Post-menopausal women with recurring UTI; $n = 10$	Four cranberry capsules per day (Natural Cranberry Extract, 400 mg with vitamin C vegetable, capsules, Solgar)	6 months	No symptomatic UTI during trial, almost all urine cultures were sterile

(continued)

Table 4.1 (continued)

Reference	Study design	Patient population	Product and daily dosage	Treatment duration	Outcome
McMurdo et al. (2009)	PRCT	Women >45 years with at least 2 antibiotics treated UTIs in previous 12 months; n = 120	Cranberry tablet: 500 mg, 100 mg of trimethoprim	6 months	Study underpowered; trimethoprim had a limited advantage over cranberry, but more adverse effects; cranberry group experienced fewer infections with E. coli
Cadkova et al. (2009)	PT	Women during perioperative period leading to gynecological surgery; n = 286	Cranberry extract capsules (equivalent to 17,000 mg of fresh fruit) twice daily, 4 days before and 5 days after the surgery	6 days	No effect on the number of post-surgical UTIs
Vidlar et al. (2009)	RCT	Men, aged 45–70 years; n = 42	1500 mg of the dried powdered cranberries or no treatment	6 months	Significant improvement in International Prostate Symptom Score, QOL, urination parameters including voiding parameters, and lower total PSA level for cranberry group
Botto and Neuzillet (2010)	OT	Asymptomatic bacteriuria in patients with an ileal enterocystoplasty; n = 15	36 mg/daily PAC A (Urell, Pharmatoka)	32.8 months (median)	Significant decrease in the number of positive urine cultures during cranberry compound treatment
Juthani-Mehta et al. (2010)	PRCT, 3 arms	Elderly men and women >60 years of age with dementia; n = 56	Cranberry capsule: 1 x 650 mg once or twice daily	6 months	No difference between the 3 groups
Essadi and Elmehashi (2010)	PCT, control water: 250 mL 4 times/ daily	Pregnant women: n = 544	Cranberry juice: 250 mL 4 times/ daily	12 months	To little information to assess

Study	Design	Population	Intervention	Duration	Outcome
Barbosa-Cesnik et al. (2011)	PRCT, control matched for flavor and color	Women 18–40 years, with UTI symptoms; n = 319	Low calorie cranberry cocktail: 240 mL twice daily	6 months	Among otherwise healthy college women with an acute UTI, drinking cranberry juice did not result in a decrease in the 6-month incidence of a second UTI
Sengupta et al. (2011)	PRCT, 3 arms	Females with a history of recurrent UTIs; n = 60	Cranberry: 500 and 1000 mg/daily, 1.5% PAC	3 months	Significant reduction ($P < 0.05$) in the subjects positive for E. coli in both the high-dose and low-dose treatment groups
Beerepoot et al. (2011)	PRCT, control trimethoprim-sulfamethoxazole 480 mg	Premenopausal women >18 years with at least 3 symptomatic UTIs in the year prior to enrollment; n = 221	Cranberry extract: 500 mg twice daily (9.1 mg/g type A PAC)	12 months	Trimethoprim-sulfamethoxazole is more effective than cranberry capsules to prevent recurrent UTIs; authors suggest that non-antibiotic therapies additionally beneficial to prevent antibiotic resistance
Uberos et al. (2012), (2015) and Fernandez-Puentes et al. (2015)	PRCT, control trimethoprim 8 mg/kg	Children aged from 1 month to 13 years, with recurrent UTI; n = 192	Cranberry syrup: 0.2 mL/kg yielding 36 mg PACs (Urell, Pharmatoka), trimethoprim	12 months	Similar efficacy between trimethoprim and cranberry
Bonetta and Di Pierro (2012)	PCT	Patients with external beam radiotherapy; n = 370	200 mg of a highly standardized cranberry extract titered as 30% proanthocyanidins	~7 weeks	Significantly fewer lower urinary tract infections in the verum group

(continued)

Table 4.1 (continued)

Reference	Study design	Patient population	Product and daily dosage	Treatment duration	Outcome
Salo et al. (2012)	PRCT	Children; $n = 255$	Cranberry juice 5 mL/kg up to 300 mL 1–2 doses daily	6 months	No significant reduction in the number of children who experienced a recurrence of UTI; reduction in actual number of recurrences
Afshar et al. (2012)	RCT	Children; $n = 40$	2 cc/kg cranberry juice containing 37% PAC (method not given) Placebo: same volume of juice with no PAC or other cranberry products	12 months	65% reduction in risk of urinary tract infection ($P = 0.045$)
Stapleton et al. (2012)	RCT	Premenopausal women with a history of recent UTI; $n = 176$	4 oz of cranberry juice, 8 oz of cranberry juice, or placebo	5.6 months (median)	Strong (though not significant) reduction in P-fimbriated E. coli; no significant reduction in UTI risk
Bianco et al. (2012)	DBRPCT, 4 arms	Elderly; $n = 80$	108, 72, and 36 mg PAC, placebo	1 month	Dose-dependent trend toward decrease in bacteriuria and pyuria
Cowan et al. (2012)	PRCT, control placebo juice	Adults >18 years with cervical or bladder cancer requiring radiation therapy; $n = 113$	Cranberry juice twice daily (volume and concentration not stated)	6 weeks	Significant decrease of UTI and urinary symptoms ($P = 0.240$)
Mutlu and Ekinci (2012)	RCTX	Children with neurogenic bladder; $n = 20$	One cranberry capsule (no further information given), placebo	12 months	Significant reduction in UTIs ($P = 0.012$); significant decrease in pyuria ($P = 0.000$)

Takahashi et al. (2013)	PRCT	Outpatients aged 20–79 years with acute exacerbation of acute uncomplicated cystitis or chronic complicated cystitis (including self-catheterization) who had a past history of multiple relapses of UTI; $n = 213$	125 mL cranberry juice or placebo	24 weeks	Significant reduction ($P = 0.0425$) in rate of relapse of UTIs in females 50 years or more
Gallien et al. (2014)	DBRCT	Multiple sclerosis patients; $n = 171$	Cranberry powder 18 mg proanthocyanidins sachets twice daily	12 months	No reduction in incidence of UTIs
Caljouw et al. (2014)	DBPCT	Elderly long-term care facility patients stratified to high- or low risk for UTIs; $n = 928$	Undisclosed cranberry preparation and dose given twice daily	12 months	26% reduction in incidence of UTIs in high risk subjects ($n = 516$); no difference in low-risk subjects
Lin et al. (2014)	OBS; control (catherized patients)	Elderly long-term care facility; patients with long-term in dwelling catheter; $n = 11$	Cranberry juice (300 mL daily; characterization not disclosed) along with 2200 mL water	6 months	No reduction in asymptomatic bacteriuria or incidence of UTIs
Mathison et al. (2014)	DBRCTX	Healthy adults; $n = 12$	Cranberry leaf extract beverage, low-calorie cranberry juice cocktail, or placebo	Single dose	Cranberry showed significant ($P < 0.05$) ex vivo anti-adhesion activity against *P*-fimbriated *E. coli* in urine compared with placebo.

(continued)

Table 4.1 (continued)

Reference	Study design	Patient population	Product and daily dosage	Treatment duration	Outcome
Foxman et al. (2015)	DBRPCT	Female adults post elective gynecologic surgery; $n = 160$	TheraCran® cranberry capsules (Theralogix) equivalent to 16 oz cranberry juice, or placebo	6 weeks after surgery	Incidence of UTI was significantly lower in the cranberry treatment group compared to the placebo group (15/80 (19%) versus 30/80 (38%); OR = 0.38; 95% CI: 0.19, 0.79; $P = 0.008$)
Barnoiu et al. (2015)	OT	Patients with in-dwelling catheter; $n = 62$	Cranberry preparation (unspecified), 120 mg, or placebo	5 days prophylactic treatment	Reduced incidence of UTI compared to control in patients with in-dwelling catheters (12.9% versus 38.75%, respectively; $P = 0.04$)
Hamilton et al. (2015)	DBRPCT	Male adults with radiation cystitis; $n = 41$	Cranberry capsules (Naturo Pharm), 72 mg PACs according to UV-VISEP/CN, or placebo	70 days	Incidence of cystitis was lower in men taking cranberry capsules (65%) compared to placebo (90%) ($P = 0.058$; severe cystitis occurred in 30% of men in the cranberry arm and 45% with placebo ($P = 0.30$)
Ledda et al. (2015), (2016), (2017)	RCT	Adolescents with history of recurrent UTI, $n = 36$	120 mg of cranberry extract (Anthocran®), standardized to 36 mg PACs	60 days	Number of UTIs in the cranberry group (0.31 ± 0.2) significantly lower than in control (2.3 ± 1.3) and compared to mean number of UTIs at baseline (1.74 ± 1.1), $P = 0.0001$ for both.

Vostalova et al. (2015)	DBRPCT, 2 arms	Female adults with 2 or more UTI episodes over 1 year, $n = 182$	500 mg cranberry fruit powder (NATUREX-DBS), or placebo	6 months	Fewer UTIs in the cranberry group (10.8% vs. 25.8%, $P=0.04$)
Vidlar et al. (2016)	DBRPCT, 3 arms	Male adults with benign prostate hyperplasia, $n = 124$	250 or 500 mg cranberry powder (Flowens™), or placebo	6 months	Lower international prostate symptoms score (IPSS) in both Flowens™ groups (-3.1 and -4.1 in the 250- and 500-mg groups, $P = 0.05$ and $P < 0.001$, respectively)
Maki et al. (2016)	DBRPCT	Female adults with a history of recurrent UTI, $n = 373$	8 oz CJC, or placebo	24 weeks	39 UTI episodes in cranberry group vs. 67 in placebo (antibiotic use–adjusted incidence rate ratio: 0.61; 95% CI: 0.41, 0.91; $P = 0.016$
Juthani-Mehta et al. (2016)	DBRPCT	Elderly females in nursing homes, $n = 185$	Two capsules of cranberry dry extract totaling 72 mg PACs (Pharmatoka), or placebo	1 year	No significant difference in the presence of bacteriuria plus pyuria between the treatment group vs the control group (29.1% vs 29.0%; OR, 1.01; 95% CI, 0.61-1.66; $P=0.98$), no significant differences in number of symptomatic UTIs (10 episodes in the treatment group vs 12 in the control group).

(continued)

Table 4.1 (continued)

Reference	Study design	Patient population	Product and daily dosage	Treatment duration	Outcome
Singh et al. (2016)	RCT	Patients with a history of UTI, $n = 72$	Cranpac™ (containing PAC-A 60 mg per capsule), 2 capsules per day, or placebo	12 weeks	bacterial adhesion scoring decreased (0.28) v. placebo (2.14) ($P < 0.001$); biofilm ($P < 0.01$) and bacterial growth ($p < 0.001$) decreased; microscopic pyuria score was 0.36 vs. 2.0 ($P < 0.001$); UTI decreased to 33.33 vs. 88.89% ($P < 0.001$); mean subjective dysuria score was 0.19 vs. 1.47.
Lee et al. (2016), Wan et al. (2016)	DBRPCT, 3 arms	Circumcised, $n = 12$) and uncircumcised, $n = 55$, boys	4 oz cranberry juice, or placebo	6 months	Incidence of bacteriuria were 25% (7/28), 37% (10/27), and 33.3% (4/12) in groups 1 (uncircumcised cranberry), 2 (uncircumcised placebo) and 3 (circumcised cranberry), respectively.
De Leo et al. (2017)	RCT	Perimenopausal women with recurrent cystitis, with or without bacteriuria ($n = 150$)	Kistinox® Forte sachets, one per the first 10 days of the month for three months or placebo	3 months	Complete remission of urinary symptoms in 92 women, excellent tolerability.
Letouzey et al. (2017)	DBRPCT	Women undergoing pelvic surgery aged 18+ ($n = 272$)	One capsule daily containing 36 mg PACs or placebo.	10 days treatment, control after 15 and 40 days	PAC did not significantly reduce the risk of bacteriuria treatment within 15 or 40 days of surgery.

Thomas et al. (2017)	OT	Males and females with long-term indwelling catheters and recurrent symptomatic UTIs (n = 22)	One capsule of cranberry daily (Ellura®, Pharmatoka) containing 36 mg PACs.	6 months	Effective in all patients, with 28% reduction in antibiotic resistance and 58.65% reduction in colony counts.
Gunnarson et al. (2017)	DBRPCT	Females aged 60 and older (n = 227) with indwelling urinary catheter post hip surgery	Two capsules of 550 mg of cranberry powder (NutriCran®) three times a day, or placebo	5 days	No statistically significant difference between treatment and placebo.
Shatkin-Margolis et al. (2018)	OBS	Patients with a catheter after pelvic reconstructive surgery (n = 167)	Cranberry capsules (TheraCran) 180 mg twice daily or placebo.	6 weeks	UTI rates were not significantly different between groups.
Bruyère et al. (2019)	DBRPCT	Women aged 18+ with recurrent acute cystitis (n = 85)	Two capsules of cranberry-propolis-zinc (DUAB®) or placebo daily	6 months	Mean number of infections, total number of cystitis episodes, and mean time to onset of the first UTI were lower in treatment group.

DB double blind, *R* randomized, *CT* controlled trial, *PCT* placebo-controlled, *PT* prospective trial, *X* crossover, *OBS* observational, *PG* parallel group, *CJC* Cranberry Juice Cocktail® (Ocean Spray®, CJC ~27% cranberry juice), *CJ* cranberry juice, *CJC* cranberry juice of unknown concentration from unknown manufacturer, *CE* cranberry extract, *CC* cranberry juice concentrate, *OT* open trial conclusions about the efficacy of cranberry when comparing these studies. While many reviews have concluded that cranberry may help prevent infections, particularly in women with recurrent UTIs (Chen et al. 2019; Mantzorou and Giaginis 2018; Luís et al. 2017; Fu et al. 2017; Jepson and Craig 2008; Wang et al. 2012), one past review concluded that there was no benefit for cranberry (Jepson et al. 2012). Risk ratios of <1.0 (calculated relative risk of developing UTIs in the treated vs control groups) were interpreted as positive outcomes by Wang et al. (2012) but not by Jepson et al. (2012) with different confidence intervals reported in each study. Compliance in some studies included in the Cochrane Review (Jepson et al. 2012) was low but may have been confounded using poor compliance measures. Most of the studies used cranberry products that were not standardized to A-type PACs and may not have had sufficient amounts of bioactive PAC to achieve clinical efficacy. Additionally, the choice of study subjects is particularly important, as the pathogenesis of UTI is specific to different patient groups. More studies completed since the Cochrane Review (Jepson et al. 2012) had positive outcomes for cranberry in preventing UTI recurrence and are included in the more recent reviews (Chen et al. 2019; Mantzorou and Giaginis 2018; Luis et al. 2017; Fu et al. 2017; Micali et al. 2014). A recent meta-analysis of 28 clinical investigations into the effects of cranberry preparations on the reduction of incidence of UTIs found that cranberry significantly reduces the incidence of UTIs (weighted risk ratio (0.6750, 95% CI 0.5516–0.7965, $P < 0.0001$). Additionally, patients at risk for UTI showed more susceptible to the effect of cranberry treatment (Luis et al. 2017). A further meta-analysis (Fu et al. 2017) assessed the efficacy of cranberry on the risk of recurrent UTI in otherwise healthy women based on 7 randomized controlled trials ($n = 1498$). Results showed that cranberry reduced the UTI risk by 26% (pooled risk ratio 0.74; 95% CI: 0.55, 0.98; $I^2 = 54\%$). In another review and meta-analysis of randomized controlled trials, Huang et al. (2017) assessed a total of 26 studies with 4709 participants for efficacy of cranberry in the prevention of UTI. Through subgroup analysis, authors found cranberry to be more effective in females (pooled risk ratio 0.73; 95% CI:0.58, 0.92; $P = 0.002$; $I^2 = 59\%$) with recurrent UTI (pooled risk ratio 0.71; 95% CI: 0.54, 0.93; $P = 0.002$; $I^2 = 65\%$).

4.7 Cranberry and Recurrent UTIs

Recent reviews of clinical studies (Fu et al. 2017; Micali et al. 2014) support the use of cranberry in the prevention of recurrent UTIs in young and middle-aged women (Takahashi et al. 2013; Vostalova et al. 2015; Afshar et al. 2012; Ferrara et al. 2009; Kontiokari et al. 2001; Salo et al. 2012; Stothers 2002; Uberos et al. 2012; Walker et al. 1997). Several studies did not demonstrate a significant effect (Barbosa-Cesnik et al. 2011; Stapleton et al. 2012). All trials recruited healthy women, ages 18–79 years, with a history of at least one UTI within the previous year. Cranberry regimens and dosing varied greatly among these studies.

Takahashi et al. (2013), provided 125 mL/day of cranberry juice compared with placebo for 24 weeks to women between 20–79 years with recurrent UTI. In the subgroup of females aged 50 years or more, there was a significant difference in the rate of relapse of UTI between groups A and P (log-rank test; $P = 0.0425$). Vostalova et al. (2015) dosed with 500 mg of cranberry whole fruit powder daily for 6 months, resulting in significantly fewer UTIs in women with a history of recurrent UTI. Walker et al. (1997) provided participants with 400 mg of encapsulated cranberry solids taken once per day for 3 months. Kontiokari et al. (2001) used cranberry-lingonberry juice made from concentrates, primarily containing cranberry: 7.5 g cranberry concentrate and 1.7 g lingonberry concentrate diluted in 50 mL water once daily for 6 months. Kontiokari et al. (2001) also compared 100 mL of a probiotic milk drink containing *Lactobacillus* for 5 days/week for 1 year to 150 women who were recruited with UTIs; an open group served as open controls. After 6 months, the women on the cranberry treatment experienced 56% fewer UTIs (defined as $>10^5$ cfu/mL) than the control group ($P = 0.02$). After 12 months, the cumulative occurrence of the first episode of UTI was still significantly different between the groups ($P = 0.048$), suggesting that cranberry juice drink was effective in preventing UTI, while the probiotic drink was not. Stothers (2002) had 2 cranberry treatment arms, administered as a juice or tablet. Participants in the juice arm consumed 240 mL of "pure, unsweetened" cranberry juice 3 times/day, and the tablet arm received a 1:30 parts concentrated cranberry juice tablet twice per day for 12 months. A double-blind, placebo-controlled, crossover study by Walker et al. (1997) found that dried cranberry powder was effective in reducing UTI occurrence. Women between the ages of 28 and 44 with a history of recurrent UTIs were recruited to take two 400-mg cranberry extract pills per day for 3 months (and 3 months of placebo). While taking cranberry pills, 7 out of the 10 women experienced fewer UTIs. Only 6 UTIs occurred among the 10 subjects on cranberry supplementation, while 15 UTIs occurred among the 10 subjects on the placebo. The authors concluded that cranberry extract pills were more effective than placebo in reducing UTI occurrences ($P < 0.005$). Participants in the Barbosa-Cesnik et al. (2011) study consumed 2–240 mL cranberry beverage per day for 6 months; these subjects entered the trial with acute UTIs. Participants in the study of Stapleton et al. (2012) consumed the same juice beverage, assigned to one 120 mL/day or 240 mL/day for 6 months. Papas et al. (1966) conducted an uncontrolled study in which 480 mL/day of cranberry juice cocktail was administered for 21 days to 60 patients (44 women and 16 men) diagnosed with acute UTI. After 3 weeks, 53% of the participants experienced fewer UTIs following cranberry juice consumption. Six weeks after discontinuation of cranberry treatment, bacteriuria returned in most subjects. Each study reported total PAC concentration but used different methods to quantify PACs, thus giving inaccurate and varied results that are inconsistent with current quantification methods.

A recent study in which women with recurrent UTI were given 42 g dried cranberries/day for 2 weeks followed by observations for 6 months showed that women taking dried cranberries had significantly lower incidence of UTI, with a mean UTI rate at 6 months decreasing from 2.4 to 1.1 compared to a historical control group enrolled in a previous vaccine control study (Burleigh et al. 2013). Those women in the dried cranberry group also had a significant reduction in *E. coli* in a rectal swab taken post-consumption. A study by Sengupta et al. (2011) found symptomatic relief and significant reduction ($P < 0.05$) in subjects positive for *E. coli* in both the high dose (1000 mg) and low dose (500 mg) treatment groups given a standardized cranberry powder for 90 days, compared to baseline evaluation in a randomized clinical trial of 60 female subjects between 18–40 years of age. Two recent clinicals found efficacy in women consuming mixtures of cranberry with propolis (Bruyère et al. 2019), which reduced UTI incidence during the first 3 months in women with recurrent UTI, and propolis plus D-mannose (DE Leo et al. 2017), which eliminated urinary symptoms of UTI in perimenopausal women.

4.8 Cranberry in Vulnerable Populations

4.8.1 The Elderly

The majority of clinical studies support efficacy of cranberry in UTI reducing bacteriuria in the elderly. One of the earliest large, double-blind, placebo-controlled randomized clinical trials evaluated a low calorie 27% cranberry juice cocktail for its effect on bacteriuria (defined as $>10^5$ cfu/mL of urine) and pyuria (white blood cells in urine) in 153 elderly women over a 6-month period (Avorn et al. 1994). Participants consumed either 300 mL/day of cranberry juice cocktail or 300 mL/day of a placebo drink. After 48 weeks of treatment, bacteriuria and pyuria were reduced by nearly 50% in the group that consumed cranberry juice cocktail with their odds of remaining bacteriuric-pyuric at only 27% of the odds of the control group ($P = 0.006$). In another study (randomized, controlled, crossover), Haverkorn and Mandigers (1994) administered 30 mL/day of cranberry juice diluted in water to 17 elderly men and women for 4 weeks. Participants consuming the cranberry treatment had fewer occurrences of bacteriuria compared to those who drank water ($P = 0.004$), confirming Avorn's findings that cranberry juice consumption reduces frequency of bacteriuria in the elderly. A double-blind, randomized, placebo-controlled pilot study aimed at identifying the optimal dose of cranberry capsules that reduced the incidence of bacteriuria plus pyuria was conducted over a 1-month period in elderly nursing home patients (Bianco et al. 2012). Subjects ($n = 80$) were given either 3 cranberry capsules (108 mg PAC determined by DMAC assay); 2 cranberry capsules (72 mg PAC) plus one placebo; or one cranberry capsule (36 mg PAC) plus 2 placebos; or 3 placebo capsules for 30 days, measuring episodes of bacteriuria and pyuria at days 7, 14, 21, and 28. In those consuming cranberry, a dose-dependent trend towards a reduction in bacteriuria and pyuria (particularly with *E. coli*) was observed, most notably in women. Cranberry did not affect bacteriuria with pathogens other than *E. coli*. The effects of the 2-capsule dose were comparable to those of the 3-capsule dose. Neither the long-term sustainability of the reduction in bacteriuria and pyuria, nor effects on clinical outcomes related to UTI (e.g., hospitalization, antibiotic therapy) was determined. Another study found no significant difference in presence of bacteriuria plus pyuria over 1 year in elderly women in nursing homes given cranberry capsules vs placebo (Juthani-Mehta et al. 2016). However, these results were confounded by certain methodological issues, namely that no description of compliance was given, and reliable clean-catch urine specimen collection was at issue. Of the 185 patients enrolled, 78% had dementia, 59% of patients had 1–8 ADL disabilities, 16% resisted care and 7% were not alert. Furthermore, there were 68% urinary incontinence and 44% bowel incontinence rates. These patients would not have been able to give reliable urine samples without contamination. Additionally, in the most recent "Clinical Practice Guideline for the Management of Asymptomatic Bacteriuria" 2019 Update by the Infectious Diseases Society of America, pyuria accompanying asymptomatic bacteriuria is not an indication for antimicrobial treatment in the elderly, because it is frequently present and is not indicative of an active UTI. Therefore, the effectiveness of a cranberry intervention in these study participants may not have been relevant for management of UTI. An uncontrolled study of 28 elderly patients in a long-term care facility found that cranberry juice was effective in preventing UTIs (Gibson et al. 1991). Participants drank 120–180 mL of cranberry juice cocktail almost daily for 7 weeks. UTIs were prevented in 19 of the 28 participants. A retrospective cross-sectional study and a longitudinal cohort study (Dignam et al. 1998) were carried out in a long-term care facility in which there was a 20-month pre-intervention period when UTI rates were recorded, and an 8-month intervention period when cranberry juice or cranberry capsules were given to participants (only 4% received the capsules instead of the juice). The cross-sectional study involved 538 elderly people (77% women and 23% men). During the 20-month pre-intervention period, UTIs were reduced significantly between these 2 periods ($P = 0.008$), with 545 UTIs compared with 164 UTIs during the 8-month intervention period when cranberry juice was consumed. In the longitudinal cohort study, 113 residents participated. There were 103 UTIs during the pre-intervention period and 84 UTIs during the intervention period, which represented a trend toward reduction in UTIs. A double-blind randomized placebo-controlled multicenter trial ($n = 928$) was conducted to determine the efficacy of cranberry (undisclosed characterization and dose taken twice daily for 12 months) in reducing the incidence of UTIs in residents of long-term care facilities (703 women, median age 84 years) in the Netherlands (Caljouw et al. 2014). Subjects were stratified by low or high UTI-risk (including long-term catheterization, diabetes mellitus, and ≥ 1 UTI in the preceding year). Of the total subjects, 516 were stratified as having a high risk for UTI; 412 were considered low risk. Compared to placebo, a 26% reduction in UTI was observed in the high-risk group, while no difference was observed in the low-risk group. One limitation of this study is that the actual incidence of UTIs was lower in the cranberry compared to placebo group. Another small trial found evidence to suggest that elderly men (>age 65) with moderate prostatic hyperplasia may benefit from cranberry intake to prevent recurrent UTIs (Ledda et al. 2016). Participants taking cranberry in encapsulated dried powder form for 60 consecutive days had significantly fewer ($p = 0.0062$) UTI episodes compared to those participants not on cranberry. The authors suggested that cranberry could help avoid some antibiotic treatments, if taken for prevention.

4.8.2 Children

A recent meta-analysis of the use of cranberry in the prevention of UTIs in children concludes that cranberry products are effective in otherwise healthy children and at least as effective as antibiotics in children with urogenital abnormalities. Dosage and frequency recommendations are confounded by the variability of products and dosages used in the trials included in this analysis (Durham et al. 2015). Recent trials in the pediatric population have demonstrated a benefit from cranberry consumption. The 7 available trials used a variety of cranberry products: 7.5 g cranberry concentrate plus 1.7 g of lingonberry concentrate diluted in 50 mL of water per day for 6 months (Ferrara et al. 2009); commercially available cranberry juice containing 8.2 g of cranberry concentrate per 200 mL water administered at 5 mL/kg body weight per day for 6 months (Salo et al. 2012); a cranberry syrup containing 36 mg PAC (measured by DMAC with A2 reference standard) administered at 5 mL per day depending on body weight (Uberos et al. 2012); cranberry juice containing 37% PACs (PAC quantitation not specified) administered at 2 mL/kg body weight for 1 year (Afshar et al. 2012); one 120 mg encapsulated dried cranberry extract standardized to 36 mg PAC taken daily for 60 days (Ledda et al. 2017); and cranberry juice (unspecified) given at 120 mL a day for 6 months (Wan et al. (2016). The primary outcomes analyzed demonstrated that cranberry consumption was efficacious in reducing UTI risk by 65% (Afshar et al. 2012) and preventing UTI recurrences (Ledda et al. 2017; Ferrara et al. 2009). UTI recurrence and bacteriuria were reduced in uncircumcised boys, ages 6–18 years (Wan et al. 2016). The primary outcome of reducing the number of children who experienced a recurrent UTI was not statistically significant in the Salo et al. (2012) trial. Cranberry treatment did, however, significantly reduce the number of recurrent UTIs and the number of days on antibiotics. In the trial of Uberos et al. (2012), cranberry prophylaxis was safe and effective having non-inferiority with respect to trimethoprim in recurrent UTI in relation to vesico-urethral reflux in 192 children ages 1 month to 13 years. In a follow-up review of this study (Uberos et al. 2015), cranberry intake was correlated with high levels of hydroxycinnamic and hydroxybenzoic acids in urine, both of which have displayed anti-adhesion activity, leading researchers to suggest these metabolites play a therapeutic role in the UTI preventive effects of cranberry in vivo, as suggested in other studies. A more recent controlled, double-blind trial was conducted to review the efficacy and safety of a cranberry extract syrup compared to the antibiotic trimethoprim in children with recurrent UTIs (Uberos et al. 2015). A secondary endpoint was to determine if there was a correlation between the excretion of phenolic acids (and their metabolites) in urine with bacterial anti-adherent activity of cranberry syrup. One group of subjects was given a 3% glucose solution of cranberry extract (Urell/Ellura®, Pharmatoka SAS, Rueil-Malmaison, France) yielding 4732 μg/mL of PACs at a dose equivalence of 5.6 mg/kg of extract; the other group was given trimethoprim in a similar syrup base at a concentration of 8 mg/mL and 0.1% CC-1000-WS (E-120) at a dose of 1.6 mg/kg. Subjects included 85 children under 1 year of age, 53 of whom were treated with trimethoprim and 32 with cranberry syrup and 107 children over 1 year of age, 64 of whom were treated with trimethoprim and 43 with cranberry syrup. There were marked differences in efficacy in children under 1 year of age and those over, as well as between treatment groups. In the trimethoprim group, rates of UTI in males and females under 1 year of age were 19% and 43%, respectively. Interestingly, gender associated efficacy was reversed in the cranberry group, the UTI rates in male and female children under 1 year of age being 46% and 17%, respectively. When adjusting for gender differences, in those under 1 year of age, the overall rates of UTI recurrence in the trimethoprim group was 28% and in the cranberry group 35%. Similarly, a reversal of the rate of efficacy was observed in children over 1 year of age, the UTI rate being 35% in the trimethoprim group and 26% in the cranberry group. These researchers concluded that overall, cranberry syrup was similar in efficacy and safety to trimethoprim, but that in children under 1 year of age trimethoprim was more effective than cranberry syrup. Conversely, cranberry was slightly more effective than trimethoprim in reducing the incidence of multi-resistant bacteria in urine culture, with 22.9% of the cranberry group displaying positive cultures compared to 33.3% in the trimethoprim group. Cranberry intake was correlated with high levels of hydroxycinnamic and hydroxybenzoic acids in urine, both of which have displayed anti-adhesion activity, leading researchers to suggest these metabolites play a therapeutic role in the UTI preventive effects of cranberry in vivo, as suggested in other studies (Uberos et al. 2015).

Use of cranberry for catheterized patients has not been sufficiently established, with mixed results in clinical trials. In a recent study by Barnoiu et al. (2015), prophylactic administration of 120 mg cranberry daily (preparation not characterized) significantly reduced the incidence of UTI as compared to a control group of patients with in-dwelling catheters (12.9% versus 38.75%, respectively; $n = 31$ in treatment and control groups; $P = 0.04$). Foxman et al. (2015), found that administration of the equivalent of two eight-ounce servings of cranberry daily in solid encapsulated dosage form for 6 weeks significantly ($P = 0.008$) reduced the incidence of UTIs in post-surgical catheterized subjects ($n = 160$) by approximately 50% as compared to placebo controls. Subjects in this study were instructed to take the capsules morning and evening with an 8-ounce glass of water beginning at time of discharge and were specifically asked to avoid any other cranberry product or vitamin C supplementation for the duration of 4–6 weeks, or until their post-operative visit. There were no differences in adverse effects, including GI upset in this study. Two studies did not find a benefit for UTI reduction with cranberry in catheterized patients following pelvic surgery (Letouzey et al. 2017; Shatkin-Margolis et al. 2018). One was a retrospective cohort study that compared two 6-months periods, one before cranberry capsules were included in patient regimes, and one after (Shatkin-Margolis et al. 2018). Catheterization lasted an average of 8 days. UTI rates were not significantly different between the two groups that totaled 167 patients. A randomized, double-blind, placebo-controlled trial of 272 catheterized women that underwent pelvic surgery did not demonstrate any significant reduction in risk of postoperative bacteriuria after cranberry capsules were given following surgery (Letouzey et al. 2017). Gunnarsson et al. (2017) also found no decrease in the incidence of postoperative UTIs following pre-operative administration of cranberry (juice concentrate) to catheterized female hip fracture patients. These negative results may be due to the fact that catheter-induced UTI is often caused by bacteria other than *E. coli*, such as *Proteus* and *Klebsiella*, which are less responsive to anti-adhesive activity of cranberry (Schmidt and Sobota 1988).

4.8.3   Pregnant Women

The use of cranberry for maintenance of urinary tract health is of interest to pregnant women because of the potentially harmful effects of antibiotics on the unborn fetus. Asymptomatic bacteriuria (ASB), defined as $>10^5$ CFUs/mL of uropathogenic bacteria in the urine, without the traditional symptoms associated with UTIs, is of particular concern in pregnant women due to their association with pre-term delivery and low birth weight (Romero et al. 1989; Sheiner et al. 2009). The first study published to investigate the effect of cranberry on ASB/UTI in pregnant women did not find a statistical difference in asymptomatic bacteriuria, UTI, or neonatal outcomes among participants who were compliant with zero, one, or two 240-mL servings of a cranberry beverage per day. The beverage treatments were reduced to 2 servings per day (Wing et al. 2008), and although this study was underpowered, women compliant with two 240-mL servings of cranberry per day experienced a 57% reduction in asymptomatic bacteriuria and 41% reduction in UTIs, indicating that cranberry may be efficacious in preventing asymptomatic bacteriuria and UTIs in pregnant women. Further studies would help solidify this area of importance for pregnant women and UTI prevention. A recent literature review of pregnant women taking cranberry supplements compared with antibiotics showed no adverse effects on the mother or infants including no increased risk of malformations nor any of the following pregnancy outcomes: stillbirth/neonatal death, preterm delivery, low birth weight, small for gestational age, low Apgar score, and neonatal infections, suggesting that cranberry consumption during pregnancy has no safety concern (Heitmann et al. 2013). Although an association was found between use of cranberry in late pregnancy and vaginal bleeding after pregnancy week 17, further sub-analyses of more severe bleeding outcomes did not support a significant risk. The 2019 Update of "Clinical Practice Guideline for the Management of Asymptomatic Bacteriuria" published by the Infectious Disease Society (Nicolle et al. 2019) recommends screening for and treating ASB in pregnant women, with the shortest and most effective course of antibiotics to reduce over-prescription that leads to resistance. Given the safety and potential efficacy of cranberry XE "Cranberries" as a prophylaxis, its use by pregnant women may be warranted.

4.9 Cranberry and Antibiotics

Cranberry utilization as a UTI prophylactic may have benefit for curtailing the overuse of antibiotics for prevention and treatment. Antibiotic over-prescribing is a world-wide issue leading to resistance development to many of the drugs used to treat UTI's. According to the World Health Organization (WHO 2014) Antimicrobial Resistance Global Report on Surveillance, resistance to fluoroquinolones for controlling UTIs is very widespread, and compared to the 1980's, resistance rates have gone from zero to 100% in many parts of the world. Cranberry products, therefore, may be a prudent nutritional therapy that can help maintain urinary tract health (Blumberg et al. 2013). Importantly, the Cochrane Review (Jepson et al. 2012) concluded that in studies comparing low-dose antibiotics to cranberry for UTI prevention, there was little difference between cranberry and antibiotic prophylaxes, with both being similarly effective. In fact, in a study conducted by Beerepoot et al. (2011), antibiotic prophylaxis resulted in TMP-SMX resistance in 86.3% of fecal and 90.5% of asymptomatic bacteriuria *E. coli* isolates after 1 month on low-dose TMP-SMX, while in the cranberry group, 23.7% of fecal and 28.1% of asymptomatic bacteriuria *E. coli* isolates were TMP-SMX resistant. These same researchers also found increased resistance rates for trimethoprim, amoxicillin, and ciprofloxacin in these *E. coli* isolates after 1 month in the TMP-SMX group. Due to the very low risk of resistant bacterial strain development, cranberry was recommended by these study authors as a viable alternative to low-dose antibiotics to prevent UTI. Since the increasing prevalence of *E. coli* resistance to first-line antimicrobials in the treatment of acute UTI in women is also a serious problem (Stapleton 2013), use of cranberry to prevent initial infections may help reduce the need for subsequent antibiotic treatments and slow the pace of resistance development.

4.10 Conclusions

Efficacy of cranberry for urinary tract health is likely due to multiple effects that include anti-adhesion activity, modulation of bacterial motility, bactericidal activity, immune modulation, and urine acidification. This assessment has recently been confirmed by the Committee for Medicinal Products for Human Use of the European Medicines Agency (2016) regarding the mode of action of proanthocyanidins. Based on available data, the committee members opine that metabolites of PACs and other cranberry constituents exhibit complex, pharmacological activity and rule out a purely mechanical mode of action, thus challenging the marketing of cranberry products as medical devices. Consequently, the European Commission issued a decision that "the group of products whose principal intended action, depending on proanthocyanidins (PAC) present in cranberry (*Vaccinium macrocarpon*), is to prevent or treat cystitis, does not fall within the definition of medical devices (European Commission 2017). In 2016, the UK Medicines and Healthcare products Regulatory Agency (MHRA) granted the first ever Traditional Herbal Registration Certificate for a traditional herbal medicinal product containing cranberry as an active ingredient (revised 2018, UKPAR 2019), thus affirming its status as an herbal medicine. This product has since been registered as a Traditional Herbal Medicine in several other European countries.

A-type cranberry PACs are predominantly associated with the anti-adhesion activity, with suggestions that 36 mg of PACs daily (as determined by DMAC with the A2 reference standard) is the target dose. This 36-mg dose is the amount of PACs in a typical 300 mL serving of cranberry juice cocktail (27% juice) which showed efficacy in prevention of bacteriuria in elderly women (Avorn 1996). Efficacy has been demonstrated for a variety of preparations including cranberry juice, cranberry juice cocktail, cranberry juice extracts, dried fruit, solid extracts, and a syrup. An important aspect of cranberry as a nutritional approach in potential prevention of UTIs is in lessening the need for conventional antibiotic therapy that leads to resistant bacterial strains.

Cranberry is generally recognized as safe (GRAS) and lacks any adverse side effects acutely when consumed as a typical part of the diet. Long-term human studies similarly indicate a generally high level of safety for cranberry and its preparations used as foods or dietary supplements.

The accumulative *in vivo* data, including numerous positive clinical studies with more than 3000 thousand subjects, along with strong mechanistic rationale, suggest efficacy of cranberry and its preparations for maintaining urinary tract health and the potential to help prevent UTIs. Conversely, a number of studies failed to show efficacy in UTI prophylaxis or treatment, likely due to methodological issues and the use of unstandardized cranberry products that were not administered at an efficacious dose. The overall consensus from recent reviews and meta-analyses is that there are beneficial effects of cranberry consumption on reducing risk of UTI recurrence. Additional studies are needed to determine the optimal dose, frequency of administration, length of consumption, subject characteristics, and product form.

References

Afshar K, Stothers L, Scott H, MacNeily AE (2012) Cranberry juice for the prevention of pediatric urinary tract infection: a randomized controlled trial. J Urol 188:1584–1587
Alfaro-Viquez E, Esquivel-Alvarado D, Madrigal-Carballo S, Krueger CG, Reed JD (2019) Proanthocyanidin-chitosan composite nanoparticles prevent bacterial invasion and colonization of gut epithelial cells by extra-intestinal pathogenic Escherichia coli. Int J Biol Macromol 135:630–636

Anger J, Lee U, Ackerman L, Chou R, Chughtai B (2019) Recurrent uncomplicated urinary tract infections in women: AUA/CUA/SUFU guideline

Avorn J, Monane M, Gurwitz JH, Glynn RJ, Choodnovskiy I, Lipsitz LA (1994) Reduction of bacteriuria and pyuria after ingestion of cranberry juice. JAMA 271:751–754

Avorn J (1996) The effect of cranberry juice on the presence of bacteria and white blood cells in the urine of elderly women. What is the role of bacterial adhesion? Adv Exp Med Biol 408:185–186

Bailey DT, Dalton C, Daugherty FJ, Tempesta MS (2007) Can a concentrated cranberry extract prevent recurrent urinary tract infections in women? A pilot study. Phytomedicine 14:237–241

Barbosa-Cesnik C, Brown MB, Buxton M, Zhang L, DeBusscher J, Foxman B (2011) Cranberry juice fails to prevent recurrent urinary tract infection: results from a randomized placebo-controlled trial. Clin Infect Dis 52:23–30

Barnoiu OS, Sequeira J, del Moral G, Sanchez-Martinez N, Diaz-Molina FL, Baena-Gonzalez V (2015) American cranberry (proanthocyanidin 120 mg): its value for the prevention of urinary tract infections after ureteral catheter placement. Actas Urol Esp 39:112–117

Beerepoot MA, ter Riet G, Nys S, van der Wal WM, de Borgie CA, de Reijke TM, Prins JM, Koeijers J, Verbon A, Stobberingh E et al (2011) Cranberries vs antibiotics to prevent urinary tract infections: a randomized double-blind noninferiority trial in premenopausal women. Arch Intern Med 171:1270–1278

Bianco L, Perrelli E, Towle V, Van Ness PH, Juthani-Mehta M (2012) Pilot randomized controlled dosing study of cranberry capsules for reduction of bacteriuria plus pyuria in female nursing home residents. J Am Geriatr Soc 60:1180–1181

Blatherwick NR (1914) The specific role of foods in relation to the composition of the urine. Arch Intern Med 14:409–450

Blatherwick NR, Long L (1923) Studies of urinary acidity. II. The increased acidity produced by eating prunes and cranberries. J Biol Chem 57:815–819

Blumberg JB, Camesano TA, Cassidy A, Kris-Etherton P, Howell A, Manach C, Ostertag LM, Sies H, Skulas-Ray A, Vita JA (2013) Cranberries and their bioactive constituents in human health. Adv Nutr 4:618–632

Bodel PT, Cotran R, Kass EH (1959) Cranberry juice and the antibacterial action of hippuric acid. J Lab Clin Med 54:881–888

Bone K, Morgan M (1999) Vaccinium macrocarpon-cranberry. MediHerb Prof Rev 72:1–4

Bonetta A, Di Pierro F (2012) Enteric-coated, highly standardized cranberry extract reduces risk of UTIs and urinary symptoms during radiotherapy for prostate carcinoma. Cancer Manag Res 4:281–286

Botto H, Neuzillet Y (2010) Effectiveness of a cranberry (Vaccinium macrocarpon) preparation in reducing asymptomatic bacteriuria in patients with an ileal enterocystoplasty. Scand J Urol Nephrol 44:165–168

Borukh IF, Kirbaba VI, Senchuk GV (1972) Antimicrobial properties of cranberry. Vopr Pitan 31:82

Brumfitt W, Percival A (1962) Adjustment of urine pH in the chemotherapy of urinary-tract infections. Lancet 279:186–190

Bruyère F, Azzouzi AR, Lavigne JP, Droupy S, Coloby P, Game X, Karsenty G, Issartel B, Ruffion A, Misrai V, Sotto A, Allaert FA (2019) A multicenter, randomized, placebo-controlled study evaluating the efficacy of a combination of Propolis and cranberry (Vaccinium macrocarpon) (DUAB®) in preventing low urinary tract infection recurrence in women complaining of recurrent cystitis. Urol Int 22:1–8

Burleigh AE, Benck SM, McAchran SE, Reed JD, Krueger CG, Hopkins WJ (2013) Consumption of sweetened, dried cranberries may reduce urinary tract infection incidence in susceptible women — a modified observational study. Nutr J 12:139

Cadkova I, Doudova L, Novackova M, Chmel R (2009) Effect of cranberry extract capsules taken during the perioperative period upon the post-surgical urinary infection in gynecology. Ceska Gynekol 74:454–458

Caljouw MA, van den Hout WB, Putter H, Achterberg WP, Cools HJ, Gussekloo J (2014) Effectiveness of cranberry capsules to prevent urinary tract infections in vulnerable older persons: a double-blind randomized placebo-controlled trial in long-term care facilities. J Am Geriatr Soc 62:103–110

Carballo SM, Haas L, Krueger CG, Reed JD (2017) Cranberry Proanthocyanidins – protein complexes for macrophage activation. Food Funct 8(9):3374–3382

Chen O, Mah E, Liska D (2019) Effect of cranberry on urinary tract infection risk: a meta-analyses (P06-116-19). Curr Dev Nutr 3(Suppl 1). pii: nzz031.P06-116-19

Cowan CC, Hutchison C, Cole T, Barry SJE, Paul J, Reed NS, Russell JM (2012) A randomised double-blind placebo-controlled trial to determine the effect of cranberry juice on decreasing the incidence of urinary symptoms and urinary tract infections in patients undergoing radiotherapy for cancer of the bladder or cervix. Clin Oncol 24:e31–e38

DE Leo V, Cappelli V, Massaro MG, Tosti C, Morgante G (2017) Evaluation of the effects of a natural dietary supplement with cranberry, Noxamicina® and D-mannose in recurrent urinary infections in perimenopausal women. Minerva Ginecol 69(4):336–341

de Llano DG, Esteban-Fernández A, Sánchez-Patán F, Martínlvarez PJ, Moreno-Arribas MV, Bartolomé B (2015) Anti-adhesive activity of cranberry phenolic compounds and their microbial-derived metabolites against uropathogenic Escherichia coli in bladder epithelial cell cultures. Int J Mol Sci 16(6):12119–12130

Di Martino P, Agniel R, David K, Templer C, Gaillard JL, Denys P, Botto H (2006) Reduction of Escherichia coli adherence to uroepithelial bladder cells after consumption of cranberry juice: a double-blind randomized placebo-controlled cross-over trial. World J Urol 24:21–27

Dignam RR, Ahmed M, Kelly KG, Denman SJ, Zayon M, Kleban M (1998) The effect of cranberry juice on urinary tract infection rates in a long-term care facility. Ann Long-Term Care 6:163–167

Durham SH, Stamm PL, Eiland LS (2015 Dec) Cranberry products for the prophylaxis of urinary tract infections in pediatric patients. Annals of Pharmacotherapy. 49(12):1349–1356

Eck P (1931) The American cranberry. Rutgers University Pr, New Brunswick. 420 p. Reprint Edition 1990

Essadi F, Elmehashi MO (2010) Efficacy of cranberry juice for the prevention of urinary tract infections in pregnancy. J Matern-Fetal Neonat Med 23:378

European Commission (2017) Commission Implementing Decision (EU) 2017/1445 of 8 August 2017 on the group of products whose principal intended action, depending on proanthocyanidins (PAC) present in cranberry (Vaccinium macrocarpon), is to prevent or treat cystitis (notified under document C(2017) 5341). Accessed November 16, 2017 at http://data.europa.eu/eli/dec_impl/2017/1445/oj

European Medicines Agency (2016) CHMP scientific opinion to DG internal market, industry, entrrepreneurship and SMEs, Unit GROW D.4. "Health Technology & Cosmetics" on the principal mode of action of proanthocyanidins intended to be used for prevention and treatment of urinary tract infections. EMA/427414/2016. Accessed November 16, 2017 at http://ec.europa.eu/DocsRoom/documents/19961

Feliciano RP, Krueger CG, Shanmuganayagam D, Vestling MM, Reed JD (2012) Deconvolution of matrix-assisted laser desorption/ionization time-of-flight mass spectrometry isotope patterns to determine ratios of A-type to B-type interflavan bonds in cranberry proanthocyanidins. Food Chem 135:1485–1493

Feliciano RP, Meudt JJ, Shanmuganayagam D, Krueger CG, Reed JD (2014) Ratio of 'A-type' to 'B-type' proanthocyanidin interflavan bonds affects extra-intestinal pathogenic Escherichia coli invasion of gut epithelial cells. Am Chem Soc 62:3919–3925

Feliciano RP, Mills CE, Istas G, Heiss C, Rodriguez-Mateos A (2017 Mar) Absorption, metabolism and excretion of cranberry (poly)phenols in humans: a dose response study and assessment of inter-individual variability. Nutrients 11:9(3)

Fellers CR, Redmon BC, Parrott EM (1933) Effect of cranberries on urinary acidity and blood alkali reserve. J Nutr 6(5):455–463

Felter HW, Lloyd JU (1905) King's American dispensatory, vol 2, 19th edn. Ohio Valley, Cincinnati, 2172 p

Fernández-Puentes V, Uberos J, Rodríguez-Belmonte R, Nogueras-Ocaña M, Blanca-Jover E, Narbona-López E (2015) Efficacy and safety profile of cranberry in infants and children with recurrent urinary tract infection. An Pediatr 82(6):397–403

Ferrara P, Romaniello L, Vitelli O, Gatto A, Serva M, Cataldi L (2009) Cranberry juice for the prevention of recurrent urinary tract infections: a randomized controlled trial in children. Scand J Urol Nephrol 43:369–372

Foda MM, Middlebrook PF, Gatfield CT, Potvin G, Wells G, Schillinger JF (1995) Efficacy of cranberry in prevention of urinary tract infection in a susceptible pediatric population. Can J Urol 2:98–102

Foo LY, Lu Y, Howell AB, Vorsa N (2000a) The structure of cranberry proanthocyanidins which inhibit adherence of uropathogenic P-fimbriated Escherichia coli in vitro. Phytochemistry 54:173–181

Foo LY, Lu Y, Howell AB, Vorsa N (2000b) A-type proanthocyanidin trimers from cranberry that inhibit adherence of uropathogenic P-fimbriated Escherichia coli. J Nat Prod 63:1225–1228

Foxman B (1990) Recurring urinary tract infection: incidence and risk factors. Am J Public Health 80:331–333

Foxman B (2002) Epidemiology of urinary tract infections: incidence, morbidity, and economic costs. Am J Med 113(Suppl 1A):5S–13S

Foxman B, Brown P (2003) Epidemiology of urinary tract infections: transmission and risk factors, incidence, and costs. Infect Dis Clin N Am 17:227–241

Foxman B, Cronenwett AEW, Spino C, Berger MB, Morgan DM (2015) Cranberry juice capsules and urinary tract infection after surgery: results of a randomized trial. Amer J Obst Gynecol 213(2):194ff

Fu Z, Liska D, Talan D, Chung M (2017) Cranberry reduces the risk of urinary tract infection recurrence in otherwise healthy women: a systematic review and meta-analysis. J Nutr 147(12):2282–2288

Gallien P, Amarenco G, Benoit N, Bonniaud V, Donzé C, Kerdraon J, de Seze M, Denys P, Renault A, Naudet F et al (2014) Cranberry versus placebo in the prevention of urinary infections in multiple sclerosis: a multicenter, randomized, placebo-controlled, double-blind trial. Mult Scler 20:1252–1259

Gibson L, Pike L, Kilbourn JP (1991) Clinical study: effectiveness of cranberry juice in preventing urinary tract infections in long-term care facility patients. J Nat Med 2:45–47

Griebling TL (2005) Urologic diseases in America project: trends in resource use for urinary tract infections in women. J Urol 173:1281–1287

Grove K, Robinson R (1931) An anthocyanin of Oxycoccus macrocarpus Pers. Biochem J 25:1706–1711

Gu L, Kelm MA, Hammerstone JF, Beecher G, Holden J, Haytowitz D, Prior RL (2003) Screening of foods containing proanthocyanidins and their structural characterization using LC-MS/MS and thiolytic degradation. J Agric Food Chem 51:7513–7521

Gu L, Kelm MA, Hammerstone JF, Beecher G, Holden J, Haytowitz D, Gebhardt S, Prior RL (2004) Concentrations of proanthocyanidins in common foods and estimations of normal consumption. J Nutr 134:613–617

Gunnarsson AK, Gunningberg L, Larsson S, Jonsson KB (2017) Cranberry juice concentrate does not significantly decrease the incidence of acquired bacteriuria in female hip fracture patients receiving urine catheter: a double-blind randomized trial. Clin Interv Aging 12:137–143

Gupta K, Chou MY, Howell A, Wobbe C, Grady R, Stapleton AE (2007) Cranberry products inhibit adherence of P-fimbriated Escherichia coli to primary cultured bladder and vaginal epithelial cells. J Urol 177:2357–2360

Hamilton K, Bennett NC, Purdie G, Herst PM (2015) Standardized cranberry capsules for radiation cystitis in prostate cancer patients in New Zealand: a randomized double blinded, placebo controlled pilot study. Support Care Cancer 23(1):95–102

Hamilton-Miller JMT, Brumfitt W (1976) Methenamine and its salts as urinary tract antiseptics: variables affecting the antibacterial activity of formaldehyde, mandelic acid, and hippuric acid in vitro. Investig Urol 14:287–291

Hannon DB, Thompson JT, Khoo C, Juturu V, Vanden Heuvel JP (2016) Effects of cranberry extracts on gene expression in THP-1 cells. Food Sci Nutr 5(1):148–159

Haverkorn MJ, Mandigers J (1994) Reduction of bacteriuria and pyuria using cranberry juice. JAMA 272:590

Heitmann K, Nordeng H, Holst L (2013) Pregnancy outcome after use of cranberry in pregnancy—the Norwegian mother and child cohort study. BMC Complement Altern Med 13:345

Henig YS, Leahy MM (2000) Cranberry juice and urinary-tract health: science supports folklore. Nutrition 16:684–687

Hess MJ, Hess PE, Sullivan MR, Nee M, Yalla SV (2008) Evaluation of cranberry tablets for the prevention of urinary tract infections in spinal cord injured patients with neurogenic bladder. Spinal Cord 46:622–626

Hollembaek H (1865) American eclectic materia medica. Hollembaek, Philadelphia, 676 p

Howell A (2013) Commentary on: Jepson RG, Williams G, Craig JC. Cranberries for preventing urinary tract infections. Cochrane Database Syst Rev 10:CD001321

Howell AB, Vorsa N, Der Marderosian A, Foo LY (1998) Inhibition of the adherence of P-fimbriated Escherichia coli to uroepithelial-cell surfaces by proanthocyanidin extracts from cranberries. New Eng J Med 339:1085–1086

Howell AB, Leahy M, Kurowska E, Guthrie N (2001) In vivo evidence that cranberry proanthocyanidins inhibit adherence of P-fimbriated E. coli bacteria to uroepithelial cells. FASEB J 15:A284

Howell AB, Reed JD, Krueger CG, Winterbottom R, Cunningham DG, Leahy M (2005) A-type cranberry proanthocyanidins and uropathogenic bacterial anti-adhesion activity. Phytochemistry 66:2281–2291

Howell AB, Botto H, Combescure C, Blanc-Potard AB, Gausa L, Matsumoto T, Tenke P, Sotto A, Lavigne JP (2010) Dosage effect on uropathogenic Escherichia coli anti-adhesion activity in urine following consumption of cranberry powder standardized for proanthocyanidin content: a multicentric randomized double blind study. BMC Infect Dis 10:94

Huang YC, Chen PS, Tung TH (2017) Effectiveness of cranberry ingesting for prevention of urinary tract infection: a systematic review and meta-analysis of randomized controlled trials. World Acad Sci, Eng Technol, Int J Med Health Sci 5(5)

Ibragimov DI, Kazanskaia GB (1981) Antimicrobial action of cranberry bush, common yarrow and Achillea biebersteinii. Antibiotiki 26:108–109

Jackson B, Hicks LE (1997) Effect of cranberry juice on urinary pH in older adults. Home Healthc Nurse 15:199–202

Jepson RG, Craig JC (2008) Cranberries for preventing urinary tract infections. Cochrane Database Syst Rev (1). Art. No.:CD001321

Jepson RG, Williams G, Craig JC (2012) Cranberries for preventing urinary tract infections. Cochrane Database Syst Rev 10:CD001321

Juthani-Mehta M, Perley L, Chen S, Dziura J, Gupta K (2010) Feasibility of cranberry capsule administration and clean-catch urine collection in long-term care residents. J Am Geriatr Soc 58:2028–2030

Juthani-Mehta M, Van Ness PH, Bianco L, Rink A, Rubeck S, Ginter S, Argraves S, Charpentier P, Acampora D, Trentalange M, Quagliarello V (2016) Effect of cranberry capsules on bacteriuria plus pyuria among older women in nursing homes: a randomized clinical trial. JAMA 316(18):1879–1887

Kahn HD, Panariello VA, Saeli J, Sampson JR, Schwartz E (1967) Implications for therapy of urinary tract infection and calculi: effect of cranberry juice on urine. J Am Diet Assoc 51:251–254

Karonen M, Leikas A, Loponen J, Sinkkonen J, Ossipov V, Pihlaja K (2007) Reversed-phase HPLC–ESI–MS analysis of birch leaf proanthocyanidins after their acidic degradation in the presence of nucleophiles. Phytochem Anal 18:378–386

Kerr KG (1999) Cranberry juice and prevention of recurrent urinary tract infection. Lancet 353:673

Kinney AB, Blount M (1979) Effect of cranberry juice on urinary pH. Nurs Res 28:287–290

Konowalchuk J, Speirs JI (1978) Antiviral effect of commercial juices and beverages. Appl Environ Microbiol 35:1219–1220

Kontiokari T, Salo J, Eerola E, Uhari M (2005) Cranberry juice and bacterial colonization in children – a placebo-controlled randomized trial. Clin Nutr 24:1065–1072

Kontiokari T, Sundqvist K, Nuutinen M, Pokka T, Koskela M, Uhari M (2001) Randomised trial of cranberry-lingonberry juice and Lactobacillus GG drink for the prevention of urinary tract infections in women. BMJ 322:1571

Kraemer RJ (1964) Cranberry juice and the reduction of ammoniacal odor of urine. Southwest Med 45:211–212

Krueger CG, Vestling MM, Reed JD (2003) Matrix-assisted laser desorption/ionization time-of-flight mass spectrometry of heteropolyflavan-3-ols and glucosylated heteropolyflavans in sorghum (Sorghum bicolor (l.) Moench). J Agric Food Chem 51:538–543

Krueger CG, Maudi JJ, Howell AB, Khoo C, Shanmuganayugari D, Reed JD (2013a) Consumption of cranberry powder shifts urinary protein profile in healthy human subjects. FASEB J 27:637.32

Krueger CG, Reed JD, Feliciano R, Howell AB (2013b) Quantifying and characterizing proanthocyanidins in cranberries in relation to urinary tract health. Anal Bioanal Chem 405:4385–4395

LaPlante KL, Sarkisian SA, Woodmansee S, Rowley DC, Seeram NP (2012) Effects of cranberry extracts on growth and biofilm production of Escherichia coli and Staphylococcus species. Phytother Res 26:1371–1374

Lavigne JP, Bourg G, Combescure C, Botto H, Sotto A (2008) In-vitro and in-vivo evidence of dose-dependent decrease of uropathogenic Escherichia coli virulence after consumption of commercial Vaccinium macrocarpon (cranberry) capsules. Clin Microbiol Infect 14:350–355

Ledda A, Bottari A, Luzzi R, Belcaro G, Hu S, Dugall M, Hosoi M, Ippolito E, Corsi M, Gizzi G, Morazzoni P (2015) Cranberry supplementation in the prevention of non-severe lower urinary tract infections: a pilot study. Eur Rev Med Pharmacol Sci 19(1):77–80

Ledda A, Belcaro G, Dugall M, Feragalli B, Riva A, Togni S, Giacomelli L (2016) Supplementation with high titer cranberry extract (Anthocran®) for the prevention of recurrent urinary tract infections in elderly men suffering from moderate prostatic hyperplasia: a pilot study. Eur Rev Med Pharmacol Sci 20(24):5205–5209

Ledda A, Belcaro G, Dugall M, Riva A, Togni S, Eggenhoffner R, Giacomelli L (2017) Highly standardized cranberry extract supplementation (Anthocran®) as prophylaxis in young healthy subjects with recurrent urinary tract infections. Eur Rev Med Pharmacol Sci 21(2):389–393

Lee BB, Haran MJ, Hunt LM, Simpson JM, Marial O, Rutkowski SB, Middleton JW, Kotsiou G, Tudehope M, Cameron ID (2007) Spinal-injured neuropathic bladder antisepsis (SINBA) trial. Spinal Cord 45:542–550

Lee WK, Ko MC, Huang CS (2016) Cranberries for preventing recurrent urinary tract infections in uncircumcised boys. Altern Ther Health Med 22(6):20

Letouzey V, Ulrich D, Demattei C, Alonso S, Huberlant S, Lavigne JP, de Tayrac R (2017) Cranberry capsules to prevent nosocomial urinary tract bacteriuria after pelvic surgery: a randomised controlled trial. BJOG 124(6):912–917

Light I, Gursel E, Zinnser HH (1973) Urinary ionized calcium in urolithiasis: effect of cranberry juice. Urology 1:67–70

Lin SC, Wang CC, Shih SC, Tjung JJ, Tsou MT, Lin CJ (2014) Prevention of asymptomatic bacteriuria with cranberries and Roselle juice in home-care patients with long-term urinary catheterization. Intern J Gerontol 8:152–156

Linsenmeyer TA, Harrison B, Oakley A, Kirshblum S, Stock JA, Millis SR (2004) Evaluation of cranberry supplement for reduction of urinary tract infections in individuals with neurogenic bladders secondary to spinal cord injury. A prospective, double-blinded, placebo-controlled, crossover study. J Spinal Cord Med 27:29–34

Luís Â, Domingues F, Pereira L (2017) Can cranberries contribute to reduce the incidence of urinary tract infections? A systematic review with meta-analysis and trial sequential analysis of
clinical trials. J Urol 198:614–621

Maki KC, Kaspar KL, Khoo C, Derrig LH, Schild AL, Gupta K (2016) Consumption of a cranberry juice beverage lowered the number of clinical urinary tract infection episodes in women
with a recent history of urinary tract infection. Am J Clin Nutr 103(6):1434–1442

Mantzorou M, Giaginis C (2018) Cranberry consumption against urinary tract infections: clinical
state of- the-art and future perspectives. Curr Pharm Biotechnol 19(13):1049–1063

Mathison BD, Kimble LL, Kaspar KL, Khoo C, Chew BP (2014) Consumption of cranberry beverage improved endogenous antioxidant status and protected against bacteria adhesion in healthy
humans: a randomized controlled trial. Nutr Res 34:420–427

Mazokopakis EE, Karefilakis CM, Starakis IK (2009) Efficacy of cranberry capsules in prevention of urinary tract infections in postmenopausal women. J Altern Complement Med 15:1155

McCall J, Hidalgo G, Asadishad B, Tufenkji N (2013) Cranberry impairs selected behaviors essential for virulence in Proteus mirabilis HI4320. Can J Microbiol 59(6):430–436

McGuinness SD, Krone R, Metz LM (2002) A double-blind, randomized, placebo-controlled trial
of cranberry supplements in multiple sclerosis. J Neurosci Nurs 34:4–7

McKay DL, Chen CY, Zampariello CA, Blumberg JB (2015) Flavonoids and phenolic acids from
cranberry juice are bioavailable and bioactive in healthy older adults. Food Chem 168:233–240

McMurdo ME, Bissett LY, Price RJ, Phillips G, Crombie IK (2005) Does ingestion of cranberry
juice reduce symptomatic urinary tract infections in older people in hospital? A double-blind,
placebo-controlled trial. Age Ageing 34:256–261

McMurdo ME, Argo I, Phillips G, Daly F, Davey P (2009) Cranberry or trimethoprim for the prevention of recurrent urinary tract infections? A randomized controlled trial in older women. J
Antimicrob Chemother 63:389–395

Micali S, Isgro G, Bianchi G, Miceli N, Calapai G, Navarra M (2014) Cranberry and recurrent
cystitis: more than marketing? Crit Rev Food Sci Nutr 54:1063–1075

Monroe J (1824) The American botanist and American physician. Jonathan Morrison, Eaton, 203 p

Mutlu H, Ekinci Z (2012) Urinary tract infection prophylaxis in children with neurogenic bladder
with cranberry capsules: randomized controlled trial. ISRN Pediatr 2012:1–4

Neto CC, Krueger CG, Lamoureaux TL, Kondo M, Vaisberg AJ, Hurta RAR, Curtis S, Matchett
MD, Yeung H, Sweeney-Nixon MI et al (2006) MALDI-TOF MS characterization of proanthocyanidins from cranberry fruit (Vaccinium macrocarpon) that inhibit tumor cell growth and
matrix metalloproteinase expression in vitro. J Sci Food Agric 86:18–25

Nickey KE (1975) Urinary pH: effect of prescribed regimes of cranberry juice and ascorbic acid.
Arch Phys Med Rehabil 56:556

Nicolle LE, Gupta K, Bradley SF, Colgan R, DeMuri GP, Drekonja D, Eckert LO,
Geerlings SE, Köves B, Hooton TM, Juthani-Mehta M, Knight SL, Saint S, Schaeffer AJ,
Trautner B, Wullt B, Siemieniuk R (2019) Clinical practice guideline for the management of
asymptomatic Bacteriuria: 2019 update by the Infectious Diseases Society of America. Clin
Infect Dis 68:e83–e110

Nowack R, Schmitt W (2008) Cranberry juice for prophylaxis of urinary tract infections – conclusions from clinical experience and research. Phytomedicine 15:653–667

Ofek I, Goldhar J, Sharon N (1996) Anti-Escherichia coli adhesin activity of cranberry and blueberry juices. In: Kahane I, Ofek I (eds) Toward anti-adhesion therapy for microbial diseases.
Plenum, New York, pp 179–183

Papas PN, Brusch CA, Ceresia GC (1966) Cranberry juice in the treatment of urinary tract infections. Southwest Med 47:17–20

Patel N, Daniels IR (2000) Botanical perspectives on health: of cystitis and cranberries. J R Soc
Promot Heal 120:52–53

Peron G, Sut S, Pellizzaro A, Brun P, Voinovich D, Castagliuolo I, Dall'Acqua S (2017) The antiad-
hesive activity of cranberry phytocomplex studied by metabolomics: intestinal PAC-A metabo-

lites but not intact PAC-A are identified as markers in active urines against uropathogenic *Escherichia coli*. Fitoterapia 122:67–75

Porter ML, Krueger CG, Wiebe DA, Cunningham DG (2001) Cranberry proanthocyanidins associate with low-density lipoprotein and inhibit in vitro Cu2+–induced oxidation. J Sci Food Agric 81:1306–1313

Rafinesque CS (1830) Medical flora, or manual of the medical botany of the United States of North America, vol 2. Samuel C Atkinson, Philadelphia, 276 p

Rajbhandari R, Peng N, Moore R, Arabshahi A, Wyss JM, Barnes S, Prasain JK (2011) Determination of cranberry phenolic metabolites in rats by liquid chromatography-tandem mass spectrometry. J Agric Food Chem 59:6682–6688

Reed JD, Krueger CG, Vestling MM (2005) MALDI-TOF mass spectrometry of oligomeric food polyphenols. Phytochemistry 66:2248–2263

Reid G (1999) Potential preventive strategies and therapies in urinary tract infection. World J Urol 17:359–363

Risco E, Miguelez C, Sanchez de Badajoz E, Rouseaud A (2010) Effect of American cranberry (Cysticlean) on Escherichia coli adherence to bladder epithelial cells. In vitro and in vivo study. Arch Esp Urol 63:422–430

Rogers J (1991) Clinical: pass the cranberry juice. Nurs Times 27:36–37

Romero R, Oyarzun E, Mazor M, Sirtori M, Hobbins JC, Bracken M (1989) Meta-analysis of the relationship between asymptomatic bacteriuria and preterm delivery/low birth weight. Obstet Gynecol 73:576–582

Salo J, Uhari M, Helminen M, Korppi M, Nieminen T, Pokka T, Kontiokari T (2012) Cranberry juice for the prevention of recurrences of urinary tract infections in children: a randomized placebo-controlled trial. Clin Infect Dis 54:340–346

Schappert SM, Rechtsteiner EA (2011) Ambulatory medical care utilization estimates for 2007. Vital Health Stat 13:1–38

Schlager TA, Anderson S, Trudell J, Hendley JO (1999) Effect of cranberry juice on bacteriuria in children with neurogenic bladder receiving intermittent catheterization. J Pediatr 135:698–702

Schmidt DR, Sobota AE (1988) An examination of the anti-adherence activity of cranberry juice on urinary and nonurinary bacterial isolates. Microbios 55(224–225):173–181

Schultz AS (1984a) Efficacy of cranberry on urinary pH. J Com Health Nursing 1:159–169

Schultz AS (1984b) Efficacy of cranberry juice and ascorbic acid in acidifying the urine in multiple sclerosis subjects. J Community Health Nurs 1:139–169

Sengupta K, Alluri KV, Golakoti T, Gottumukkala GV, Raavi J, Kotchrlakota L, Sigalan SC, Dey D, Gosh S, Chatterjee A (2011) A randomized, double blind, controlled dose dependent clinical trial to evaluate the efficacy of a proanthocyandines standardized whole cranberry (Vaccinium macrocarpon) powder on infections of the urinary tract. Curr Bioac Comp 7:39–46

Shanmuganayugari D, Johnson RE, Meudt JJ, Felciano RP, Kohlman KL, Nechyporenko AV, Heinz J, Krueger CG, Reed JD (2013) A-type proanthocyanidins from cranberry inhibit the ability of extraintestinal pathogenic E. coli to invade gut epithelial cells and resist killing by macrophages. FASEB J 27:637.16

Shatkin-Margolis A, Warehime J, Pauls RN (2018) Cranberry supplementation does not reduce urinary tract infections in patients with indwelling catheters after pelvic reconstructive surgery. Female Pelvic Med Reconstr Surg 24(2):130–134

Sheiner E, Mazor-Drey E, Levy A (2009) Asymptomatic bacteriuria during pregnancy. J Matern Fetal Neonatal Med 22:423–427

Singh I, Gautam LK, Kaur IR (2016) Effect of oral cranberry extract (standardized proanthocyanidin-A) in patients with recurrent UTI by pathogenic E. coli: a randomized placebo-controlled clinical research study. Int Urol Nephrol 48(9):1379–1386

Sobota AE (1984) Inhibition of bacterial adherence by cranberry juice: potential use for the treatment of urinary tract infections. J Urol 131(5):1013–1016

Stapleton AE (2013) Cranberry-containing products are associated with a protective effect against urinary tract infections. Arch Intern Med 172:988–996

Stapleton AE, Dziura J, Hooton TM, Cox ME, Yarova-Yarovaya Y, Chen S, Gupta K (2012) Recurrent urinary tract infection and urinary Escherichia coli in women ingesting cranberry juice daily: a randomized controlled trial. Mayo Clin Proc 87:143–150

Stearns S (1801) The American herbal or materia medica. David Carlisle, Walpole, 360 p

Sternlieb P (1963) Cranberry juice in renal disease. New Engl J Med 268:57

Stothers L (2002) A randomized trial to evaluate effectiveness and cost effectiveness of naturopathic cranberry products as prophylaxis against urinary tract infection in women. Can J Urol 9:1558–1562

Swartz JH, Medrek TF (1968) Antifungal properties of cranberry juice. Appl Microbiol 16(1524):1527

Takahashi S, Hamasuna R, Yasuda M, Arakawa S, Tanaka K, Ishikawa K, Kiyota H, Hayami H, Yamamoto S, Kubo T et al (2013) A randomized clinical trial to evaluate the preventive effect of cranberry juice (UR65) for patients with recurrent urinary tract infection. J Infect Chemother 19:112–117

Tempera G, Corsello S, Genovese C, Caruso FE, Nicolosi D (2010) Inhibitory activity of cranberry extract on the bacterial adhesiveness in the urine of women: an ex-vivo study. Int J Immunopathol Pharmacol 23:611–618

Thomas D, Rutman M, Cooper K, Abrams A, Finkelstein J, Chughtai B (2017) Does cranberry have a role in catheter-associated urinary tract infections? Can Urol Assoc J E421(E424):11

Uberos J, Nogueras-Ocana M, Fernandez-Puentes V, Rodriguez-Belmonte R, Narbona-Lopez E, Molina-Carballo A, Munoz-Hoyos A (2012) Cranberry syrup vs trimethoprim in the prophylaxis of recurrent urinary tract infections among children: a controlled trial. Open Access J Clin Trials 4:31–38

Uberos J, Rodríguez-Belmonte R, Rodríguez-Pérez C, Molina-Oya M, Blanca-Jover E, Narbona-Lopez E, Muñoz-Hoyos A (2015) Phenolic acid content and antiadherence activity in the urine of patients treated with cranberry syrup (Vaccinium macrocarpon) vs. trimethoprim for recurrent urinary tract infection. J Funct Foods 18:608–616

Ujvary I, Orlik J, Racz G, Donath (1961) On the fungistatic effect of the cranberry's and cranberry products. Univ Mass Agr Exp (Vaccinium Vitis Idoea L.) crop-extract. Orv Szemle 4:406–409

UKPAR (2019) Ellura capsules. Medicines and Healthcare products Regulatory Agency (MHRA) 2019. Accessed August 10, 2019 at http://www.mhra.gov.uk/home/groups/par/documents/websiteresources/con745794.pdf

Upton R, Brendler T (Eds.) (2016). Cranberry fruit. American herbal pharmacopoeia. Monograph revision

USP 24-NF 19 (1999) United States pharmacopeia-national formulary. US Pharmacopeial Convention Inc, Rockville. 2729 p

USP-NF PF (2019) United States pharmacopeia-national formulary pharmaceutical forum, vol 46(6). US Pharmacopeial Convention Inc., Rockville

Valentova K, Stejskal D, Bednar P, Vostalova J, Cihalik C, Vecerova R, Koukalova D, Kolar M, Reichenbach R, Sknouril L et al (2007) Biosafety, antioxidant status, and metabolites in urine after consumption of dried cranberry juice in healthy women: a pilot double-blind placebo-controlled trial. J Agric Food Chem 55:3217–3224

Vidlar A, Vostalova J, Ulrichova J, Student V, Stejskal D, Reichenbach R, Vrbkova J, Ruzicka F, Simanek V (2009) Beneficial effects of cranberries on prostate health: evidence from a randomized controlled trial. Eur Urol Suppl 8:660

Vidlar A, Student V, Vostalova J, Fromentin E, Roller M, Simanek V (2016) Cranberry fruit powder (Flowens™) improves lower urinary tract symptoms in men: a double-blind, randomized, placebo-controlled study. World J Urol 34(3):419–424

Vostalova J, Vidlar A, Simanek V, Galandakova A, Kosina P, Vacek J, Vrbkova J, Zimmermann BF, Ulrichova J, Student V (2015) Are high proanthocyanidins key to cranberry efficacy in the prevention of recurrent urinary tract infection? Phytother Res 29(10):1559–1567

Waites KB, Canupp KC, Armstrong S, DeVivo MJ (2004) Effect of cranberry extract on bacteriuria and pyuria in persons with neurogenic bladder secondary to spinal cord injury. J Spinal Cord Med 27:35–40

Walker EB, Barney DP, Mickelsen JN, Walton RJ, Mickelsen RA (1997) Cranberry concentrate: UTI prophylaxis. J Fam Pract 45:167–168

Wan KS, Liu CK, Lee WK, Ko MC, Huang CS (2016) Cranberries for preventing recurrent urinary tract infections in uncircumcised boys. Altern Ther Health Med 22(6):20–23

Wang CH, Fang CC, Chen NC, Liu SS, Yu PH, Wu TY, Chen WT, Lee CC, Chen SC (2012) Cranberry-containing products for prevention of urinary tract infections in susceptible populations: a systematic review and meta-analysis of randomized controlled trials. Arch Intern Med 172:988–996

Wing DA, Rumney PJ, Preslicka CW, Chung JH (2008) Daily cranberry juice for the prevention of asymptomatic bacteriuria in pregnancy: a randomized, controlled pilot study. J Urol 180:1367–1372

Zafriri D, Ofek I, Adar R, Pocino M, Sharon N (1989) Inhibitory activity of cranberry juice on adherence of type 1 and type P fimbriated Escherichia coli to eucaryotic cells. Antimicrob Agents Chemother 33:92–98

Zhou Y, Zhuang W, Hu W, Liu GJ, Wu TX, Wu XT (2011) Consumption of large amounts of Allium vegetables reduces risk for gastric cancer in a meta-analysis. Gastroenterology 141:80–89

Chapter 5
Medicinal Attributes of American Elderberry

Andrew L. Thomas, Patrick L. Byers, P. Leszek Vincent, and Wendy L. Applequist

Abstract American elderberry (*Sambucus nigra* subsp. *canadensis*) is a common fruiting shrub native to much of Eastern North America. While the fruit and flowers have been used for eons as food and medicine by both early and contemporary North Americans, its use is seeing a resurgence. This has resulted in a renewed interest in horticultural development and cultivation of elderberry, with numerous new products developed especially in the form of dietary supplements. Recent scientific research continues to underscore the health-benefitting attributes of both elderberry fruit and flowers, and is further fueling the development of a significant elderberry value chain from production, to processing, marketing, and consumption.

Keywords *Sambucus* · Dietary supplement · Anti-oxidant · Elderflower

A. L. Thomas (✉)
Division of Plant Sciences, Southwest Research Center, University of Missouri, Mt. Vernon, MO, USA
e-mail: ThomasAL@missouri.edu

P. L. Byers
Cooperative Extension Service, University of Missouri, Marshfield, MO, USA
e-mail: ByersPL@missouri.edu

P. L. Vincent
Department of Biochemistry, and Division of Plant Sciences, University of Missouri, Columbia, MO, USA
e-mail: Leszek@missouri.edu

W. L. Applequist
Missouri Botanical Garden, St. Louis, MO, USA
e-mail: wendy.applequist@mobot.org

© Springer Nature Switzerland AG 2020   119
Á. Máthé (ed.), *Medicinal and Aromatic Plants of North America*, Medicinal and Aromatic Plants of the World 6,
https://doi.org/10.1007/978-3-030-44930-8_5

5.1 Introduction

American elderberry is a common deciduous woody shrub native to much of Eastern North America, ranging from Florida to Québec, and west to Manitoba and New Mexico (Allen et al. 2002; Hosie 1979; Small et al. 2004). It is also found in Mexico and Central America, where it may have been introduced by humans (e.g., Bolli 1994; Mejicanos and Ziller 1994; Pedraza and Williams-Linera 2003). The taxonomy of the widespread elderberry genus, *Sambucus* L., is much-debated. While *Sambucus* was formerly placed in the family Caprifoliaceae, molecular phylogenetic data have shown that it is best placed in a smaller segregate family (e.g., Donoghue et al. 2001; Angiosperm Phylogeny Group et al. 2016), which was formerly known as Adoxaceae but is now properly called Viburnaceae (Wilson 2016; Turland et al. 2017). Most botanists now concur that American elderberry is best considered a subspecies of the European elderberry (*Sambucus nigra* L., now best called *Sambucus nigra* subsp. *nigra*) and is most correctly called *Sambucus nigra* L. subsp. *canadensis* (L.) Bolli (Applequist 2015). Even the USDA National Plant Germplasm System (2019) officially changed the name of American elderberry accordingly in 2013. Nevertheless, American elderberry is still frequently called *Sambucus canadensis* L. in both popular and scientific literature.

Other North American *Sambucus* taxa include red elderberry (*S. racemosa* L., a very widespread species when broadly circumscribed, the eastern North American populations being *S. racemosa* subsp. *pubens* (Michx.) Hultén or *S. pubens* Michx.), and blue or Mexican elderberry (*S. cerulea* Raf., whose synonyms include *S. nigra* subsp. *cerulea* (Raf.) Bolli). Red elderberry fruit is considered poisonous but was nevertheless widely consumed by Native Americans after cooking and removal of seeds (Losey et al. 2003). Blue elderberry is commonly consumed by humans as both a food and medicine, and was tentatively proposed as a horticultural crop by the well-known plant breeder Luther Burbank (Hummer et al. 2012). This report focuses on the North American taxon commonly called American elderberry (*Sambucus nigra* subsp. *canadensis*), which will often be referred to herein simply as "elderberry".

American elderberry typically occurs at the edge of wooded areas, in fence-rows, in openings near streams and waterways, and on the upper edges of road-side ditches where water is abundant but does not stand for extended periods. Elderberry thrives in areas that are mowed annually, such as roadsides, where competing woody vegetation is kept to a minimum. Occasionally elderberry is found in open dry areas where weedy vegetation is moderated by either animal grazing or mowing. While elderberry can grow in shade, it rarely flowers or fruits abundantly therein. Once established, American elderberry can vigorously re-grow from ground-level buds, and produce flowers and fruit in one season (Thomas et al. 2009), whereas European elderberry does not flower on new seasonal growth. American elderberry tends to sucker and spread as a multi-stemmed shrub if conditions allow, usually reaching heights of 2–3 m, and sometimes taller in southern climates. It produces abundant, large, flat corymbs of creamy white, perfect flowers (Figs. 5.1 and 5.2) typically in

Fig. 5.1 Cultivated American elderberry in full bloom, Southwest Missouri, USA

June in the Midwest, followed by dark, glossy purple to black berries (Fig. 5.3) from late July through early September. The juicy berries are small, generally ranging from 70–90 mg, with several hundred berries produced per corymb (Finn et al. 2008). Pollination of elderberry has not been well-studied, but appears to combine wind pollination (likely the major factor) with some assistance from insects (Way 1981). While small solitary bees are often seen on elderflowers, honeybees are infrequent visitors (A.L.T. pers. obs. 22 years). Elderberries are of great benefit to wildlife, providing excellent cover and habitat for small mammals, birds, and beneficial insects, and nutritious food for birds and deer. The berries are edible and considered highly nutritious and beneficial to humans due to their high content of anthocyanins, polyphenols, and vitamins. The nutritious flowers are also consumed either cooked or more often in beverages, teas, and liqueurs. Both fruit and flowers are also of increasing interest for use as natural food colorants. During the last 15 years, American elderberry has emerged as an important specialty crop primarily grown for fruit used in health-promoting dietary supplements; a trip to most any North American health store will reveal an abundance of elderberry dietary supplement products for sale, most at a high price. However, the majority of those products are still coming from Europe.

Fig. 5.2 American elderberry corymbs at full bloom

5.2   Traditional and Historic Use as Food and Medicine

The close biological relationship between European and American elderberry should prompt the reader to incorporate records of elderberry use as a food and medicine in Europe (e.g., Knudsen and Kaack 2015a, b) to approximate the history of human elderberry use in North America. European records cite the human use of European elder since ancient times for a plethora of food, medicinal, and spiritual purposes (Salamon and Grulova 2015). American elderberry has a similar and extensive history of use among nearly 100 North American indigenous nations and tribes (Moerman 1998), and subsequently among the peoples who migrated more recently to North America from Europe. The latter group's long-established use of European elderberry as food and medicine was likely and logically perpetuated upon encountering the closely related North American subspecies. Nevertheless, despite this history, American elderberry remains relatively unknown to the North American public and the agricultural industry alike. While many of the reported medicinal effects still lack adequate scientific validation, there is an increasing number of studies supporting important medicinal or therapeutic properties associated with American elderberry, as described herein.

The documented medicinal uses of elderberry by indigenous Americans include the following areas: emetic, laxative, anti-rheumatic, cathartic, dermatological, burn

Fig. 5.3 American elderberry infructescence with fully ripe fruit

dressing, diaphoretic, diuretic, pediatric aid, liver aid, breast treatment, blood puri-
fier, gastrointestinal aid, analgesic, tonic, febrifuge, heart medicine, measles, venereal aid, gynecological aid, sedative, toothache, ceremonial medicine, urinary aid,
herbal steam, anti-diarrheal, anti-hemorrhagic, tuberculosis, and cold remedy
(Moerman 1998). Note that the several species and varieties cited by Moerman
(1998) are now mostly recognized as being members of *S. nigra* subsp. *canadensis*;
the variation being considered more representative of ecotype formation. While the
use of elderberry fruits is most frequently cited in traditional medicine practice,
there is evidence of use of the flowers (and leaves) for pain relief, swelling, inflammation, and diuresis (urine production), and as a diaphoretic or expectorant (Ulbricht
et al. 2014). Readers are also directed to the condensed coverage of traditional
medicinal uses of American elderberry in Duke (2002), the PDR (Physicians' Desk
Reference) for Herbal Medicines (Thomson Healthcare 2007), and the evidence-
based systematic review of elderberry and elderflower by the Natural Standard
Research Collaboration (Ulbricht et al. 2014).

Food uses by both indigenous and neo-Americans include juice and beverages,
fresh and dried fruit, pie, bread, cake, jam, jelly, sauce, relish, and in yoghurt, syrup,
and ice cream. The flowers are dried for use in making beverages or teas, or consumed fresh – often battered and fried in the form of "fritters". The flowers are also
used to flavor alcoholic spirits and in soaps for their medicinal and fragrant attributes. Excellent wines can be made from both the fruit and flowers; numerous North

American wineries now produce and market a variety of dry and sweet wine products that are popular with North American consumers.

5.3 Modern Science-Based Health Claims

Clinical and preclinical research to confirm the traditional uses of elderberry is limited, and the large majority has been done in Europe with the European subspecies, *Sambucus nigra* subsp. *nigra*. The two subspecies are chemically distinct but contain the same major compound classes (e.g., Lee and Finn 2007), and there is no reason to presume that the American subsp. *canadensis*, which is widely used medicinally and for similar purposes, lacks the bioactivities of the European subspecies. Both subspecies contain chlorogenic acid and rutin as "major" polyphenolic compounds, with American elderberry also containing substantial amounts of neochlorogenic acid and isorhamnetin 3-rutinoside (Johnson et al. 2017; Lee and Finn 2007). In one *in vitro* study, extracts of fruits of American elderberry demonstrated a greater number of anticancer bioactivities than European elderberry (Thole et al. 2006). Similarly, flowers of both subspecies contain numerous acyl spermidines, though they have distinct fingerprints (Kite et al. 2013).

A second confounding issue is that most published clinical trials are of multispecies products. It is common in herbal medicine, both in formalized professional modalities such as Traditional Chinese Medicine and in traditional herbalism, for multiple plant species to be used in formulas. Such formulas are believed to have greater efficacy, and sometimes also tolerability, than a high dose of a single species, much as a botanical extract containing multiple compounds may have a better effect profile than any one chemical from that extract. This creates challenges for the evaluation of those formulas. The insistence that only one species at a time should be tested is not rationally justifiable, as it would mean that botanical medicines as they were really used would not be tested, and their full potential benefit might never be revealed. However, when a multicomponent product is tested, it is not yet possible to know to what degree the observed benefits are derived from the individual activity of any one component or synergistic activities requiring that component's presence. Therefore, single-species products should not be presumed to have the same efficacy as tested multispecies products that contain those species. The use of "omics" technology is promoted as a means of elucidating putative synergistic molecular activity in the context of animal-based physiological models (Khakimov et al. 2012).

The most common and best-supported medicinal use of elderberry is for upper respiratory infections (URIs). Clinical trials of elderberry alone for treatment of URI symptoms have been favorable, but most have been small. A meta-analysis of trials of elderberry for URIs reported a "large mean effect size" (Hawkins et al. 2019); however, the trials included in that analysis had only 180 participants. At least two placebo-controlled human trials have tested the standardized extract Sambucol® (Razei Bar, Jerusalem, Israel) for influenza; recovery was up to 4 days faster in elderberry users (Zakay-Rones et al. 1995, 2004). Elderberry is also effective against influenza in mice (Kinoshita et al. 2012).

A larger placebo-controlled trial gave elderberry or a placebo prophylactically to 312 airline passengers traveling economy class from Australia to overseas destinations (Tiralongo et al. 2016). Though the number of colds was not significantly reduced in those who consumed a standardized elderberry extract (12 vs. 17 in the placebo group), the number of days of illness was halved (57 vs. 117, P = 0.02) and average symptoms suffered by those who were ill also substantially lowered (P = 0.05).

Perhaps the best-tested elderberry-containing product is Echinaforce Hotdrink® (A. Vogel Bioforce AG, Roggwil, Switzerland). This syrup contains 240 mg of a standardized hydroethanolic extract of *Echinacea purpurea* (L.) Moench (echinacea) and 276.5 mg elderberry extract per mL; it is made into a hot drink by adding 5 mL to hot water. In a large clinical trial, 473 people with early influenza were randomized to take Echinaforce Hotdrink or the anti-influenza drug oseltamivir (Rauš et al. 2015). Blinding was created by having each participant take both a hot beverage, five times a day for 3 days, then three times a day for 7 days, and capsules twice a day for 10 days. The Echinaforce group received Echinaforce Hotdrink syrup plus placebo capsules; the oseltamivir group received a placebo syrup and 5 days of drug capsules followed by 5 days of placebo capsules. The two treatments were similar in efficacy, with 1.5% vs. 4.1% reporting recovery (absent or only mild symptoms) after 1 day, 50.2% vs. 48.8% after 5 days, and 90.1% vs. 84.8% after 10 days. The incidence of complications was near-significantly lower with Echinaforce (2.46% vs. 6.45%; P = 0.076), and nausea and vomiting were five times more frequent with oseltamivir, sometimes necessitating discontinuation.

Rauš et al. (2015) technically demonstrated only that Echinaforce Hotdrink is not inferior to oseltamivir in efficacy. However, the additional considerations of oseltamivir's lesser accessibility, greater cost, and neurological and psychiatric side effects (Ishii et al. 2008; Jefferson et al. 2014) would support a consumers'-perspective conclusion that Echinaforce Hotdrink is superior to the pharmaceutical alternative when all costs and outcomes are considered. Most other Echinaforce® branded products contain only echinacea, which has been subjected to far more clinical trials than elderberry, and has been well demonstrated to reduce incidence and severity of URIs when correctly used (Shah et al. 2007; Schapowal et al. 2015). Perhaps because echinacea is so much better known, Rauš et al. (2015: 66) referred to the tested product as "*Echinacea purpurea*-based." However, the likely contribution of elderberry, with its well-demonstrated anti-influenza mechanisms of action (see below), to the efficacy of this and similar formulae should not be minimized.

Bark of American and European elderberry has been used as a purgative, while twigs and fruits are used for a laxative effect and flower infusions used for stomach upset and given to babies with colic (Moerman 1998). There are no clinical trials of single-species products for these effects. A placebo-controlled crossover trial of a four-species formula for chronic constipation found that the formula reduced mean colonic transit time from 42.3 h during placebo treatment to 15.7 h during active treatment (Picon et al. 2010). Frequency of elimination and patient perception of bowel function also improved significantly. That botanical formula, sold commercially in Brazil since 1926, contains one-third elderflower by weight, with the remainder flowers of *Cassia angustifolia* Vahl (= *Senna alexandrina* Mill., senna) and fruits of *Pimpinella anisum* L. (anise) and *Foeniculum vulgare* Mill. (fennel). Because senna leaf is a well-known laxative, it may be suspected of being the most potent part of this formula, which does not mean that the other species have no benefit; they could serve to support efficacy while mitigating side effects.

American and European elderberry have occasionally been used as diuretics; the inner bark is sometimes the part used (Moerman 1998), but sometimes the fruit is used. European elderberry has been shown to be a true diuretic in rodents, leading to increased sodium excretion as well as increased urinary output (Beaux et al. 1999). A European formulation of four botanicals including elderflower (Herbensurina®; Deiters Laboratories, Badalona, Spain) has been reported to reduce induced calcium oxalate crystal formation, microcalcifications, and kidney fibrosis in male rats, with the middle tested dosage of 125 mg/kg being most effective (Crescenti et al. 2015). However, there are no human trials.

PerioPatch® (Izun Pharmaceuticals Ltd., Jerusalem, Israel, and New York City, USA) is a topical herbal patch containing elderberry, echinacea, and *Centella asiatica* (L.) Urb. (gotu kola), which is used for treatment of oral wounds and ulcerations. A clinical trial in humans with gingivitis (Grbic et al. 2011) found that use for 3 days reduced the gingival index, compared to users of a placebo patch, on days 4 and 15, though not on day 8. The trial was intended to be a crossover trial, but because gingival inflammation in some users of the active patch had not rebounded to the baseline level by day 15, results from the second half were not reported. A study in rats compared PerioPatch to a placebo patch and no patch for treatment of oral surgical wounds. After 12 days, the herbal patch group had smaller epithelial gaps and greater collagen deposition and angiogenesis at the flap site (Chaushu et al. 2015). In a human trial, an herbal mouth rinse, HM-302, developed by the same company and containing the same species, reduced the development of gingivitis significantly better than a water rinse and nonsignificantly better than essential oil or cetylpyridinium rinses (Samuels et al. 2012).

5.4 Mechanisms of Bioactivity

Sambucus nigra has direct antiviral activity, especially against a wide range of influenza viruses, in which hemagglutination is reduced and replication is inhibited (Zakay-Rones et al. 1995; Krawitz et al. 2011). Strains against which the commercial extract Sambucol is effective *in vitro* include multiple H1N1, H3N2, and type B strains, an H5N1 strain, and animal strains from European swine and turkeys. Elderberry also inhibits an H9N2 strain in human epithelial cell culture (Shahsavandi et al. 2017), possibly preventing viral spread between cells by interfering with lipid raft association. Flavonoids from elderberry bind to H1N1 virions, inhibiting their ability to penetrate host cells (Roschek et al. 2009). In one study, the IC_{50} for the direct H1N1 inhibitory activity of fruit extract was 252 ± 34 µg/mL, while two iso-lated flavonoids displayed inhibitory activity at concentrations comparable to osel-tamivir and amantadine respectively (Roschek et al. 2009). Contrarily, Kinoshita et al. (2012) reported that *in vitro* anti-influenza activity of elderberry was relatively weak, compared with its benefits in mice *in vivo*, which were therefore attributed primarily to immune stimulation.

Elderberry is also directly active against infectious bronchitis virus, a coronavi-rus causing illness in chickens; treatment with *S. nigra* extracts reduced viral titers in Vero cells by four to six orders of magnitude (Chen et al. 2014). The standardized extract Rubini® (BerryPharma AG, Leichlingen, Germany), which, like Sambucol, is made from the European subspecies, inhibits growth of four bacterial species that can cause upper respiratory tract infections, including *Streptococcus pyogenes*, group C and G *Streptococci*, and *Branhamella catarrhalis* (Krawitz et al. 2011). Elderberry therefore might reduce complications of influenza in part by impeding development of secondary infections.

Some studies have reported that elderberry and elderflower reduce inflammation both *in vitro* and *in vivo*. Nitric oxide (NO) inhibition and complement fixing activi-ties were demonstrated *in vitro* (e.g., Ho et al. 2017a). In obese mice, the inflamma-tory marker tumor necrosis factor-α (TNF-α) was significantly reduced by elderberry extract (Farrell et al. 2015b). However, Sambucol products have been reported to be pro-inflammatory *in vitro* (Barak et al. 2001; Waknine-Grinberg et al. 2009), increasing production of both anti- and pro-inflammatory cytokines and TNF-α (Barak et al. 2001, 2002). Mice that consumed Sambucol both before and after infection with malaria had an increased incidence of cerebral malaria, which is thought to be caused by inflammatory response, although Sambucol protected mice from leishmaniasis (Waknine-Grinberg et al. 2009). Another study reported that elderberry extract increased TNF-α in a diabetic rat model, which was inexplicably reported as indicating "anti-inflammatory effects" (Badescu et al. 2015).

Elderberry has not been shown to be pro-inflammatory (or anti-inflammatory) in humans, and it is not clear whether the manufacturing of Sambucol may affect its bioactivities. Some symptoms of influenza stem from the inflammatory response to the illness, so a product that did increase cytokine production might be expected to worsen symptoms rather than improve them. Hypothetically, this might provide a good rationale for the herbal practice of combining elderberry with echinacea. Echinacea is well demonstrated to reduce production of pro-inflammatory cyto-kines (e.g., Sharma et al. 2009a, b; Vimalanathan et al. 2017) so in addition to add-ing a different direct anti-influenza activity (Pleschka et al. 2009), it might counteract any potential side effects of the elderberry.

Though anthocyanins or flavonoids are sometimes presumed to be "the" active compounds in elderberry, pectic polysaccharides in elderberry and elderflower pro-vide complement fixing and macrophage stimulating activities (Ho et al. 2015, 2016). In mice, also, the elderberry fraction containing polysaccharides was reported to be the strongest in reducing viral load and increasing antibody production (Kinoshita et al. 2012), but other fractions also evidently had some activity. Whole elderberry products or crude extracts are therefore likely to be preferable to products that isolate a single subset of constituents.

5.5 Emerging Health Benefits

Recent scientific studies provide data suggesting that elderberry might be useful not only in treating acute conditions, but in preventing or ameliorating some of the chronic degenerative diseases associated with the Western diet and lifestyle. These include dementia, diabetes, cardiovascular disease (CVD), liver disease, and/or obesity, which often plays a causal role in the development of these chronic diseases. Though the evidence for prophylactic benefits of elderberry is far from definitive, it is certainly enough to demonstrate that further research is called for.

In particular, an ever-increasing amount of research indicates that consumption of berry fruits is beneficial for brain health. Elderberry has been relatively little studied in this context. However, promising preclinical research exists. Mice fed diets containing American elderberry suffered substantially less disability after 24 h and less neuronal cell death after 3 days in response to global cerebral ischemia (effectively, an induced stroke) than non-supplemented mice (Chuang et al. 2014). Flowers of *Sambucus cerulea*, which was once treated as a subspecies of *S. nigra* though it is morphologically distinct (Applequist 2015), are used by members of the North American Lumbee tribe to alleviate symptoms of Parkinson's disease, and *in vitro* the extract has neuroprotective activities, including improvement of mitochondrial function in neurons and protection against rotenone neurotoxicity (de Rus Jacquet et al. 2017).

Methanolic extracts, especially of elder bark and leaf, reduce convulsions in mice induced by pentylenetetrazole and electroshock, and protect against mortality as well as diazepam (Ataee et al. 2016). Fruit and leaf extracts demonstrate substantial activity in mouse models of depression (Mahmoudi et al. 2014); results of that study are not relevant to humans because acute doses of up to 1200 mg/kg body weight were injected intraperitoneally.

An extract of *Sambucus nigra* had antioxidant activity in isolated neurons *in vitro* (Spagnuolo et al. 2016). Ethanolic extracts of American elderberry inhibit microglial activation (Simonyi et al. 2015) and sometimes reduce the production of reactive oxygen species (ROS) in microglial cells. However, when a variety of juice extracts were tested, some actually increased NO production (Jiang et al. 2015), so the antioxidant activity may not be ubiquitous.

Elderflower tea is reported to have particularly high antioxidant potential, suggesting that it might be beneficial for mitigating many potential health problems caused by chronic oxidative stress (e.g., Loizzo et al. 2016; Viapiana and Wesolowski 2017). Elderflower aqueous extracts can also inhibit the pro-inflammatory activities of bacteria responsible for gum disease (Harokopakis et al. 2006), perhaps contributing to the reported anti-gingivitis effects.

European elderberry is sometimes used traditionally to treat diabetes, but there is no relevant human research to date. *In vitro*, elderberry and elderflower increase glucose uptake in human skeletal muscle and liver carcinoma cells; multiple components, including flavonoids and phenolic acids, and their metabolites inhibit 15-lipoxygenase, α-amylase, and α-glucosidase, the latter two better than acarbose (Ho et al. 2017b, c). In isolated mouse muscles, an aqueous extract increases 2-deoxy-glucose transport, glucose oxidation, glycogenesis, and insulin secretion; activities derive from multiple constituents (Gray et al. 2000). Increased glucose uptake in porcine myotube cultures is attributed at least partly to phenolic compounds, including naringenin and 5-O-caffeoylquinic acid (Bhattacharya et al. 2013). Elderflower extracts can activate peroxisome proliferator-activated receptor-gamma (PPAR-γ); three constituents known to have this activity, including naringenin and fatty acids, are present but are not wholly responsible for the activity (Christensen et al. 2010).

In vivo studies have shown inconsistent effects. In one study of type 2 diabetic rats, a polar extract reduced hyperglycemia and a lipophilic extract reduced insulin secretion; both lowered insulin resistance (Salvador et al. 2016). In obese mice fed a high-fat diet, an elderberry extract reduced insulin resistance (Farrell et al. 2015a). However, in a different mouse model no significant benefit was reported for elder berry (Swanston-Flatt et al. 1989). In hyperlipidemic mice deprived of the apolipoprotein E gene, elderberry reduced fasting glucose (Farrell et al. 2015b). In a diabetic mouse model of neuropathic pain, aqueous extract of elderflower, administered by intraperitoneal injection, reduced blood sugar better than a pharmaceutical control and reduced pain (Raafat and El-Lakany 2015). Elderberry may also improve bone density in diabetic rats (Badescu et al. 2012).

In a rat model of hypertension, a polyphenol-rich European elderberry extract reduced blood pressure; also, antioxidant activity was improved by the combination of elderberry extract and the pharmaceutical aliskiren better than by either alone (Ciocolu et al. 2016). In hamsters with hyperlipidemia due to fish oil feeding, elderberry decreased plasma and hepatic lipids and lipid peroxidation (Dubey et al. 2012). In mice fed a high-fat diet, elderberry extract reduced lipid synthesis (Farrell et al. 2015b). Two studies in mice deprived of the apolipoprotein E gene have had partly contradictory results. In one, in which the mice were fed a low-fat diet, elderberry supplementation increased non-HDL cholesterol (Millar et al. 2018). However, HDL function was improved, as were measures of liver inflammation and atherosclerotic plaque formation. In the other, serum lipids did not change, but aspartate transaminase and cholesterol content in the aorta (a measure of atherosclerosis progression) were reduced (Farrell et al. 2015b).

In a placebo-controlled human study, a product equivalent to 5 mL elderberry juice per day, taken for 2 weeks by young subjects, reduced cholesterol only nonsignificantly (from a mean of 199 to 190 mg/dL, vs. a slight increase in the placebo group) (Murkovic et al. 2004). However, that study used a very small dose and a very short treatment period. A 12-week placebo-controlled trial in healthy postmenopausal women tested the effects of 500 mg/day extracted anthocyanins as cyanidin glycosides, equivalent to the amount contained in 25 g/day elderberries, on biomarkers of CVD risk (Curtis et al. 2009). No statistically significant differences were seen in any biomarkers for CVD risk, including blood pressure, serum lipids or glucose, markers of inflammation, or platelet reactivity, or markers of liver and kidney function. However, the authors acknowledged that other studies testing anthocyanin-rich fruits containing a broader array of constituents, rather than one isolated fraction, in people with higher baseline inflammation levels have had positive results. No human trial, therefore, has yet effectively tested the potential benefits of elderberry, as consumers might use it, for CVD.

The possible effect of elderberry on body fat or weight is unclear. Elderflower extract reduces fat accumulation in *Caenorhabditis elegans* (a nematode worm), with phenolic compounds partly responsible for bioactivity (Bhattacharya et al. 2013). In a human study, the combination of elderberry and asparagus products was reported to result in improvements in weight and blood pressure (Chrubasik et al. 2008); unfortunately, the study design was inadequate to allow clear assessment of causality.

5.6   Negative Perceptions or Possible Side Effects from Elderberry Consumption

Conventional wisdom suggests that American elderberry contains cyanide, or specifically cyanogenic glycosides, and that consuming leaves, stems, green berries, and even unprocessed ripe berries can be harmful. Similar claims suggest that cooking, processing, or fermenting elderberries or elderberry juice mitigates such putative cyanide occurrence. This perception has been immortalized by a well-known incident in 1983 where a group of worshipers in California consumed a beverage made from raw elderberry fruits, leaves, and stems (likely *S. cerulea*), along with other ingredients, resulting in emergency hospitalization of eight people (Centers for Disease Control 1984). While cyanide poisoning was initially suspected as the cause, all victims had normal serum cyanide levels and recovered quickly. Nevertheless, this incident is frequently cited as a basis for the alleged occurrence of cyanide in elderberries. Countless websites state that elderberry contains cyanide or cyanogenic glycosides, and that the fruit should not be consumed raw; however, the scientific literature supporting such claims is scant.

Cyanogenic glycosides are naturally occurring organic compounds that are found in more than 2600 plant species (Irchhaiya et al. 2015), as well as in some insects, especially Lepidoptera (Zagrobelny et al. 2004). They are nitrogenous secondary metabolites consisting of an aglycone and a sugar moiety, with a nitrile functional group (R-CN). Cyanogenic glycosides are believed to play roles in plant defense mechanisms against herbivory, and in both pest and disease resistance and response (Nahrstedt 1985). The cyanogenic glycosides in plant tissues are generally considered safe and stable, but when such tissues are disrupted by herbivory or infection, endogenous enzymes can be activated resulting in the release of toxic hydrogen cyanide from these compounds (Bak et al. 2006).

Dellagreca et al. (2000) and Senica et al. (2016) have documented the occurrence of notable and possibly toxic levels of cyanogenic glycosides in European elderberry (subsp. *nigra*) leaves, which are not considered edible, and are rarely used medicinally. Senica et al. (2017) quantified cyanogenic glycosides in leaf, flower, and fruit of European elderberry; levels of sambunigrin (a common cyanogenic glycoside) ranged from 29–210 µg/g in leaves, 7–19 µg/g in flowers, and from 0.08–0.59 µg/g in fruits (all generally increasing with altitude where plants were growing). Using the picrate paper test to evaluate American elderberry (subsp. *canadensis*) leaves, Buhrmester et al. (2000) found consistent detectable levels of evolved hydrogen cyanide (HCN) in only one population of wild elderberry collected in Illinois (USA), whereas eight other populations either tested negative or were inconsistent in detectable amounts of evolved HCN. Little research has been published on the occurrence of cyanogenic glycosides in American elderberry fruit. Bolarinwa et al. (2015) reported levels of 1–7 µg cyanide equivalents/g in commercial apple juice, and 10–40 µg/g in fresh apple juice pressed from 15 varieties. Appenteng et al. (manuscript in preparation) documented levels of cyanogenic glycosides in American elderberry juice at or below levels found in concurrent analyses of commercial apple juice. Frozen and canned commercial apple juice produced 0.16–0.21 µg cyanide equivalents/g juice in picrate paper tests, whereas dissected American elderberry fruits from two cultivars produced levels of 0.03–0.24 µg/g in juice, 0.01–0.24 µg/g in seeds, 0.01–0.64 µg/g in fruit skins, and 0.05–0.54 µg/g in pedicels (small green stems attaching the berry to the corymb). Further study found mean levels of 0.25 and 0.37 µg cyanide equivalents/g of green (unripe) berries and pedicels, respectively, thus suggesting that unripe elderberries and associated stems have slightly higher levels of cyanogenic glycosides compared with commercial apple juice. These levels are still well below published numbers for *S. nigra* fresh and processed juice (18.8 and 10.6 µg/g, respectively) (Senica et al. 2016). Elderberry processors already eschew incorporating unripe berries and stems in their products because they degrade quality, taste, and appearance; therefore, American elderberry products are very unlikely to contain any appreciable amount of cyanogenic glycosides or cyanide.

Thermal processing and fermenting of foods containing cyanogenic glycosides have indeed been shown to render them safe to consume. For example, cassava (*Manihot esculenta* Crantz), one of the world's most important staple food crops, is highly toxic in its unprocessed state due to cyanogenic glycosides, but is safe to consume after processing or cooking (Montagnac et al. 2009). Senica et al. (2016) documented a 44% reduction in sambunigrin content in heat-treated subsp. *nigra* juice, and a 96% reduction in processed liqueur and fruit spread products.

The internet also abounds with claims that elderberries contain alkaloids, and suggesting that the purported alkaloids therein are toxic. For example, the USDA-NRCS Plant Guide for American Elderberry (2003) states "The active alkaloids in elderberry plants are hydrocyanic acid and sambucine. Both alkaloids will cause nausea so care should be observed with this plant." However, there is no scientific citation to support this statement, nor can a definitive corroborating published work be found. Nevertheless, such assertions are widely perpetuated throughout the internet and popular media. In fact, hydrocyanic acid is NOT considered an alkaloid, but is a weak acid formed by dissolving HCN (a gas) in water [National Research Council (US) Subcommittee on Acute Exposure Guideline Levels 2002]. As stated above, the potential occurrence of HCN in American elderberry fruit is very low; therefore, the formation of harmful amounts of hydrocyanic acid in either raw or processed elderberry products is inherently unlikely. Similarly, sambucine (also known as sambucin, antirrhinin, and cyanidin 3-O-rutinoside) is also not an alkaloid, but rather an anthocyanin (red pigment) that can act as an anti-oxidant in the human body upon consumption (Lila 2004). In high doses, however, sambucine may react in the human body in the presence of certain enzymes to produce an alkaloid that would be considered toxic (Kamsteeg et al. 1978).

The authors are aware of some rare anecdotal incidents where people who consumed large amounts of raw elderberries experienced upset stomach. The cause of this may be related to individual sensitivities to elderberry fruits in general, or possibly to one or more specific compounds in elderberry. The bottom line is that more research is needed to understand not only the health-benefitting metabolites and attributes of elderberry, but also to scientifically confirm or debunk the widespread unsubstantiated claims regarding the supposed toxicity and potential negative side-effects of elderberry consumption.

5.7 Horticulture and Developing Markets

Historically, much of the American elderberry crop was harvested from wild plants under minimal management. While commercial-scale European elderberry production is well-established in Europe, American elderberry remained a largely undeveloped crop until interest in use of the fruit and flowers in dietary supplements surged in the early 2000s (Charlebois 2007). American elderberry is well-suited to horticultural production in and near its native environment, provided adequate management is employed. The rapidly growing need for standardized quality parameters in fruit and flower production intended for commercial processed products has given rise to the development of improved elderberry cultivars and recommended production practices (Byers et al. 2014; Thomas et al. 2015a). Additionally, the importance of sustainable and profitable production has helped drive farmers and elderberry researchers into increasingly intensive management of American elderberry. Production practices developed in the twentieth century focused primarily on maximizing raw fruit production for processing markets (jelly, jam, juice, wine). More recently, as interest in the medicinal attributes of American elderberry has grown, emphasis has shifted to production practices that not only emphasize productivity, but also influence a range of elderberry bioactive compounds found in fruit and flowers. An overview of modern production practices is detailed in Charlebois et al. (2010).

Among the many production practices available to farmers, cultivar selection has perhaps the greatest influence on not only profitability, but also fruit and flower quality parameters. Cultivar selection with American elderberry is a recent development (Charlebois et al. 2010), and, in part, reflects the need among elderberry farmers and processors for uniform raw materials. Selecting cultivars with regional adaptation can help drive elderberry production in new areas; the success of the Missouri (USA) industry is due in part to the release of cultivars adapted to that region (Byers et al. 2010; Byers and Thomas 2011). While genotype has a demonstrated effect on the spectrum and levels of bioactive compounds found in American elderberry cultivars (Thomas et al. 2015b), establishing a planting that consistently produces fruit or flowers that meet quality standards is confounded by the interaction of genotype with environment (Finn et al. 2008; Perkins-Veazie et al. 2015; Thomas et al. 2008; Thomas et al. 2013).

Production practices that enhance productivity have a role to play in sustainable American elderberry production (Byers et al. 2014). Pruning and fertility management practices that influence plant growth and harvest efficiency by promoting uniform ripening have the potential to improve the consistency of processed products. For example, research at the University of Missouri/Missouri State University (Thomas et al. 2009) demonstrated that, for many American elderberry genotypes, removal of plants to the ground during the winter resulted in regrowth the following season that produced a fruit crop that ripened uniformly. This approach to pruning, while widely adopted by elderberry farmers in the central USA, may have limitations in production areas with shorter growing seasons. Elderberry fertility management has an important influence on plant growth and quality parameters in flowers and fruit. Initial studies by Byers et al. (2015) suggest that elemental content of elderberry plant parts could help guide a fertility management plan, thus giving farmers a useful management tool. Additional research is needed to determine optimum nutrient application rates and timing. While much of the American elderberry industry targets fruit production, elderberry flowers are of growing interest as a medicinal product. McGowan et al. (2019) demonstrated that management practices can be implemented that offer the potential of both a fruit and flower harvest from the same plant.

Pest and disease issues have the potential to reduce profitability of elderberry production and influence flower and fruit quality, but little research has been published exclusively on American elderberry. Warmund et al. (2019) confirmed that elderberry rust (*Puccinia sambuci*) requires an alternate host (sedge; *Carex* sp.) to complete its life cycle, and documented potential economic losses and reduction to fruit quality at various levels of infection; for even moderate incidences of rust, control measures may be warranted. Microscopic Eriophyid mites are a well-known pest of European elderberry in Europe (Vaněčková-Skuhravá 1996), and are also problematic in North America on American elderberry. Warmund and Amrine (2015) studied two mite species (*Phyllocoptes wisconsinensis* and a new, unnamed species of *Phyllocoptes*) on American elderberry and determined that the majority of over-wintering female protogynes shelter under leaf buds 14–24 cm from the terminus of the stem. Thus, they propose that pruning and removal of over-wintering stems may provide adequate mite control. Keller et al. (2015) surveyed viruses in American elderberry samples from Missouri and identified two new Carlaviruses in most of the samples. The economic impact of these viruses, if any, is unknown. The researchers were able to use meristem culture to eliminate the viruses from several cultivars. Insects noted to be of particular concern with American elderberry include spotted wing drosophila (*Drosophila suzukii*), Japanese beetle (*Popillia japonica*), elder shoot borer (*Achatodes zeae*), elderberry sawfly (*Tenthrado grandis*), and elderberry longhorn beetle (*Desmocerus palliatus*) (Byers et al. 2014). Even with native crops such as elderberry, when plants are brought together into monocultures for efficient and economic production, pests and diseases must be monitored and addressed.

References

Allen GM, Bond MD, Main MB (2002) 50 common native plants important in Florida's ethnobotanical history. Circ. 1439, University of Florida IFAS Extension

Angiosperm Phylogeny Group, Chase MW, MJM C et al (2016) An update of the Angiosperm Phylogeny Group classification for the orders and families of flowering plants: APG IV. Bot J Linn Soc 181(1):1–20

Appenteng MK, Krueger R, Johnson MC, Ingold H, Bell R, Thomas AL, Greenlief CM (in preparation) Cyanogenic glycosides analysis in American elderberry: picrate paper and LC MS/MS method development and validation. Manuscript in preparation

Applequist WL (2015) A brief review of recent controversies in the taxonomy and nomenclature of *Sambucus nigra* sensu lato. Acta Hort 1061:25–33

Ataee R, Falahati A, Ebrahimzadeh MA, Shokrzadeh M (2016) Anticonvulsant activities of Sambucus nigra. Eur Rev Med Pharmacol Sci 20(14):3123–3126

Badescu L, Badulescu O, Badescu M, Ciocoiu M (2012) Mechanism by *Sambucus nigra* extract improves bone mineral density in experimental diabetes. Evid Based Complement Alternat Med 2012:848269

Badescu M, Badulescu O, Badescu L, Ciocoiu M (2015) Effects of *Sambucus nigra* and *Aronia melanocarpa* extracts on immune system disorders within diabetes mellitus. Pharm Biol 53(4):533–539

Bak S, Paquette M, Morant M et al (2006) Cyanogenic glycosides: a case study for evolution and application of cytochromes P450. Phytochem Rev 5(2–3):309–329

Barak V, Halperin T, Kalickman I (2001) The effect of Sambucol, a black elderberry-based, natural product, on the production of human cytokines: I. Inflammatory cytokines. Eur Cytokine Netw 12(2):290–296

Barak V, Birkenfeld S, Halperin T, Kalickman I (2002) The effect of herbal remedies on the production of human inflammatory and anti-inflammatory cytokines. Isr Med Assoc J 4(11 Suppl):919–922

Beaux D, Fleurentin J, Mortier F (1999) Effect of extracts of Orthoxiphon stamineus Benth, *Hieracium pilosella* L., *Sambucus nigra* L. and *Arctostaphylos uva-ursi* (L). Spreng. in rats. Phytother Res 13(3):222–225

Bhattacharya S, Christensen KB, Olsen LC et al (2013) Bioactive components from flowers of Sambucus nigra L. increase glucose uptake in primary porcine myotube cultures and reduce fat accumulation in *Caenorhabditis elegans*. J Agric Food Chem 61(46):11033–11040

Bolarinwa IF, Orfila C, Morgan MRA (2015) Determination of amygdalin in apple seeds, fresh apples and processed apple juices. Food Chem 170:437–442

Bolli R (1994) Revision of the Genus Sambucus. Dissertationes Botanicae vol 223. J. Cramer, Berlin, Stuttgart

Buhrmester RA, Ebinger JE, Seigler DS (2000) Sambunigrin and cyanogenic variability in populations of Sambucus canadensis L. (Caprifoliaceae). Biochem Syst Ecol 28:689–695

Byers PL, Thomas AL (2011) 'Bob Gordon' elderberry. J Am Pomol Soc 65(2):52–55

Byers PL, Thomas AL, Millican M (2010) 'Wyldewood' elderberry. HortScience 45(2):312–313

Byers PL, Thomas AL, Gold MA, Cernusca, MM, Godsey LD (2014) Growing and marketing elderberries in Missouri, University of Missouri Center for Agroforestry. http://www.center-foragroforestry.org/pubs/2014GrowingElderberryGuide.pdf

Byers PL, Thomas AL, Nathan MP (2015) Effect of genotype, environment, growth stage, and foliage type on American elderberry leaf elemental status. Acta Hortic 1063:183–189

Centers for Disease Control (1984) Poisoning from elderberry juice – California. Morb Mortal Wkly Rep 33(13):173–174

Charlebois D (2007) Elderberry as a medicinal plant. In: Janick J, Whipkey A (eds) Issues in new crops and new uses. Proceedings of the sixth national symposium creating markets for economic development of new crops and new uses. ASHS Press, Alexandria

Charlebois D, Byers PL, Finn CE, Thomas AL (2010) Elderberry: horticulture, botany, potential. In: Janick J (ed) Horticultural reviews, vol 37. Wiley, Hoboken, pp 213–280

Chaushu L, Weinreb M, Beitlitum I, Moses O, Nemcovsky CE (2015) Evaluation of a topical herbal patch for soft tissue wound healing: an animal study. J Clin Periodontol 42:288–293

Chen C, Zuckerman DM, Brantley S et al (2014) Sambucus nigra extracts inhibit infectious bronchitis virus at an early point during replication. BMC Vet Res 10:24. https://doi.org/10.1186/1746-6148-10-24

Christensen KB, Petersen RK, Kristiansen K, Christensen LP (2010) Identification of bioactive compounds from flowers of black elder (Sambucus nigra L.) that activate the human peroxisome proliferator-activated receptor (PPAR) gamma. Phytother Res 24(Suppl 2):S129–S132

Chrubasik C, Maier T, Dawid C et al (2008) An observational study and quantification of the actives in a supplement with Sambucus nigra and Asparagus officinalis used for weight reduction. Phytother Res 22(7):913–918

Chuang DY, Cui J, Simonyi A et al (2014) Dietary Sutherlandia and elderberry mitigate cerebral ischemia-induced neuronal damage and attenuate p47phox and phospho-ERK1/2 expression in microglial cells. ASN Neuro 6(6):175909141455494. https://doi.org/10.1177/1759091414554946

Ciocolu M, Badescu M, Badulescu O, Badescu L (2016) The beneficial effects on blood pressure, dyslipidemia and oxidative stress of Sambucus nigra extract associated with renin inhibitors. Pharm Biol 54(12):3063–3067

Crescenti A, Puiggròs F, Colomé A et al (2015) Antiurolithiasic effect of a plant mixture of Herniaria glabra. Agropyron repens, Equisetum arvense and Sambucus nigra (Herbensurina®) in the prevention of experimentally induced nephrolithiasis in rats. Arch Esp Urol 68(10):739–749

Curtis PJ, Kroon PA, Hollands WJ et al (2009) Cardiovascular disease risk biomarkers and liver and kidney function are not altered in postmenopausal women after ingesting an elderberry extract rich in anthocyanins for 12 weeks. J Nutr 139:2266–2271

de Rus Jacquet A, Timmers M, Ma SY et al (2017) Lumbee traditional medicine: neuroprotective activities of medicinal plants used to treat Parkinson's disease-related symptoms. J Ethnopharmacol 206:408–425

Dellagreca M, Fiorentino A, Monaco P, Previtera L, Simonet AM (2000) Cyanogenic glycosides from Sambucus nigra. Nat Prod Lett 14:175–182

Donoghue MJ, Eriksson T, Reeves PA, Olmstead RG (2001) Phylogeny and phylogenetic taxonomy of Dipsacales, with special reference to Sinadoxa and Tetradoxa (Adoxaceae). Harv Pap Bot 6(2):459–479

Dubey P, Jayasooriya AP, Cheema SK (2012) Fish oil induced hyperlipidemia and oxidative stress in BioF1B hamsters is attenuated by elderberry extract. Appl Physiol Nutr Metab 37(3):472–479

Duke JA (2002) Handbook of medicinal herbs, 2nd edn. CRC Press, Boca Raton

Farrell NJ, Norris GH, Ryan J, Porter CM, Jiang C, Blesso CN (2015a) Black elderberry extract attenuates inflammation and metabolic dysfunction in diet-induced obese mice. Br J Nutr 114(8):1123–1131

Farrell N, Norris G, Lee SG, Chun OK, Blesso CN (2015b) Anthocyanin-rich black elderberry extract improves markers of HDL function and reduces aortic cholesterol in hyperlipidemic mice. Food Funct 6(4):1278–1287

Finn CE, Thomas AL, Byers PL, Serçe S (2008) Evaluation of American (Sambucus canadensis) and European (*S. nigra*) elderberry genotypes grown in diverse environments and implications for cultivar development. HortScience 43(5):1385–1391

Gray AM, Abdel-Washab YH, Flatt PR (2000) The traditional plant treatment, *Sambucus nigra* (elder), exhibits insulin-like and insulin-releasing actions in vitro. J Nutr 130(1):15–20

Grbic J, Wexler I, Celenti R, Altman J, Saffer A (2011) A phase II trial of a transmucosal herbal patch for the treatment of gingivitis. J Am Dent Assoc 142(10):1168–1175

Harokopakis E, Albzreh MH, Haase EM, Scannapieco FA, Hajishengallis G (2006) Inhibition of proinflammatory activities of major periodontal pathogens by aqueous extracts from elder flower (Sambucus nigra). J Peridontol 77(2):271–279

Hawkins J, Baker C, Cherry L, Dunne E (2019) Black elderberry (*Sambucus nigra*) supplementation effectively treats upper respiratory symptoms: a meta-analysis of randomized, controlled clinical trials. Complement Ther Med 42:361–365

Ho GT, Ahmed A, Zou YF, Aslaksen T, Wangensteen H, Barsett H (2015) Structure-activity relationship of immunomodulating pectins from elderberries. Carbohydr Polym 125:314–322

Ho GT, Zou YF, Wangensteen H, Barsett H (2016) RG-I regions from elderflower pectins substituted on GalA are strong immunomodulators. Int J Biol Macromol 92:731–738

Ho GT, Wangensteen H, Barsett H (2017a) Elderberry and elderflower extracts, phenolic compounds, and metabolites and their effect on complement, RAW 264.7 macrophages and dendritic cells. Int J Mol Sci 18(3):E584

Ho GT, Kase ET, Wangensteen H, Barsett H (2017b) Phenolic elderberry extracts, anthocyanins, procyanidins, and metabolites influence glucose and fatty acid uptake in human skeletal muscle cells. J Agric Food Chem 65(13):2677–2685

Ho GT, Kase ET, Wangensteen H, Barsett H (2017c) Effect of phenolic compounds from elderflowers on glucose- and fatty acid uptake in human myotubes and HepG2-cells. Molecules 22(1):90. https://doi.org/10.3390/molecules22010090

Hosie R (1979) Native trees of Canada, 8th edn. Canadian Forestry Service, Dept. Environment, Ottawa

Hummer K, Pomper KW et al (2012) Emerging fruit crops. In: Badenes ML, Byrne DH (eds) Fruit breeding. Springer, New York

Irchhaiya R, Kumar A, Yadav A, Gupta N, Kumar S, Gupta N, Kumar S, Yadav V, Prakash A, Gurjar H (2015) Metabolites in plants and its classification. World J Pharm Pharm Sci 4:287–305

Ishii K, Hamamoto H, Sasaki T et al (2008) Pharmacologic action of oseltamivir on the nervous system. Drug Discov Ther 2(1):24–34

Jefferson T, Jones MA, Joshi P et al (2014) Neuraminidase inhibitors for preventing and treating influenza in healthy adults and children. Cochrane Database Syst Rev 2014(4):CD008965. https://doi.org/10.1002/14651858.CD008965.pub4

Jiang JM, Zong Y, Chuang DY et al (2015) Effects of elderberry juice from different genotypes on oxidative and inflammatory responses in microglial cells. Acta Hort 1061:281–288

Johnson MC, Tres MDL, Thomas AL, Rottinghaus GE, Michael Greenlief C (2017) Discriminant analyses of the polyphenol content of American elderberry juice from multiple environments provide genotype fingerprint. J Agric Food Chem 65:4044–4050

Kamsteeg J, van Brederode J, van Nigtevecht G (1978) Identification, properties, and genetic control of UDP-glucose: Cyanidin-3-rhamnosyl-(1 leads to 6)-glucoside-5-O-glucosyltransferase isolated from petals of the red campion (*Silene dioica*). Biochem Genet 16(11–12):1059–1071

Keller KE, Mosier NJ, Thomas AL, Quinto-Avila DF, Martin RR (2015) Identification of two new carlaviruses in elderberry. Acta Hortic 1063:161–164

Khakimov B, Amigo JM, Bak S, Engelsen SB (2012) Plant metabolomics: resolution and quantification of elusive peaks in liquid chromatography-mass spectrometry profiles of complex plant extracts using multi-way decomposition methods. J Chromatogr A 1266:84–94

Kinoshita E, Hayashi K, Katayama H, Hayashi T, Obata A (2012) Anti-influenza virus effects of elderberry juice and its fractions. Biosci Biotechnol Biochem 76(9):1633–1638

Kite GC, Larsson S, Veitch NC, Porter EA, Ding N, Simmonds MS (2013) Acyl spermidines in inflorescence extracts of elder (*Sambucus nigra* L., Adoxaceae) and elderflower drinks. J Agric Food Chem 61(14):3501–3508

Knudsen BF, Kaack KV (2015a) A review of traditional herbal medicinal products with disease claims for elder (Sambucus nigra) flower. Acta Hort 1061:109–119

Knudsen BF, Kaack KV (2015b) A review of human health and disease claims for elderberry (*Sambucus nigra*) fruit. Acta Hort 1061:121–131

Krawitz C, Mraheil MA, Stein M et al (2011) Inhibitory activity of a standardized elderberry liquid extract against clinically-relevant human respiratory bacterial pathogens and influenza A and B viruses. BMC Complement Altern Med 11:16

Lee J, Finn CE (2007) Anthocyanins and other polyphenolics in American elderberry (Sambucus canadensis) and European elderberry (S. nigra) cultivars. J Sci Food Agric 87(14):2665–2675

Lila MA (2004) Anthocyanins and human health: an in vitro investigative approach. J Biomed Biotechnol 2004(5):306–313. https://doi.org/10.1155/S111072430440401X

Loizzo MR, Pugliese A, Bonesi M et al (2016) Edible flowers: a rich source of phytochemicals with antioxidant and hypoglycemic properties. J Agric Food Chem 64(12):2467–2474

Losey RJ, Stenholm N, Whereat-Phillips P, Vallianatos H (2003) Exploring the use of red elderberry (*Sambucus racemosa*) fruit on the southern northwest coast of North America. J Archaeol Sci 30(6):695–707

Mahmoudi M, Ebrahimzadeh MA, Dooshan A, Arimi A, Ghasemi N, Fathiazad F (2014) Antidepressant activities of *Sambucus ebulus* and *Sambucus nigra*. Eur Rev Med Pharmacol Sci 18(22):3350–3353

McGowan KG, Byers PL, Jose S, Gold M, Thomas AL (2019) Flower production and effect of flower harvest on berry yields within six American elderberry genotypes. Acta Hortic 1265:99–105

Mejicanos G, Ziller J (1994) Observaciones sobre la producción y calidad de la biomasa de Sauco amarillo (*Sambucus canadensis*), Engorda ganado (?) y Chompipe (Bomarea nirtella) en San Marcos, Guatemala. In: Benavides JE (ed) Arboles y arbustos forrajeros en América Central. Vol. 1. Centro Agronomico Tropical de Investigacion y Enseñanza Catie, Informe Técnico No. 236, Turrialba, Costa Rica

Millar CL, Norris GH, Jiang C et al (2018) Long-term supplementation of black elderberries promotes hyperlipidemia, but reduces liver inflammation and improves HDL function and atherosclerotic plaque stability in apolipoprotein E-knockout mice. Mol Nutr Food Res 62(23):e18004004. https://doi.org/10.1002/mnfr.201800404

Moerman DE (1998) Native American ethnobotany. Timber Press, Portland

Montagnac JA, Davis CR, Tanumihardjo SA (2009) Processing techniques to reduce toxicity and antinutrients of cassava for use as a staple food. Compr Rev Food Sci Food Saf 8:17–27

Murkovic M, Abuja PM, Bergmann AR et al (2004) Effects of elderberry juice on fasting and postprandial serum lipids and low-density lipoprotein oxidation in healthy volunteers: a randomized, double-blind, placebo-controlled study. Eur J Clin Nutr 58(2):244–249

Nahrstedt A (1985) Cyanogenic compounds as protecting agents for organisms. Plant Syst Evol 150:35–47

National Research Council (US) Subcommittee on Acute Exposure Guideline Levels (2002) Acute exposure guideline levels for selected airborne chemicals. Vol. 2(5). Hydrogen cyanide: acute exposure guideline levels. National Academies Press (US), Washington, DC. https://www.ncbi. nlm.nih.gov/books/NBK207601/

Pedraza RA, Williams-Linera G (2003) Evaluation of native tree species for the rehabilitation of deforested areas in a Mexican cloud forest. New For 26:83–99

Perkins-Veazie P, Thomas AL, Byers PL, Finn CE (2015) Fruit composition of elderberry (*Sambucus* spp.) genotypes grown in Oregon and Missouri, USA. Acta Hortic 1063:219–224

Picon PD, Picon RV, Costa AF et al (2010) Randomized clinical trial of a phytotherapic compound containing *Pimpinella anisum*, *Foeniculum vulgare*, *Sambucus nigra*, and *Cassia augustifolia* [sic] for chronic constipation. BMC Complement Altern Med 10:17

Pleschka S, Stein M, Schoop R, Hudson JB (2009) Anti-viral properties and mode of action of standardized Echinacea purpurea extract against highly pathogenic avian Influenza virus (H5N1, H7N7) and swine-origin H1N1 (S-OIV). Virol J 6:197. https://doi.org/10.1186/1743-422X-6-197

Raafat K, El-Lakany A (2015) Acute and subchronic in-vivo effects of Ferula hermonis L. and *Sambucus nigra* L. and their potential active isolates in a diabetic mouse model of neuropathic pain. BMC Complement Altern Med 15:257

Rauš K, Pleschka S, Klein P, Schoop R, Fisher P (2015) Effect of an echinacea-based hot drink versus oseltamivir in influenza treatment: a randomized, double-blind, double-dummy, multi-center, noninferiority clinical trial. Curr Ther Res 77:66–72

Roschek B Jr, Fink RC, McMichael MD, Li D, Alberte RS (2009) Elderberry flavonoids bind to and prevent H1N1 infection in vitro. Phytochemistry 70(10):1255–1261

Salamon I, Grulova D (2015) Elderberry (*Sambucus nigra*): from natural medicine in ancient times to protection against witches in the middle ages – a brief historical overview. Acta Hort 1061:35–39

Salvador ÂC, Król E, Lemos VC et al (2016) Effect of elderberry (*Sambucus nigra* L.) extract supplementation in STZ-induced diabetic rats fed with a high-fat diet. Int J Mol Sci 18(1). https://doi.org/10.3390/ijms18010013

Samuels N, Grbic JT, Saffer AJ, Wexler ID, Williams RC (2012) Effect of an herbal mouth rinse in preventing periodontal inflammation in an experimental gingivitis model: a pilot study. Compend Contin Educ Dent 33(3):204–206. 208–211

Schapowal A, Klein P, Johnston SL (2015) Echinacea reduces the risk of recurrent respiratory tract infections and complications: a meta-analysis of randomized controlled trials. Adv Ther 32(3):187–200

Senica M, Stampar F, Veberic R, Mikulic-Petkovsek M (2016) Processed elderberry (*Sambucus nigra* L.) products: a beneficial or harmful food alternative? LWT-Food Sci Technol 72:182–188

Senica M, Stampar F, Veberic R, Mikulic-Petkovsek M (2017) The higher the better? Differences in phenolics and cyanogenic glycosides in *Sambucus nigra* leaves, flowers and berries from different altitudes. J Sci Food Agric 97:2623–2632

Shah SA, Sander S, White CM, Rinaldi M, Coleman CI (2007) Evaluation of echinacea for the prevention and treatment of the common cold: a meta-analysis. Lancet Infect Dis 7(7):473–480

Shahsavandi S, Ebrahimi MM, Hasaninejad Farahani A (2017) Interfering with lipid raft association: a mechanism to control influenza virus infection by Sambucus Nigra. Iran J Pharm Res 16(3):1147–1154

Sharma M, Anderson SA, Schoop R, Hudson JB (2009a) Induction of multiple pro-inflammatory cytokines by respiratory viruses and reversal by standardized Echinacea, a potent antiviral herbal extract. Antivir Res 83(2):165–170

Sharma M, Schoop R, Hudson JB (2009b) Echinacea as an antiinflammatory agent: the influence of physiologically relevant parameters. Phytother Res 23(6):863–867

Simonyi A, Chen Z, Jiang J et al (2015) Inhibition of microglial activation by elderberry extracts and its phenolic components. Life Sci 128:30–38

Small E, Catling PM, Richer C (2004) Poorly known economic plants of Canada—41. American elder [*Sambucus nigra* subsp. canadensis (L.) R. Bolli] and blue elderberry [S. nigra subsp. cerulea (Raf.) R. Bolli]. Can Bot Assoc Bul 37:20–28

Spagnuolo C, Napolitano M, Tedesco I, Moccia S, Milito A, Russo GL (2016) Neuroprotective role of natural polyphenols. Curr Top Med Chem 16(17):1943–1950

Swanston-Flatt SK, Day C, Flatt PR, Gould BJ, Bailey CJ (1989) Glycaemic effects of traditional European plant treatments for diabetes. Studies in normal and streptozotocin diabetic mice. Diabetes Res 10(2):69–73

Thole JM, Kraft TF, Sueiro LA et al (2006) A comparative evaluation of the anticancer properties of European and American elderberry fruits. J Med Food 9(4):498–504

Thomas AL, Byers PL, Finn CE, Chen Y-C, Rottinghaus GE, Malone AM, Applequist WL (2008) Occurrence of rutin and chlorogenic acid in elderberry leaf, flower, and stem in response to genotype, environment, and season. Acta Hortic 765:197–206

Thomas AL, Byers PL, Ellersieck MR (2009) Productivity and characteristics of American elderberry in response to various pruning methods. HortScience 44(3):671–677

Thomas AL, Perkins-Veazie P, Byers PL, Finn CE, Lee J (2013) A comparison of fruit characteristics among diverse elderberry genotypes grown in Missouri and Oregon. J Berry Res 3:159–168

Thomas AL, Byers PL, Avery JD Jr, Kaps M, Gu S (2015a) Horticultural performance of eight American elderberry genotypes at three Missouri locations. Acta Hortic 1063:237–244

Thomas AL, Byers PL, Gu S, Avery JD Jr, Kaps M, Datta A, Fernando L, Grossi P, Rottinghaus GE (2015b) Occurrence of polyphenols, organic acids, and sugars among diverse elderberry genotypes grown in three Missouri (USA) locations. Acta Hortic 1063:147–154

Thomson Healthcare (2007) PDR for herbal medicines, 4th edn. Thomson Healthcare, Montvale

Tiralongo E, Wee SS, Lea RA (2016) Elderberry supplementation reduced cold duration and symptoms in air-travellers: a randomized, double-blind placebo-controlled clinical trial. Nutrients 8(4):182. https://doi.org/10.3390/nu8040182

Turland NJ, Wiersema JH, Monro AM, Deng YF, Zhang L (2017) XIX international botanical congress: report of congress action on nomenclature proposals. Taxon 66(5):1234–1245

Ulbricht C, Basch E, Cheung L, Goldberg H, Hammerness P, Isaac R, Khalsa KPS, Romm A, Rychlik I, Varghese M, Weissner W, Windsor RC, Wortley J (2014) An evidence-based systematic review of elderberry and elderflower (Sambucus nigra) by the Natural Standard Research Collaboration. J Diet Suppl 11(1):80–120. https://doi.org/10.3109/19390211.2013.859852

USDA – National Plant Germplasm System (2019) For example: https://npgsweb.ars-grin.gov/gringlobal/accessiondetail.aspx?1684794

USDA-NRCS Plant Guide (2003) Common elderberry. https://plants.usda.gov/plantguide/pdf/cs_sanic4.pdf

Vaněčková-Skuhravá I (1996) Life cycles of five eriophyid mites species (Eriophyoidea, Acari) developing on trees and shrubs. J Appl Entomol 120:513–517

Viapiana A, Wesolowski M (2017) The phenolic contents and antioxidant activities of infusions of Sambucus nigra L. Plant Foods Hum Nutr 72(1):82–87

Vimalanathan S, Schoop R, Suter A, Hudson J (2017) Prevention of influenza virus induced bacterial superinfection by standardized Echinacea purpurea, via regulation of surface receptor expression in human bronchial epithelial cells. Virus Res 233:51–59

Waknine-Grinberg JH, El-On J, Barak V, Barenholz Y, Golenser J (2009) The immunomodulatory effect of Sambucol on leishmanial and malarial infections. Planta Med 75(6):581–586

Warmund MR, Amrine JW Jr (2015) Eriophyid mites inhabiting American elderberry. Acta Hortic 1063:155–160

Warmund MR, Mihail JD, Hensel K (2019) Puccinia sambuci infection of American elderberry plants. HortScience 54:880–884

Way RD (1981) Elderberry culture in New York State. New York's Food and Life Sciences Bull 91:1–4. New York State Agricultural Experiment Station, Geneva

Wilson KL (2016) Report of the general committee: 14. Taxon 65(4):878–879

Zagrobelny M, Bak S, Rasmussen AV, Jørgensen B, Naumann CM, Møller BL (2004) Cyanogenic glucosides and plant–insect interactions. Phytochemistry 65:293–306

Zakay-Rones Z, Varsano N, Zlotnik M et al (1995) Inhibition of several strains of influenza virus in vitro and reduction of symptoms by an elderberry extract (Sambucus nigra L.) during an outbreak of influenza B Panama. J. Altern Complement Med 1(4):361–369

Zakay-Rones Z, Thom E, Wollan T, Wadstein J (2004) Randomized study of the efficacy and safety of oral elderberry extract in the treatment of influenza A and B virus infections. J Int Med Res 32(2):132–140

Chapter 6
History, Conservation, and Cultivation of American Ginseng, North America's Most Famous Medicinal Plant

Jennifer L. Chandler and James B. McGraw

Abstract Asian ginseng played an integral role in traditional Chinese medicine for millennia, but overexploitation of the plant, coupled with habitat loss, contributed to steep declines in natural populations throughout China. The discovery of the closely related medicinal herb, American ginseng, in Canada in the early 1700s ignited three centuries of trade between North America and China. Profits made from the export of American ginseng to China facilitated western expansion in the United States, and the fortunes of some early Americans were built on the American ginseng trade. American ginseng has been harvested continuously in North America following its initial discovery, punctuated by regional spurts of intense harvest in response to economic needs of rural residents. The ongoing extraction of American ginseng from deciduous forests throughout its range have resulted in modern populations that contain fewer individuals and individuals of smaller stature relative to historic populations, yet thousands of these remnant populations remain. In fact, concerns about overharvest of American ginseng led to its placement on Appendix II of the Convention on the International Trade in Endangered Species of Wild Fauna and Flora (CITES) in 1975. Overexploitation, however, is not the only threat to wild populations of American ginseng. Climate change and browse by overabundant white-tailed deer also negatively affect American ginseng and pose a threat to the future viability of populations. Conservation efforts are needed to ensure that populations of wild American ginseng persist. Population stewardship is a low-labor conservation strategy where individuals promote the growth of wild ginseng populations by strategically planting seeds and removing the aboveground portion of large plants just prior to the onset of harvest season to prevent potential harvesters from locating the adult plants. Another conservation strategy, conservation through cultivation, involves shifting harvest pressure from wild American ginseng to cultivated ginseng. Wild-simulated cultivation is a low-labor technique where seeds are

J. L. Chandler (✉)
Department of Biology, West Chester University of Pennsylvania, West Chester, PA, USA

J. B. McGraw
Department of Biology, West Virginia University, Morgantown, WV, USA

© Springer Nature Switzerland AG 2020
Á. Máthé (ed.), *Medicinal and Aromatic Plants of North America*, Medicinal and Aromatic Plants of the World 6,
https://doi.org/10.1007/978-3-030-44930-8_6

141

planted into desirable habitat and are left to grow with little intervention until the roots are ready for harvest years later. Wild-simulated cultivation produces roots that are virtually indistinguishable from true wild roots, which makes this type of cultivation a desirable option for the conservation through cultivation approach. Additional cultivation techniques include woods-cultivation, which is ginseng grown in prepared beds under natural forest canopies, and artificial shade cultivation, which is the most labor-intensive cultivation technique and requires large initial investments in shade structures and site preparation. The last two cultivation techniques can produce large roots and abundant seeds in a short period of time. However, the roots lack morphological features desired in traditional Chinese medicine, and are less valuable than both wild-simulated and true wild roots. The abundance and size of wild American ginseng populations are decreasing range-wide, and this negative trajectory is certain to continue unless efforts are made to both improve the management of wild populations and to increase conservation efforts.

Keywords Conservation through cultivation · Cultivation · Harvest · Non-timber forest product · *Panax ginseng* · *Panax quinquefolius* · Stewardship · Sustainable harvest

6.1   Introduction

Harvesting to supply the Asian traditional medicine market made ginseng North America's most harvested wild plant for two centuries. Asian ginseng has played an integral role in traditional Chinese medicine (TCM) for millennia. The subsequent discovery of the closely related medicinal herb, American ginseng, in Canada in the early 1700s ignited three centuries of trade between North America and China.

6.1.1   Asian Ginseng (**Panax ginseng**)

For thousands of years, practitioners of traditional Chinese medicine consumed Asian ginseng (*Panax ginseng*) as a tonic to maintain vital energy, or qi (Taylor 2006). Asian ginseng is considered an adaptogen, a plant that is capable of maintaining and restoring a body's daily balance (Manget 2012; Taylor 2006). Traditionally, Asian ginseng was combined with numerous other medicinal herbs and prescribed on an individual basis to increase one's qi and to promote well-being (Taylor 2006). The mechanisms through which Asian ginseng affected the body were not understood, although Li Shizhen indicated that the herb could treat illness throughout the body due to its expression of yang in his Great Compendium of Herbs, which was published posthumously in 1596 (Taylor 2006).

The harvest, trade, and use of the root of Asian ginseng prompted a period of extreme overexploitation in China. Overharvesting of Asian ginseng in central China, as well as habitat loss, led to near extinction by the end of the fifteenth century, at which point harvesters were forced to focus their hunting efforts far north in Manchuria (Taylor 2006; McGraw et al. 2013). Nurhaci, emperor of the Qing dynasty, sent thousands of harvesters into the Long White Mountains near Korea to harvest Asian ginseng under an imperial monopoly (Taylor 2006). European interest in medicinal herbs increased in the seventeenth century, with Asian ginseng being regularly exported from the port of Canton by the East India Company, further increasing demand (Taylor 2006). The high demand for Asian ginseng root enticed poachers to harvest illegally (Taylor 2006). Realizing that root theft could not be controlled, and that overharvest was inevitable, the imperial court amplified harvest efforts (Taylor 2006). Eventually, wild Asian ginseng was extirpated from many forests throughout China. Although the supply of Asian ginseng was depleted, the demand remained high due to its continued use in traditional medicine and to an ever-increasing Chinese population (Taylor 2006).

6.1.2   American Ginseng (Panax quinquefolius)

In the early eighteenth century, members of the Jesuit order served as missionaries to the Mohawk tribe of the Iroquois in Quebec, in what would later be Canada (Taylor 2006). One of those early missionaries was Joseph-Francois Lafitau (Taylor 2006). In 1715, Lafitau came across an article written by Father Pierre Jartoux, a Jesuit who had been contracted in 1702 to produce an Atlas of China (Taylor 2006; Davis and Persons 2014). The article described the reputation of Asian ginseng as a restorative in Chinese medicine, included details about its preparation and traditional use, and even recounted Father Jartoux's personal use of Asian ginseng, along with his opinion that the herb did have restorative properties (Taylor 2006). The article further provided a botanical description of the plant and a description of the habitat in which the plant was found (Taylor 2006). Of particular importance, however, was Father Jartoux's suggestion that the plant may be found in similar mountainous habitats in the New World (Taylor 2006).

The following spring, Lafitau and his Mohawk guides searched diligently for the herb about which he had read using little more than a physical description of the plant and a hope that the plant would be found in North America (Taylor 2006). Lafitau spent months searching the forest to no avail until that fall when he stumbled upon a plant near his homestead that bore bright red berries (Figs. 6.1 and 6.2, Taylor 2006). With the help of the Mohawk peoples, Lafitau confirmed the taxonomic similarity of this New World *American* ginseng plant to the Asian ginseng described in Jartoux's writings. As it turns out, Native Americans had harvested and used American ginseng medicinally, and although the importance of the herb varied

Fig. 6.1 Bright red berries
of a large, adult American
ginseng (*Panax
quinquefolius*) plant in
southern Virginia, U.S.A.

Fig. 6.2 Four-prong adult
American ginseng in a
West Virginia, U.S.A. forest

among regions and tribes (Pritts 1995), American ginseng on its own was likely not of great importance in traditional Native American medicine (Johannsen 2006; McGraw et al. 2013).

Demand for Asian ginseng in China still outpaced supply when Lafitau confirmed the identification of American ginseng in Canada. Asian ginseng and American ginseng are taxonomically similar, but are thought to have different medicinal properties in Chinese medicine, with Asian ginseng producing a stronger stimulant effect and American ginseng a more calmative effect. Nevertheless, the abundance of American ginseng from this New World source ignited the North American ginseng industry (Taylor 2006; McGraw et al. 2013). Natives and European colonists throughout Montreal set out to the woods the year following Lafitau's discovery to dig American ginseng to sell to French merchants that were making shipments to Asia (Taylor 2006). The initial rush for ginseng resulted in swift extirpation of the plant from the forests surrounding Montreal, and hunters were forced to expand their search range (Taylor 2006; McGraw et al. 2013). By the end of the eighteenth century, the forests of Canada were virtually devoid of the wild resource that had made many people a great deal of money. The hunt for ginseng had slowed in Canada, but ginseng prospects were high in the British colonies to the south (Taylor 2006).

Wild American ginseng was first discovered and harvested in the forests of New England in the early 1700s (Davis and Persons 2014), though the harvesters did not stay in one locale for long. Ginseng harvest helped pave the way for the western expansion of settlers in America (Taylor 2006; Davis and Persons 2014). For those individuals that could identify the plant among the myriad of other forest herbs, the harvest and trade of American ginseng provided the immediate income needed to build homesteads and to support families in lean times (Davis and Persons 2014), a service that is paralleled in contemporary Appalachia, particularly when coupled with other seasonal harvest activities (Bailey 1999). Indeed, there were two major exports from early America, furs and ginseng, and the same individuals often dealt in both commodities (Davis and Persons 2014). The fur trade is typically credited as being the backbone that built early fortunes, while ginseng has inexplicably been ignored as a major economic player of the time (Davis and Persons 2014). Although often overlooked, ginseng played a meaningful role in the lives of some well-known early Americans. John Jacob Astor, the fur trader who founded the American Fur Company, built an early portion of his empire on the ginseng trade (Davis and Persons 2014). The famous pioneer, Daniel Boone, made his fortune exporting barrels full of ginseng roots beginning in 1788, one year after he lost an entire shipment of ginseng to the Ohio River after his boat capsized (Taylor 2006; Davis and Persons 2014).

The importance of the ginseng trade was solidified in early America, and harvest and exports continued in earnest throughout the next century. Harvest progressed westward with the frontier expansion in a pattern of discovery-depletion. Harvesters would move from location to location extracting the majority of the ginseng from previously untouched forests before moving to areas of higher ginseng density (Davis and Persons 2014). Harvest and export of ginseng from America remained high through the mid 1800s, with ginseng exports regularly exceeding 300,000 pounds (Davis and Persons 2014). One estimate suggests that at least 64 million roots were exported from North America in 1841 alone (McGraw et al. 2013).

For many families living in the mountains of southern Appalachia, ginseng harvest was a vital source of income in the decades surrounding the American Civil War. Historian Luke Manget (2012) paints a picture of the ginseng harvest in western North Carolina in the mid 1800s. Men, women, and even children would venture into the woods to harvest ginseng. During the early years of harvest, individuals rarely had to travel far from their homes to find substantial amounts of ginseng, as the abundance of forested land and the lack of both people and infrastructure meant that this common resource was untapped (Manget 2012). Nevertheless, as time passed and larger increments of the forest were scoured for wild ginseng, exports began to decline. Such declines were likely due to decreased ginseng abundance near the farming settlements, and due to the fact that local farmers were prospering without added income from ginseng sales (Manget 2012).

The Civil War produced an extended economic depression in the badly-damaged southern states (Manget 2012). Southern Appalachian families who once prospered were now forced to take to the mountains to harvest wild ginseng in order to feed their families, and so exports once again increased, and southern Appalachia became the epicenter of ginseng harvest (Manget 2012). Approximately 1,400,000 pounds of ginseng were exported from the United States in the 3 years following the war, a quantity nearly equaling the total ginseng exports throughout the 1850s (Manget 2012). United States ginseng exports in 1888 alone brought in approximately $656,817 (Davis and Persons 2014), equivalent to over $17,000,000 in 2018 dollars. The quantity of roots harvested and exported in the 1800s was far beyond anything observed in the more recent past (McGraw et al. 2013). What's more, photographic evidence along with data collected from herbaria throughout ginseng's range suggest that plants being harvested at the end of the first ginseng boom in 1900 were large compared to plants obtained in 2000 (McGraw 2001).

As with many industries that rely on natural resources for profit, with a strong boom comes a rapid bust. In the post-war South, the increased reliance on the harvest and sale of ginseng, and the rate at which ginseng was being pulled from the mountains, led to concerns that ginseng would soon be extirpated from forests throughout the region (Manget 2012). In response to this concern, several states including North Carolina, Georgia, and West Virginia established harvest regulations in the 1860s and 1870s (Manget 2012). These early regulations dictated that ginseng could only be dug during a pre-specified harvest season (Manget 2012). By the end of the 1800s, easily accessible ginseng populations full of large plants had been severely overexploited, and the export of ginseng from the United States decreased substantially, reaching levels as low as 200,000 pounds annually by 1895 (Davis and Persons 2014).

By the late 1800s and early 1900s, another industry was sweeping through the forests of eastern North America. The ancient trees that once provided habitat for American ginseng were now being removed by large-scale clearcut logging operations (Wyatt and Silman 2014; Chandler and McGraw 2015). In mere decades, the vast majority of the old growth forests that supported American ginseng were irrevocably lost. In the eastern deciduous forest, less than 1% of the original area of old growth remains today, and this decrease is attributed largely to early timber harvest (Wyatt and Silman 2014), but also to the conversion of forests to agricultural land. The rate of land conversion from forest to farmland peaked in the mid-1800s, and resulted in degraded soil profiles and decreased forest regeneration due to constant grazing and trampling by livestock (McGraw et al. 2013; Davis and Persons 2014). The prior overharvest of ginseng, coupled with timber harvest and land use conversion, resulted in smaller populations that were scarcer across ginseng's wide range. In the early 1900s, farmlands located on steep, hard to access slopes throughout Appalachia were being abandoned at a greater rate than forests were being cleared, especially following widespread outbreak of chestnut blight. Decreases in timber harvest coupled with farm abandonment resulted in a net increase in forest cover that persisted throughout the twentieth century (McGraw et al. 2013). Nevertheless, recolonization of ginseng into these secondary forests has been slow, especially in reclaimed farmlands, and harvest has continued, albeit at a slower pace than in the 1800s. Despite the precipitous decline in American ginseng abundance from the 1700s to the present, the demand for ginseng's root in China has not wavered, and so people still harvest ginseng root every fall. Indeed, just as the harvest and sale of ginseng had supported rural families during lean times in the past, the modern annual harvest of ginseng provides millions of dollars of supplemental income to people throughout Appalachia.

6.2 Threats to American Ginseng Populations

6.2.1 Harvest

McGraw et al. (2010) studied the harvest rates of thirty populations of American ginseng spread across seven states in ginseng's natural range for a period of 5–11 years. In any given year, 1.3% of the total number of individuals contained in the thirty populations were confirmed harvested, and 15% of these populations showed harvest evidence. Of the thirty populations, 43% showed signs of harvest at least once during the entire study period, and six of those were harvested multiple times. Additionally, Van der Voort and McGraw (2006) indicated that non-compliant harvesting, which is harvesting that breaks regulations set forth by each state, reduces population growth by 15% annually. The same study found that minimally compliant harvest, where harvest was delayed until the season began on August 15 (the start of the harvest season in West Virginia during the time of the study), resulted in an 8% reduction in population growth (Van der Voort and McGraw 2006). Only voluntary 'stewardship' methods, whereby harvesters showed significant self-restraint and encouraged regeneration through seed planting, resulted in a sustainable harvest practice.

Continued wild harvest, and fear that overharvest would lead to the extinction of natural populations, prompted the inclusion of American ginseng on Appendix II of the Convention on the International Trade in Endangered Species of Wild Fauna and Flora (CITES) in 1975. The purpose of CITES is to identify species whose survival in the wild may be jeopardized by international trade, and then monitor and control the trade to protect future viability of the species (CITES.org). Appendix II of CITES includes species that are not currently threatened with extinction, but that require trade regulations to avoid such a trajectory. In the United States, the monitoring requirements outlined by CITES fall under the responsibilities of the U.S. Fish and Wildlife Service (USFWS), who assess ginseng viability and trade on a state-by-state basis and determine whether the export of ginseng roots is detrimental to species survival (CITES.org, Davis and Persons 2014). The USFWS mandates that states institute formal ginseng management programs that dictate harvest seasons, harvest rules, and export records, including certification that roots are either wild or cultivated (McGraw et al. 2013; Davis and Persons 2014). All states require plants to be at least 5 years old (age assessed by counting abscission scars on the rhizome) and require that they have at least three leaves, though Illinois requires that plants be at least 10 years old with at least four leaves (United States Fish and Wildlife Service). Although harvest season varied dramatically when CITES regulations were first implemented in the U.S., harvest seasons for all states that export ginseng now begin on September 1, with the exception of Illinois, whose season begins on the first Saturday of September. The late starting date for the harvest season is important, as it allows ginseng berries to ripen, thus increasing the probability of germination (McGraw 2017). Populations in Quebec, Canada were essentially exhausted by overexploitation, and as a result, the export of wild American ginseng from Quebec has been banned outright since 1973, with the rest of Canada following suit in 1989 (Davis and Persons 2014).

6.2.2 Deer Browse

White-tailed deer (*Odocoileus virginianus*) densities are now on average two to four times higher than pre-settlement densities, and the over-abundance of these herbivores has been shown to negatively affect the growth of herbaceous species throughout their range, including American ginseng (Anderson 1994; Rooney and Dress 1997; Augustine and Frelich 1998; Fletcher et al. 2001; Rooney and Gross 2003; Knight 2004; Füredi and McGraw 2004; Füredi 2004; McGraw and Furedi 2005; Farrington et al. 2008). In contrast to the harvest of American ginseng for its root, initial damage caused by deer browse is concentrated only on aboveground portions of the plant, and doesn't directly remove portions of the root. However, deer browse can severely stunt a plant's leaf area, stem height, and reproduction (McGraw et al. 2013). If browse is repeated 2 or 3 years in a row, root mortality follows as carbohydrate supplies are depleted (Füredi 2004).

A recent study in West Virginia compared ginseng performance in plants exposed to deer browse on the forest floor to plants protected from deer browse on a rock refugium (McGraw and Chandler 2018). This study suggested that deer browse negatively affected ginseng growth, and that the portion of the population exposed to deer browse exhibited a high proportion of smaller, juvenile plants when compared to plants on the rock refugium. Additionally, while the subpopulation protected from deer browse increased in size over the study period, the subpopulation exposed to browse decreased by 4.5% per year. Similarly, a separate study found that ginseng population growth rate decreased by 2.7% when exposed to deer browse (McGraw and Furedi 2005). Another study examined the interactive effect of deer browse and root harvest on population growth (Farrington et al. 2008). This study indicated that while deer browse and harvest negatively affected population growth separately, population growth rate in the presence of both deer browse and harvest was no lower than populations exposed to harvest alone. The most likely explanation for this is that deer tend to browse the larger plants that are targeted by harvesters, and the browse of the aboveground portion of these plants precludes detection by harvesters.

6.2.3 Climate Change

Global temperatures are predicted to climb 1.3–6.1 °C above preindustrial levels by the year 2100, with a projected rise of 2.3–5.6 °C above preindustrial levels by the year 2100 for the eastern region of the United States (Rogelj et al. 2012; McGraw et al. 2013). Additionally, modelers predict that future temperatures and precipitation patterns will be more variable, leading to abnormal frequencies of both drought and flooding, as well as increases in extreme weather events. Souther and McGraw (2011) and Souther et al. (2012) investigated the effects that climate change might have on populations of ginseng found in central Appalachia. The field study showed that regardless of their site of origin, populations are locally-adapted to the temperature regime in which they evolved. A separate growth chamber study suggested that the relationship between ginseng population performance and temperature change is complex, however, and is not readily explained by direct temperature effects alone. The response is mediated by indirect effects on abiotic and biotic properties such as pollinator success, disease, and soil moisture (Souther and McGraw 2011; Souther et al. 2012). Nevertheless, when growing temperatures were experimentally increased 3–6 °C above home-site temperatures, ginseng photosynthesis and seed production decreased significantly, and leaf senescence began earlier than is typical (Jochum et al. 2007; Souther et al. 2012).

Smaller populations of ginseng are more susceptible to extirpation than larger populations under future climate scenarios (Souther 2011). This fact is problematic, as the majority of natural ginseng populations throughout the United States are small, and are therefore at a higher risk of local extinction. For the species as a whole, the large number of populations provides some insurance against total extinction in the short-term. The ability of ginseng to disperse northward or up in elevation could keep climatic ecotypes within their appropriate climate envelopes and decrease the likelihood of decline (Souther 2011). Evidence suggests that wood thrushes (*Hylocichla mustelina*) are capable of dispersing ginseng seeds nearly 100 m, so it is reasonable to infer that at least a portion of seeds produced may disperse away from the original population (Elza et al. 2016). A separate study which documented the ingestion and regurgitation of ginseng berries by a captive wood thrush further supports the assertion that wood thrush may be potential long distance dispersers of American ginseng seeds (Hruska et al. 2014). Hruska et al. (2014) found that 15% of the seeds offered to the wood thrush were ingested and that viable seeds cleaned of their pulp were regurgitated between 15 and 37 min following ingestion. It seems reasonable, then, that wood thrushes feeding in forests may similarly ingest ginseng berries, travel a relatively long distance, and regurgitate the seed far from the parent plant. Nevertheless, 90% of gravity-dispersed ginseng seeds remain within two meters of the parent plant (Van der Voort 2005; Hackney 1999), and so it is somewhat unlikely that populations will be able to track their optimal climate naturally, but instead may require assisted migration.

While harvest, deer browse, and climate change are the most pressing threats to the viability of wild ginseng populations, they are by no means the only threats. McGraw et al. (2013) detail several additional negative forces with the potential to move ginseng populations down the extinction vortex, including inbreeding depression, suburban sprawl, and surface mining. Nevertheless, not all forces acting upon ginseng are negative. Positive forces such as potential exponential growth and population stewardship may help some wild populations move up the extinction vortex toward viability.

6.2.4 Conservation

Centuries of harvest have caused American ginseng to exhibit an unusual type of rarity. Ginseng is somewhat cosmopolitan, as it can grow in a wide variety of habitats throughout temperate forests of North America (McGraw et al. 2013). One rough calculation suggests that West Virginia alone may contain approximately 87.8 million ginseng plants (McGraw et al. 2003). Nevertheless, populations tend to be sporadic and quite small, typically containing fewer than 150 individuals (McGraw et al. 2013). Further, ginseng exhibits a "slow" life history pattern, which is characterized by slow growth and a long pre-reproductive period (McGraw et al. 2013). Small population sizes coupled with slow growth and reproduction puts many ginseng populations at a high risk of extirpation. In order to conserve wild ginseng, we must shift the way we think about ginseng, its cultivation and stewardship, and the way in which it is harvested. In this section, we advocate two activities that support wild American ginseng conservation: stewardship of wild populations and conservation through cultivation.

6.3   Stewardship

With the existence of many thousands of small populations over a broad area, an opportunity exists to conserve ginseng through assisted population growth by con-servation minded landowners. We call this approach 'stewarding' natural popula-tions. For a landowner with ginseng on their property, the formal procedure begins with producing a rough map of existing plants in the population, using fixed objects such as rocks and large trees as reference points. Then, just prior to the onset of harvest season (end of August), the map is used to re-find plants each year. Red, or large green fruits are planted 2 cm (0.5–1 in.) deep in proximity (but not under) par-ent plants. Then, the tops are removed on adults with 3 or more leaves to prevent harvesters from observing the plants, a process that does not injure the plant, but ensures its survival. Utilizing this procedure from 2007 to 2018, a small population of 55 plants increased to 262 plants, a nearly five-fold increase over that time span (McGraw, unpublished data, 2019).

6.4   Conservation of American Ginseng Through Cultivation

Much attention has surrounded the concept of reducing the harvest pressure on wild American ginseng by shifting collection efforts from wild harvest to a more sustain-able model of cultivation and stewardship using a model known as "conservation through cultivation" (Burkhart 2011). In his research which focused on forest farm-ing ginseng in Pennsylvania, U.S.A., Burkhart (2011) suggested that this conserva-tion through cultivation model may be a valuable tool in ginseng conservation efforts.

Burkhart and Jacobson (2009) performed financial analyses for eight wild-harvested medicinal plants in the eastern deciduous forest to determine if a transi-tion from wild harvest to forest cultivation is economically sustainable. They found that among all eight medicinal plants, American ginseng was the only one for which agroforestry could consistently be considered profitable; cultivation of goldenseal (*Hydrastis canadensis* L.) was profitable under specific circumstances (Burkhart and Jacobson 2009). Since wild-simulated roots are virtually indistinguishable from wild roots, the monetary payoff for individuals that choose to grow wild-simulated ginseng may be high, particularly if niche markets for ethically produced roots are developed and adopted by consumers (Burkhart and Jacobson 2009). The high pay-off potential relative to labor in wild-simulated cultivation, the well-establish mar-ket for these roots, and the easy accessibility to buyers to whom growers can sell roots all make the idea of a ginseng-based "conservation through cultivation" palat-able (Burkhart 2011).

The cultural roots of ginseng growing and harvest run deep throughout much of Appalachia, and some growers and harvesters may have the desire and motivation to steward existing populations or to establish wild-simulated populations to ensure the same culture persists for generations to come. Nevertheless, efforts to steward wild populations or to institute the large-scale substitution of wild harvested roots by wild-simulated roots will be tempered by the lingering fear that a harvester may steal the plants that were so carefully cultivated or stewarded. Indeed, the tragedy of the commons (Hardin 1968), wherein individuals deplete a common resource for their own gain without regard for long-term consequences and well-being, will still exist in these scenarios. Additionally, the continued collection of wild roots will impact conservation efforts, as wild harvest is less economically risky, less expensive, and less labor intensive than both stewardship and wild-simulated cultivation. As such, it is vital that potential growers and harvesters are provided adequate education that clearly explains the ecological, economic, and cultural benefits of these conservation methods, and that sets forth clear guidelines for implementation (Burkhart and Jacobson 2009).

6.5 Cultivation of American Ginseng

This section provides a very brief overview of the standard methods for ginseng cultivation, summarized largely from Davis and Persons (2014). For an extensive description of the various farming techniques used to grow American ginseng and other non-timber forest products, we suggest the book "Growing and Marketing Ginseng, Goldenseal, and Other Woodland Medicinals" by Jeanine Davis and W. Scott Persons.

6.5.1 Artificial Shade Cultivation

Dwindling population sizes and increased difficulty locating large quantities of wild roots prompted attempts to cultivate American ginseng, a feat that was long considered an impossibility (Davis and Persons 2014). Many early attempts at cultivating ginseng in the late 1800s and early 1900s failed, as potential entrepreneurs were attempting to grow ginseng using standard methods typical in crop farming (Davis and Persons 2014). However, several early growers realized that standard methods were inadequate, as the growing conditions required by ginseng stand in stark contrast to the wide open fields used to raise cereals and other more traditional crops (Davis and Persons 2014). The first successful commercial grower, George Stanton, developed a method whereby he sowed seed into open, tilled fields, and then built a system of wooden laths to provide shade to the growing plants (Davis and Persons 2014). While many growers followed his lead, others chose to plant seed under the shade of a natural forest canopy (Davis and Persons 2014). Ginseng farming was being touted as a way in which to earn good, quick money, and many people wanted in on the potential profits. Ginseng farms sprouted up throughout the eastern half of the United States and in parts of Canada, namely Ontario (Davis and Persons 2014).

Unfortunately, these same individuals could not have envisioned outbreaks of fungus and disease that would claim entire stands of ginseng. Loss of portions of ginseng crops due to the spread of disease and fungal infections in the high-density beds was, and still is, a constant challenge for farmer's growing ginseng under artificial shade. Virtually all ginseng grown in monoculture under shade cloth require routine fungicide treatments as well as periodic treatments using other pesticides including insecticide and herbicides, the latter of which is often necessary before the initial planting and prior to ginseng emergence in the spring (Davis and Persons 2014). While several early large-scale ginseng farmers were able to make a profit from the sale of cultivated roots and seed, particularly in Wisconsin and Canada, the majority of small farms could not turn consistent profits, and farming efforts were ultimately abandoned (Davis and Persons 2014).

By the mid-late 1900s, commercial-scale ginseng farms were producing large quantities of roots, most of which were grown as monocultures using artificial shade. These plants were grown in tilled beds at high densities, under shade provided by either wooden lath or the newly-developed polypropylene shade cloth (Davis and Persons 2014). This method of growth was enticing, as 'mature' roots could be ready for harvest in as little as 3–4 years (Davis and Persons 2014). However, the roots cultivated under artificial shade lack morphological characteristics typical of wild-harvested ginseng, because the rich growing environment provides little resistance to root growth, so roots tend to grow quickly and have a smooth appearance, unlike gnarly, wrinkled roots of wild plants (Davis and Persons 2014). Today, the carrot-like roots grown under artificial shade are used in energy supplements, drinks, and lotions. They are also sold into the Asian traditional medicine market at a much lower price than wild roots.

6.5.2 Woods Cultivation

Woods cultivation is a technique where ginseng is grown at high density in prepared beds under the natural shade of the forest canopy. This labor-intensive form of ginseng farming begins by selecting a site that (a) has two degrees of slope at minimum, but optimally greater than five degrees to ensure well-drained moist soil, (b) provides adequate shade while still allowing sunflecks (transient flecks of high intensity sunlight reaching the forest floor) into the understory, and (c) has mild summer temperatures (Davis and Persons 2014). Biotic characteristics of the site, including the presence of "companion species", must then be assessed. Companion species commonly observed by growers include the canopy trees sugar maple (*Acer saccharum*) and tulip poplar (*Liriodendron tulipifera*), as well as understory indicators such as naturally-occurring American ginseng, maidenhair fern (*Adiantum* sp.), rattlesnake fern (*Botrypus virginianus*), blue cohosh (*Caulophyllum thalictroides*), and black cohosh or doll's eyes (*Actaea* sp.) (Burkhart 2011; Davis and Persons 2014; Turner and McGraw 2015). Care must be taken when relying on companion species for site selection, however, as one study provided contradicting evidence that four commonly-identified indicator species: wild sarsaparilla (*Aralia nudicaulis*), red maple (*Acer rubrum*), sweet birch (*Betula lenta*), and spicebush (*Lindera benzoin*) were actually contraindicators, and that ginseng performance was lower in the presence of these species (Turner and McGraw 2015). Once a site is selected, the area must be cleared of downed logs, saplings, briars, stumps, and any other obstacles that may prevent tillers or tractors from accessing the site (Davis and Persons 2014). The removal of these obstacles and subsequent opening of the understory also increases air flow within the cultivation site, reducing the threat of fungal disease. Seeds are then sown in tilled beds ideally at a density far lower than plants grown under artificial shade, and the sown seeds are immediately mulched with various possible materials, including leaf litter and straw (Davis and Persons 2014).

Woods cultivated ginseng face a myriad of threats, including Alternaria blight, which also wreaks havoc in plants grown in high density under artificial shade. Damping-off and root rot are other common fungal attacks that can quickly spread in the dense woods cultivated ginseng and jeopardize the entire crop (Davis and Persons 2014). Though not as frequent or widespread as fungal diseases, insects, slugs, and various rodents pose a threat to both the aboveground and belowground portion of the plant. Fungicides and insecticides are often used in woods cultivation, especially in sites planted at higher densities (Davis and Persons 2014). The majority of pesticides used are non-organic commercial products, however, some growers are exploring natural alternatives such as compost tea, which, if certified, may increase the value of their roots on niche, organic markets (Davis and Persons 2014).

The roots produced by woods cultivation are decidedly not wild, and are readily distinguishable from wild-harvested roots at the point of sale. Nevertheless, the value of woods cultivated roots exceed that of shade cultivated roots, and the cost to produce these roots is lower since costly shade cloth and lath systems are not required. W. Scott Persons reported that woods cultivated roots sell for ca. 30–70% the price of wild roots (Davis and Persons 2014). Nevertheless, the potential profit per unit area is high, and this type of cultivation can produce abundant seed stock (Davis and Persons 2014). There are, however, additional risks associated with woods cultivation, including high occurrence of deer browse and theft.

6.5.3 Wild-Simulated Cultivation

The third method, wild-simulated cultivation, is unique in that the roots produced are virtually indistinguishable from wild American ginseng. Of all the ginseng cultivation methods, wild-simulated is the most hands-off and cheapest option. Additionally, profits from wild-simulated cultivation can be substantial (Davis and Persons 2014), but can also be fairly unpredictable due to unknowns such as fluctuating weather patterns, intermittent heavy herbivory by large ungulates, and loss of plants due to theft. Site selection for wild-simulated plots is similar to that for woods-cultivated, and an extensive list of companion species can be found in Davis and Persons (2014). Methods for sowing seeds into wild-simulated plots vary. While some growers simply scatter seeds on top of the leaf litter, others rake leaf litter aside, sow seeds at a low density, and mulch with the previously-removed litter (Davis and Persons 2014). Other growers choose to till the seed into the top couple of inches of soil (Davis and Persons 2014). Regardless of method, little needs to be done in the population once the seeds are sown. The majority of labor associated with wild-simulated cultivation comes in several years when the roots are ready to be harvested (Davis and Persons 2014).

When planted in low densities as is common in wild-simulated cultivation, disease and fungi are not a significant concern, though insects, slugs, and rodents can present a challenge to the grower in some situations. Perhaps the most serious risk that growers face is the security of their plants. The potential for harvesters to steal plants is high in both woods cultivated and wild simulated growing, because unlike ginseng grown in open fields under artificial shade, more remote forests provide cover for individuals that dig and steal the ginseng roots. Security is a near constant issue in both wild-simulated and woods cultivation. One large-scale woods cultivator in Maryland often invites ginseng researchers to his farm to give tours of his growing operation and to provide ginseng seed for research. The fences surrounding his complex hold numerous no trespassing signs and several field cameras line his wooded property, with dozens more just out of sight. Theft has been a real problem on his land, such a problem that his family has not taken a vacation together in years, because at least someone must be at the property at all times to protect their ginseng. Many techniques can be used to dissuade a potential poacher from digging plants on your woodlot, though the efficacy of different techniques is likely situational. Theft deterrents range from simple no trespassing signs and fake cameras to microchip tagging devices initially produced to reduce poaching of wild ginseng in the Great Smoky Mountains National Park, TN, U.S.A. (Davis and Persons 2014).

During early ginseng farming efforts, the smooth, light colored, carrot-like cultivated ginseng roots produced when plants were grown in gardens sold for the same price as the gnarly, dark colored, stress-worn roots harvested from the wild (Davis and Persons 2014). Indeed, wild and cultivated roots were often mixed by ginseng buyers prior to export (Davis and Persons 2014). However, today's Asian consumer prefers the gnarly, wild-harvested American ginseng root that closely resemble the root of the Asian ginseng (*Panax ginseng*) that is so important in their traditional medicine (Davis and Persons 2014). As a result, the value of wild-harvested American ginseng, and the virtually identical wild simulated ginseng, far exceeds that of ginseng grown under artificial shade or woods cultivation.

6.6 Comparison of Cultivation Approaches

Ginseng is unusual in that there exists a gradient of cultivation approaches ranging from intensive farming (artificial shade cultivations) to cultivation under natural forest canopies (woods cultivation) to minimal inputs other than seed planting (wild-simulated cultivation). In truth, the methods vary continuously from intense cultivation to wild simulated, with the intensity of cultivation being a function of site conditions, knowledge base of the cultivator, and willingness to assume risk for greater profit returns. Farmers with the resources to invest in expensive shade houses, soil amendments, sprays, and manpower to implement intensive cultivation methods are rewarded by a short time-to-maturity of plants, large roots, high productivity per square meter of soil, and relatively low risk of crop loss to weather and theft. The costs of this approach are in the expense associated with all elements of cultivation and a lower price/lb. of root sold. At the opposite end of the cultivation spectrum, financial outlays can be limited to initial seed supply, labor associated with planting, and labor associated with harvesting and drying. The negative aspects of wild simulated cultivation are the larger land area required of appropriate forest type, a long time-to-harvest and the substantial risk of theft during this extended growth period. The reward could be a tenfold increase in price/lb. received for wild-looking roots.

6.7   Conclusions

Asian ginseng played an integral role in traditional Chinese medicine for millennia, but overexploitation of the plant, coupled with habitat loss, contributed to steep declines in natural populations throughout China. The discovery of the closely related medicinal herb, American ginseng, in Canada in the early 1700s ignited three centuries of trade between North America and China. Profits made from the export of American ginseng to China facilitated western expansion in the United States, and the fortunes of some early Americans were built on the American ginseng trade. The ongoing extraction of American ginseng from deciduous forests throughout its range have resulted in modern populations that contain fewer individuals and individuals of smaller stature relative to historic populations. Overharvest led to its placement on Appendix II of the Convention on the International Trade in Endangered Species of Wild Fauna and Flora (CITES), in 1975. Climate change and browse by overabundant white-tailed deer also negatively affect American ginseng and pose a threat to the future viability of populations. Conservation efforts are needed to ensure that populations of wild American ginseng persist. Possible methods include both population stewardship, as low-labor conservation, and conservation through cultivation. Wild-simulated cultivation is a low-labor technique where seeds are planted into desirable habitat and are left to grow with little intervention until the roots are ready for harvest years later. Additional cultivation techniques include woods-cultivation, where ginseng is grown in prepared beds under natural forest canopies, and artificial shade cultivation, which is the most labor-intensive cultivation technique and requires large initial investments in shade structures and site preparation. Roots that lack the morphological features desired in traditional Chinese medicine are less valuable than both wild-simulated and true wild roots. The abundance and size of wild American ginseng populations are decreasing range-wide, and this negative trajectory is certain to continue unless efforts are made to both improve the management of wild populations and to increase conservation efforts.

References

Anderson RC (1994) Height of white-flowered trillium (*Trillium grandiflorum*) as an index of deer browse intensity. Ecol Appl 4:104–109

Augustine DJ, Frelich LE (1998) Effects of white-tailed deer on populations of an understory forb in fragmented deciduous forests. Conserv Biol 12:995–1004

Bailey B (1999) Social and economic impacts of wild harvested products. Ph.D. dissertation, Morgantown, West Virginia

Burkhart EP (2011) Conservation through cultivation: Economic, sociopolitical and ecological considerations regarding the adoption of ginseng forest farming in Pennsylvania. Doctor of Philosophy, The Pennsylvania State University, State College

Burkhart EP, Jacobson MG (2009) Transitioning from wild collection to forest cultivation of indigenous medicinal forest plants in eastern North America is constrained by lack of profitability. Agrofor Syst 76:437–453

Chandler JL, McGraw JB (2015) Variable effects of timber harvest on the survival, growth, and reproduction of American ginseng (Panax quinquefolius L.). For Ecol Manag 344:1–9

Davis J, Persons WS (2014) Growing and marketing ginseng, goldenseal, and other woodland medicinals. New Society Publishers, Gabriola Island, 508 pp

Elza MC, Slover CS, McGraw JB (2016) Analysis of wood thrush (*Hylocichla mustelina*) movement patterns to explain the spatial structure of American ginseng (*Panax quinquefolius*) populations. Ecol Res 31(2):195–201

Farrington SJ, Muzika R-M, Drees D, Knight TM (2008) Interactive effects of harvest and deer herbivory on the population dynamics of American ginseng. Conserv Biol 23:719–728

Fletcher JD, Shipley LA, McShea WJ, Shumway DL (2001) Wildlife herbivory and rare plants: the effects of white-tailed deer, rodents, and insects on growth and survival of Turk's cap lily. Biol Conserv 101:229–238

Füredi MA (2004) Effects of herbivory by white-tailed deer (*Odocoileus virginianus* Zimm.) on the population biology of American ginseng (*Panax quinquefolius* L.). Ph.D. dissertation, West Virginia University, Morgantown

Füredi M, McGraw JB (2004) White-tailed deer: dispersers or predators of American ginseng seeds? Am Midl Nat 152:268–276

Hackney EE (1999) The effects of small population size, breeding system, and gene flow on fruit and seed production in American ginseng (*Panax quinquefolius* L., Araliaceae). Masters thesis, West Virginia University, Morgantown

Hardin G (1968) The tragedy of the commons. Science 162(3859):1243–1248

Hruska AM, Souther S, McGraw JB (2014) Songbird dispersal of American ginseng (Panax quinquefolius). Ecoscience 21(1):46–55

Jochum GM, Mudge KW, Thomas RB (2007) Elevated temperatures increase leaf senescence and root secondary metabolite concentrations in the understory herb *Panax quinquefolius* (Araliaceae). Am J Bot 94:819–826

Johannsen K (2006) Ginseng dreams. The secret world of America's Most valuable plant. The University Press of Kentucky, Lexington, 225 pp

Knight TM (2004) The effects of herbivory and pollen limitation on a declining population of *Trillium grandiflorum*. Ecol Appl 14:915–928

Manget L (2012) Sangin' in the mountains: the ginseng economy of the southern Appalachians, 1865–1900. Appalachian J 40(1/2)

McGraw JB (2001) Evidence for decline in stature of American ginseng plants from herbarium specimens. Biol Conserv 98:25–32

McGraw JB (2017) Taking the broad view: How are wild ginseng populations faring and when does conservation policy need to change? In: Ormsby A, Leopold S (eds) Proceedings the future of ginseng and forest botanicals. United Plant Savers, pp. 90–102

McGraw JB (2019) Unpublished Data

McGraw JB, Chandler JL (2018) Demographic hallmarks of an overbrowsed population state in American ginseng. Global Ecol Conserv 15:e00435

McGraw JB, Furedi MA (2005) Deer browsing and population viability of a forest understory plant. Science 307:920–922

McGraw JB, Sanders SM, Van der Voort M (2003) Distribution and abundance of *Hydrastis canadensis* L. (Ranunculaceae) and *Panax quinquefolius* L. (Araliaceae) in the central Appalachian region. J Torrey Bot Soc 130(2):62–69

McGraw JB, Souther S, Lubbers A (2010) Rates of harvest and compliance with regulations in natural populations of American ginseng (*Panax quinquefolius* L.). Nat Areas J 30:202–210

McGraw JB, Lubbers AE, Van der Voort M et al (2013) Ecology and conservation of ginseng (*Panax quinquefolius*) in a changing world. Ann N Y Acad Sci:62–91. https://doi.org/10.1111/nyas.12032

Pritts KD (1995) Ginseng: how to find, grow and use America's forest gold. Stackpole, Mechanicsburg, 150 pp

Rogelj J, Meinshausen M, Knutti R (2012) Global warming under old and new scenarios using IPCC climate sensitivity range estimates. Nat Clim Chang:1–6

Rooney TP, Dress WJ (1997) Species loss over sixty-six years in the ground layer vegetation of Heart's content, an old-growth forest in Pennsylvania, USA. Nat Areas J 17:297–305

Rooney TP, Gross K (2003) A demographic study of deer browsing impacts on *Trillium grandiflorum*. Plant Ecol 168:267–277

Souther S (2011) Demographic response of American ginseng (*Panax quinquefolius* L.) to climate change. Doctor of Philosophy, West Virginia University, Morgantown, WV

Souther S, McGraw JB (2011) Evidence of local adaptation in the demographic response of American ginseng to interannual temperature variation. Conserv Biol 25:922–931

Souther S, McGraw JB, Lechowicz MJ (2012) Experimental test for adaptive differentiation of ginseng populations to temperature. Ann Bot 110:829–837

Taylor DA (2006) Ginseng, the divine root: the curious history of the plant that captivated the world. Algonquin Books of Chapel Hill, Chapel Hill

Turner JB, McGraw JB (2015) Can putative indicator species predict habitat quality for American ginseng? Ecol Indic 57:110–117

Van der Voort M (2005) An ecological study of *Panax quinquefolius* in central Appalachia: Seedling growth, harvest impacts and geographic variation in demography. Doctor of Philosophy in Biology, West Virginia University, Morgantown

Van der Voort ME, McGraw JB (2006) Effects of harvester behavior on population growth rate affects sustainability of ginseng trade. Biol Conserv 130:505–516

Wyatt JL, Silman M (2014) Long-term effects of clearcutting in the southern Appalachians. In: Gilliam FS (ed) The herbaceous layer in forests of Eastern North America, 2nd edn. Oxford University Press, New York, pp 412–437

Chapter 7
Indian Tobacco (*Lobelia inflata* L.)

Ákos Máthé

Abstract *Lobelia inflata*, this old species of the New World, has retained its importance as a resource of chemical compounds suitable to treat various maladies. The
herb lobelia originally used by Native Americans in the New England region was
subsequently popularized by Samuel Thomson (1769–1843). Its English name,
"Indian tobacco", refers to the saga according to which the dried leaves of *Lobelia
inflata* were originally smoked by native Americans (Penobscot tribes), in the New
England region, as an alternative/substitute to tobacco. The genus Lobelia comprises ca. 360–400 species, with a sub-cosmopolitan distribution. *Lobelia inflata*
L. is native to several states of North America. It is found in open woods or occasionally in gardens, as weed, from the West Coast to Minnesota, south to Georgia
and Mississippi. *Lobelia inflata* has a milky sap containing piperidine/pyridine
alkaloids that suffuse all parts of the plant. The alkaloid fraction is rich in piperidine
alkaloids and has a great potential for the treatment of disorders of the Central
Nervous System. In addition, they have demonstrated antitumor and anti-inflammatory activities. Biological and chemical studies of *Lobelia inflata* alkaloids and, in
particular, (–)-lobeline, have increased over the last few years. Lobeline might serve
as a useful lead for the development of new therapeutic agents that act on nAChR
(nicotinic acetyl-choline receptors) in a novel fashion.

Lobelia inflata from open field production. Most of the commercially available
crude drugs is sourced from either wild populations, or take their origin from
imports (India and China being the main exporting countries). Recently, (Máthé
et al. Introduction of *Lobelia inflata* L. to hungary: performance of *in vitro* and generatively propagated plants. In: 2006 Abstracts 27th international horticultural congress & exhibition. August 13–19, 2006. International Society for Horticultural
Science – Korean Society for Horticultural Science, Seoul, pp 88–88, 2006) studied
the introduction into open field conditions of both generatively and *in vitro* propagated *Lobelia inflata* L. with special regards to mineral nutrition and alkaloid production in the course of plant development. Based on the survey of ecological
requirements a simplified production protocol for field cultivation is given.

Á. Máthé (✉)
Faculty of Agriculture and Food Science, Széchenyi István University,
Mosonmagyaróvár, Hungary
e-mail: akos.mathe@upcmail.hu

© Springer Nature Switzerland AG 2020   159
Á. Máthé (ed.), *Medicinal and Aromatic Plants of North America*, Medicinal
and Aromatic Plants of the World 6,
https://doi.org/10.1007/978-3-030-44930-8_7

Keywords *Lobelia inflata* · Botany · Chemistry · Ecological requirements · Traditional and modern uses · Collection · Cultivation · Veterinary uses

7.1 Introduction

This chapter briefly characterizes the fascinating North American plant species, Indian tobacco (*Lobelia inflata*) that has become a potentially important medicinal plant for the treatment of several important ailments, including the drug and alcohol abuse.

The recorded history of *L. inflata* can be traced back to 1817, when in his comprehensive book on American medicinal plants, Bigelow (Dwoskin and Crooks 2002)(Bigelow 1817) wrote that "In the United States there are many species of Lobelia, which are interesting for their beauty, singularity or use. We have few plants more elegant than the cardinal flower, and few more curious in structure than the *Lobelia dortmanna*. In his detailed description of Lobelias, Bigelow also mentions:" When chewed, it communicates to the mouth a burning, acrimonious sensation, not unlike the taste of the green tobacco."…The bitterness of the leaves and capsules, combined with a narcotic property, appears to be the foundation of their medicinal power."

Two decades later, in 1837, the medicinal values of *Lobelia inflata* were already praised in an Inaugural Essay on the botanical history, properties, medical history, chemical history of this remarkable species of the Bellflower Family (Procter Jr 1837). "Among the many medicinal agents of the vegetable kingdom, indigenous to this country, perhaps few have higher claims to the consideration of the physician, than the *Lobelia inflata*. Possessed of powerful medical qualities and capable of making the most decided impressions upon the human system, we have every reason to believe that in time, under the cognizance of the skillful practitioner, it will be numbered among our most valuable remedies" (Procter Jr 1837).

The herb lobelia originally used by Native Americans in the New England region was subsequently popularized by Samuel Thomson (1769–1843), the founder of an idiosyncratic form of medicine called Thomsonianism. The enduring popularity of lobelia is one of the legacies of this nineteenth century enthusiasm. The Thomsonians claimed that lobelia could relax muscles and nerves. On this basis, they used it for anxiety, epilepsy, kidney stones, insomnia, menstrual cramps, muscle spasms, spastic colon, and tetanus.

According to Felpin and Lebreton (Felpin and Lebreton 2004a) there are only a few medicinal plants with a history as rich as *Lobelia inflata* L.

Originally used by the native American Indians for its reputed medicinal and healing properties, to-date, *L. inflata* it is still known – mainly – as a relaxing remedy.

7.2 The Name of Indian Tobacco

7.2.1 Common Names

Lobelia inflata is known under various common names: asthma weed, bladder pod, bladder pod, bladder-podded lobelia, emetic herb, emetic herb, emetic weed, eyebright, field lobelia, gagroot, Indian tobacco, lobelia, low belia, obelia, pukeweed, tobacco lobelia, vomitwort, wild tobacco.

Its English name, "Indian tobacco", refers to the saga according to which the dried leaves of *Lobelia inflata* were originally smoked by native Americans (Penobscot tribes), in the New England region, as an alternative/substitute to tobacco. According to records, it was also used in native traditional medicine as a relaxing remedy (Felpin and Lebreton 2004a).

In the international special literature *Lobelia inflata* can be found under the following vernacular names:

- Indian-tobacco (Source: Dict Rehm) – English
- lobélie enflée (Source: Dict Rehm) – French
- tabac indien (Source: Dict Rehm) – French
- Indianer-Tabak (Source: Zander Ency) – German
- indianischer Tabak (Source: Dict Rehm) – German
- Lobelie (Source: Dict Rehm) – German
- lobelia (Source: Dict Rehm) – Portuguese
- tabaco-indiano (Source: Dict Rehm) – Portuguese
- hierba del asma (Source: Dict Rehm) – Spanish
- tabaco indio (Source: Dict Rehm) – Spanish
- läkelobelia (Source: Kulturvaxtdatabas) – Swedish
- Indián dohány, porhonrojtfű – Hungarian

After: (USDA Agricultural Research Service 2019)

7.2.2 Scientific Names

The generic name *Lobelia* of Indian tobacco, puke weed was given by Plumier in honor of the English botanist Matthias de L'Obel (1538–1616). The genus comprises 415 species distributed worldwide (Folquitto et al. 2019).

The specific name "inflata" denotes the inflation of the ripening seed capsules that clearly distinguishes this plant from the large number of other species of the genus *Lobelia*.

7.3   Botanical/Morphological Characteristics of Indian Tobacco

7.3.1   Life Cycle

Lobelia is a summer annual poisonous plant of the Bellflower Family (Campanulaceae) that contains milky juice.

The life cycle of *Lobelia inflata* is different from those of the other species of the Section Lobelia (Bowden 1959). It is ordinarily annual, shows a trace of the ability to produce secondary basal rosettes.

Lobelia inflata has been observed occasionally to express a secondary reproductive episode in the field (Hughes and Simons 2014). By experimentally manipulating the effective season length in each of 3 years of an experiment, Hughes and Simons (2014) experimentally tested whether secondary reproduction were a form of adaptive plasticity consistent with the "deferral hypothesis". They have found that although alternative adaptive explanations for secondary reproduction cannot be precluded, the characteristics of secondary reproduction found in *L. inflata* are consistent with the predictions of incomplete or transitional evolution to annual semelparity.

7.3.2   Morphological Characteristics

The plant is up to 1 m. high with branched, hairy stems. Stems are smooth above, while the lower part is rough and hairy.

The leaves are 2.54–7.5 cm long, sessile or sub-sessile and toothed. They are ovate-oblong to oblong-obovate. The lower leaves, which are about 5 cm in length, are borne on stalks, while the upper, smaller ones are stalk-less.

The pale blue flowers that appear from summer until frosts are numerous, but very small and inconspicuous. The flower has both an upper and lower lip. The latter is divided into three lobes, while the upper one into two. The irregular, two-lipped flowers are blue or white, 6–8 mm.

The flowers appear in racemes terminating the branches. The flowers are situated in the axils of alternate leaves.

The fruiting hypanthium is inflated. The inflated seed capsules are nearly round and contain very numerous extremely minute, dark-brown seeds (Sievers 1930).

Flowering period: July to September.

Table 7.1 *Lobelia inflata* L. in the Taxonomic hierarchy of Plant Kingdom

Kingdom	Plantae– plantes, planta, vegetal, plants
Subkingdom	Viridiplantae– Green plants
Infrakingdom	Streptophyta– Land plants
Superdivision	Embryophyta
Division	Tracheophyta– Vascular plants, tracheophytes
Subdivision	Spermatophytina– Spermatophytes, seed plants, phanérogames
Class	Magnoliopsida
Superorder	Asteranae
Order	Asterales
Family	Campanulaceae– Harebells, campanules
Genus	Lobelia L.
Species	*Lobelia inflata* L. – Indian tobacco, Indian-tobacco

7.3.3   Taxonomic Position

Recent scientific evidence as available from the on-line database of Integrated Taxonomic Information System (*Lobelia inflata* to the Family Campanulaceae, according to the taxonomic hierarchy) is illustrated by Table 7.1.

The bellflower family (Campanulaceae Juss.) is cosmopolitan in its distribution. It comprises a total of ca. 2300 species, ranked into 84 genera. The genus *Lobelia* contains 415 species (Lammers 2011).

The family Campanulaceae is divided into five subfamilies: Campanuloideae Burnett, Lobelioideae Burnett, Nemacladoideae Lammers, Cyphioideae (A. DC.) Walp., and Cyphocarpoideae Miers (Lammers 1990).

In the subfamily Lobelioideae Brunett (Lammers 2007), *Lobelia* is the second largest genus out of the 84 genera. It is structured by sections and subgenera that are mostly confined to particular geographic regions.

L. inflata has been placed in the Section Lobelia, along with most other eastern North American *Lobelia* species (Lammers 2011) (Hughes and Simons 2015).

Lobelia inflata is the only species in this section that is semelparous (P. Hughes and Simons 2014).

Walsh et al. (2010) have reported problems in establishing the species limits in Australia and there seem to be also considerable problems in defining the genera, both in Australia and worldwide.

In an attempt to resolve the generic limits within the subfamily Lobelioideae, Knox et al. (2008) have sequenced taxa that are representative of both all genera in the subfamily and all geographic regions in which they occur. The resultant phylogram shows that many of the genera are polyphyletic. As a result, either the segregate genera need to be defined or many of the currently recognized genera should be incorporated into a more broadly circumscribed Lobelia.

7.3.4   *Area of* Lobelia Inflata *L.*

The genus *Lobelia* comprises ca. 360–400 species, with a sub-cosmopolitan distribution. Lobelia species are mainly tropical herbs, but species growing in South Africa, North America and East Asia can also be found among them. There are about 60 species in Australia, most of them are endemic. Although Lobelias occur primarily in tropical to warm temperate regions of the world, a few species extend into cooler temperate regions.

Lobelia inflata is found in open woods or occasionally in gardens, as weed, from the West Coast to Minnesota, south to Georgia and Mississippi.

According to a most recent Database by the United States Department of Agriculture (USDA) Natural Resources Conservation Service, *Lobelia inflata* is native to several states of North America (USDA Agricultural Research Service 2019): AL, AR, CT, DC, DE, GA, IA, IL, IN, KS, KY, LA, MA, MD, ME, MI, MN, O, S, NC, NE, NH, NJ, NY, OH, OK, PA, RI, SC, TN, VA, VT, WI,WV and across southern Canada; south to Georgia; west to Arkansas and eastern Kansas: NB,NS, PE (Fig. 7.1).

7.4   Chemical Composition and Biological Activities

Lobelia inflata contains some 80 chemicals with varying biological activities. A comprehensive list of these, with reference to the highest/lowest quantities in the pant organs from which they have been isolated, has been published by Duke (1992) in his renowned Dr. Duke's Phytochemical and Ethnobotanical Databases (also accessible in the Internet).

Lobelia has a milky sap containing piperidine/pyridine alkaloids that suffuse all parts of the plant (Evans 2009) (Huber 2012).

The pharmacological effects of lobelia are attributed primarily to the piperidine-type alkaloids, in particular to lobeline, as the main alkaloid.

In addition to the piperidine-alkaloids lobelia also contains other biologically active chemicals: a bitter glycoside (lobelacrin), chelidonic acid, lipids (fats), gum, resin, a pungent volatile oil (lobelianin), chlorophyll, lignin, salts of lime and potassium, with ferric oxide.

Another important active agent in the plant with possible effects on the central nervous system (CNS), β-amyrin-palmitate, has been studied for its antidepressant activity (ANAS Subarnas et al. 1993a, b).

The scientific literature on the phytochemical composition of Lobelia seems to be insufficient, therefore Nagananda and co-workers' screening attempt of *Lobelia inflata* L. for alkaloids, carbohydrates, saponins, phytosterols, proteins, amino acids and gums in the various parts of the plants, can be regarded as a novel approach (Nagananda et al. 2012).

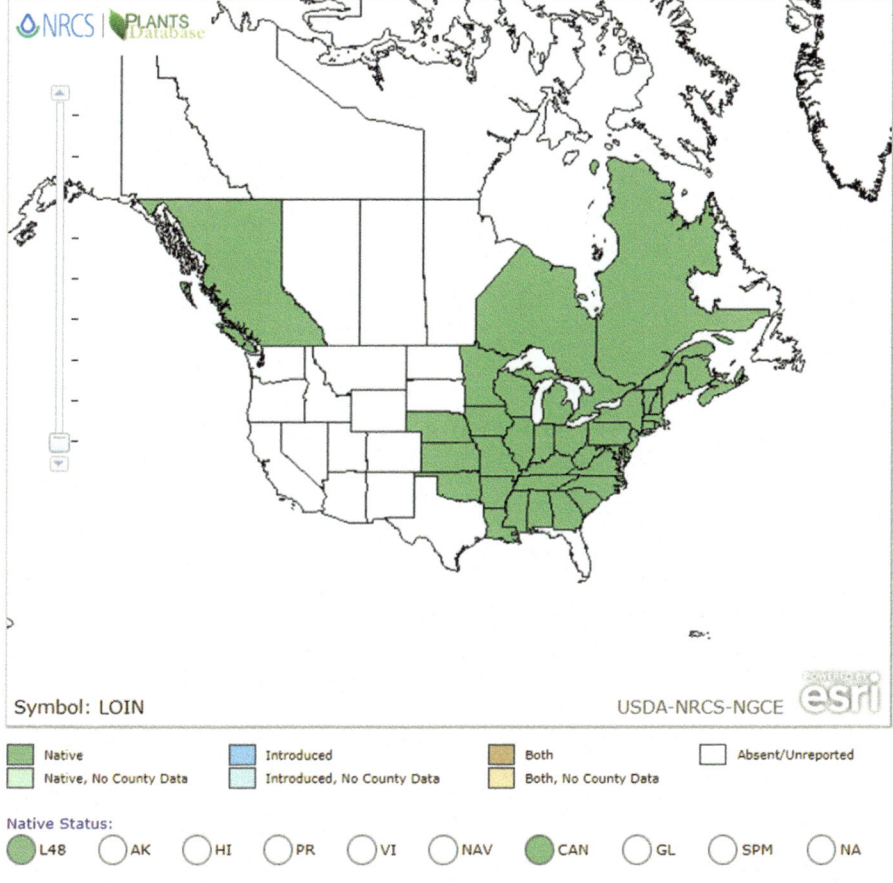

Fig. 7.1 Distribution of *Lobelia inflata* in North America. (Source: USDA Agricultural Research Service 2019)

Fig. 7.2 (+/−)-Lobeline
(Felpine and
Lebreton 2004)

7.4.1   Lobeline

In 1921, the piperidine alkaloid lobeline was isolated as the main active component of Lobelia (Wieland 1921). Its absolute stereochemistry was, however, determined only later, in 1965. (Schöpf et al. 1965).

According to our present knowledge, the main biologically active principles of *Lobelia* species have remained the alkaloids, in particular lobeline (Fig. 7.2).

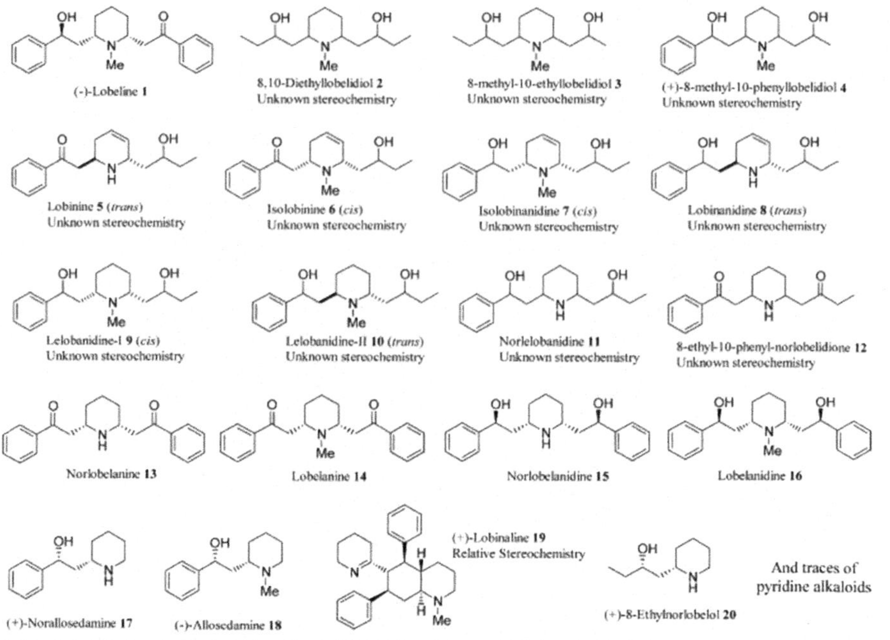

Fig. 7.3 Lobeline and farther main alkaloids of Lobelia inflata (Felpin and Lebreton 2004)

A comprehensive survey on the chemistry and biology of the alkaloids from *Lobelia inflata* was published by Felpin and Lebreton, in 2004. (Fig. 7.3).

Lobeline has a stimulatory effect on the respiratory centre and it is applied in the cases of asthma, gas- and narcotic-poisoning. Lobeline is also used in anti-smoking preparations (Krochmal et al. 1970). Antidepressant effect of beta-amirin-palmitate isolated from leaves of *L. inflata* was published by Subarnas et al., in 1993. (Subarnas et al. 1993a, b).

Farther important alkaloid constituents are lobelanine, norlobelanine, lobelanidine, and radicamine. *Lobelia inflata* is also a source of several flavonoids, and isolated compounds, like terpenes, alkynes, coumarins, and mineral elements (Duke 1992).

Biological investigations over the last 25 years have shown that many of the medicinal properties empirically discovered by the native Americans have a scientific basis. Each of the major effects: respiratory stimulant, drug deterrent (drug abuse), cognition enhancement in neurological (CNS) disorders, as well as farther minor biological properties reported for lobeline, have been discussed with regards to their mode of action and the structure-activity relationship studies (Dwoskin and Crooks 2002).

Due to the extension of life expectancy in industrial countries and the prevalence of neurological disorders like Alzheimer's disease or Parkinson's disease, the discovery of effective agents for the treatment of these pathologies is one of the major challenges in medicine for the future. Lobelia alkaloids and in particular (−)-lobeline have been studied in this regards and it has been established that lobeline might serve as a useful lead for the development of new therapeutic agents that act on nAChR (nicotinic acetyl-choline receptors) in a novel fashion. In addition the non-n-AChR-mediated pharmacological effects, lobeline might provide new opportunities for a better understanding of neuronal disorders (McCurdy et al. 1999).

Common characteristics in the biosynthesis of Lobelia alkaloids is that they are elaborated from the lysine derived Δ1-piperideine, coupling either to an aliphatic- (from acetyl-CoA precursor, e.g., pelletierine and co-alkaloids), or an aromatic part (from cinnamoyl-CoA precursor, e.g., piperine and other amides; lobeline, lobelanine and related alkaloids) (Szőke et al. 2013).

Lobelia alkaloids can be ranked into three groups: Group I: includes alkaloids with piperidine side chains without phenyl groups, Group II: comprises alkaloids with one phenyl group on one of the side chain and Group III: with two phenyl groups.

In the course of plant development, the lobeline content increases very rapidly in the young plant, reaching a first maximum shortly before blooming, a second maximum occurs at the end of blooming, followed by a rapid decrease, as the plant withers. Lobeline is distributed unequally in the different parts of the plant (Marion 1950): the blossoming apex contains 0.9–1.1%, the unripe capsule 0.88–1.05%, the leaves 0.42–0.43%, the stems 0.35–0.38%, and the roots 0.54–0.56%. According to certain authors its highest concentrations were measured in the seeds of the plant (Dwoskin and Crooks 2002).

Folquitto et al. (2019) summarize the alkaloids identified from the aerial parts of *L. inflata* as follows: lobeline, lobelanine, lobelanidine, norlobelanidine, norlobelanine, norlobeline, 8-propyl-10-phenyl lobelionol, 3-hydroxy-3-phenyl-propanoic norallosedamine, and 3-hydroxy-3-phenyl-propanoic allosedamine. Among them, lobeline appears to produce the highest yield and is most commonly found also in other *Lobelia* species (*L. urens*, *L. portoricensis*, *L. inflata*, *L. nicotianaefolia* and *L. syphilitica*).

7.4.2 Polyacetylenes

The C14-polyacetylene derivatives isolated from *L. inflata* (Fig. 7.4) by Japanese researchers (Ishimaru et al. 1991), have been investigated also as potential chemosystematic markers (Ishimaru et al. 1991).

Recent studies indicated the chemo-attractant character of *Lobelia inflata* extracts (Bae et al. 2010). Lobeline has been shown to have chemo-repellent effects (Paz et al. 2018) and currently, it is has become subject of renewed interest for the treatment of drug abuse and neurological disorders, like Alzheimer's or Parkinson's disease, which pose an important public health problem in industrialized countries (Fisher et al. 1998). Lobeline is a competitive nicotinic receptor antagonist and is

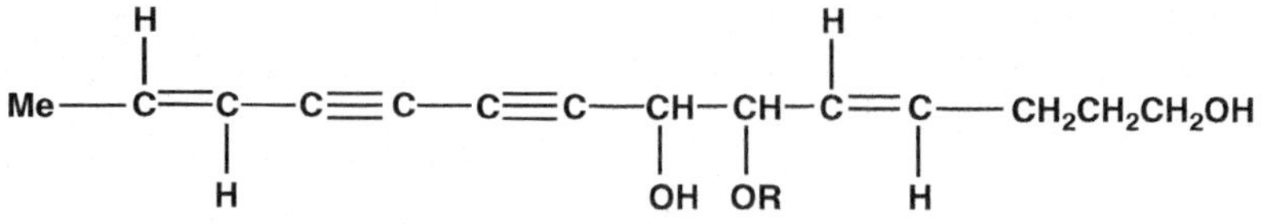

1 (lobetyol) : R=-H
2 (lobetyolin) : R=-glc
3 (lobetyolinin) :R=-glc^6 - ^{1}glc

Fig. 7.4 Polyacetylenes from *Lobelia inflata* (After Ishimaru et al. 1991)

Table 7.2 Pharmacological effects of extracts and isolated compounds	
	Respiratory stimulation,
	Anti-inflammatory activity
	Anti-microbial activity
	Anticonvulsant/neuroprotective activity
	Antitumor activity
	Immuno-modulating activity and inhibitors of alfa-glycosidases
	Analgesic activity and anti-venom activity against cobra venom
	After (Folquitto et al. 2019)

still commercialized in antismoking preparations. Interest in Lobelia alkaloids, and in particular (−)-lobeline, has increased in recent years due to their activity on the central nervous system (CNS) and the multidrug-resistance (MDR) (Ma and Wink 2008).

Beyond the above briefly discussed chemical constituents, *Lobelia inflata* has been described to contain numerous biologically active – chemical compounds (Felpin and Lebreton 2004a). The main known and potential pharmacological effects of extracts as summarized by Folquitto et al. (2019) are shown in Table 7.2.

7.4.3 Accumulation and Distribution of Lobeline in Plant Organs

Above ground plant parts of *Lobelia inflata* are generally harvested by collectors when the plant flowers and bears already seeds. According to Krochmal et al. (1970) pharmacopoeias refer to the crude drug of lobelia as dried areal parts or tops, without reference to flowers. In the British Pharmacopoeia a minimum alkaloid quantity of the drug, calculated as lobeline, is specified at 0.3%.

It is also from the researches of Krochmal et al. (1970), that we have first-hand evidence on the changing of alkaloid production during the ontogenetic development of *L. inflata*. They found that lobeline concentration decreased with plant maturity: as a contrast, total plant lobeline content increased with maturity. The average values of lobeline content were the following: mature flowering plants (0.76%), adolescent plants (1.46%) and juvenile plants in rosette stage (1.95%).

The lobeline content of different above ground plant organs of mature plants was recorded by Krochmal et al. (1970) as follows: leaves (0.38%), stems (0.58%), flowers (3.03%).

7.4.4   Alkaloid Production of In Vitro Lobelia Inflata Cultures

In the 1970ies, several laboratories reported the production of Lobelia alkaloids in tissue cultures, including *L. inflata* (Krajewska et al. 1987); (Wysokinska and Chmiel 1997). It was established as an apparent advantage of hairy roots cultures that they generally grow faster than the corresponding untransformed roots or cell suspension cultures. Moreover, they have been also observed to be potentially able to produce the root derived products of original plants (Hamill and Chandler 1994). Hairy root cultures were observed to be genetically stable with respect to the production of secondary metabolites (Aird et al. 1988). Hence, it appear both desirable and possible to produce Lobelia alkaloids in the hairy root cultures of *L. inflata*.

A comprehensive survey on the *in vitro* culture and the production of lobeline and other related secondary metabolites was published by Szöke (1994).

Studies on the alkaloid production of genetically transformed and nontransformed cultures of *Lobelia inflata* L. by Szőke et al. (2001) proved that these cultures are able to synthesize the characteristic piperidine alkaloids of the intact plant. Alkaloid precursor amino acids (Phe, Lys) and plant growth regulators affect not only the growth and differentiation of tissue cultures but also their secondary metabolism.

7.5   Ecological Requirements

The ecological requirements of *Lobelia inflata* plants can be traced back to the ecological conditions of its natural populations where it grows in weedy fields, roadsides, woods, and in partial shade (Krochmal et al. 1970). As an adaptable by nature species it can be found in the "waste" places created by humans and natural occurrences such as flooding. In its wild growing populations, it normally behaves as an annual plant and self-sows from its multitude of minuscule seeds (Geller 2013), provided common sense wildcrafting procedures are followed.

The plant profile for *Lobelia inflata*, in the database of USDA, can be found under the symbol LOIN: *L. inflata* has the native status L48 (N), CAN (N). It is facultative in the Atlantic and Gulf Coastal Plain, the Great Plains, while a facultative upland species in the Eastern Mountains and Piedmont, the Midwest and the Northcentral & Northeast (Fig. 7.1).

7.5.1   Soil and Mineral Nutrients

According to Hecht (1931) *L. inflata* prefers the loose, humus, clay or clay-sand soils. (Krochmal and Wilken 1970) have found the plants growing on soils with pHs ranging from 4.1 to 5.8, and concluded that L. inflata would thrive best on acid soils.

Influence of mineral nutrients on the growth and development of *L. inflata* were studied by Hoffmann (1949) who observed that in comparison to soils with low N content, N-rich soils enhanced the formation of alkaloids. As a conclusion, for *L. inflata* she proposed soils rich in Humus, Clay Ton, Bündner-slate und Jura, occasionally also chalk und sand. These findings are in harmony with Flück (1963), who concluded that the formerly reported controversial findings by Esdorn (1940) and Barner (1941) (cited in Krochmal et al. 1970), made with cultivated plants, can be explained by the large doses of N, supplied in the experiments, in contrast to the much lower N-level in natural soils.

In a comprehensive compilation, with reference to earlier authors (Mascré and Génot 1933) and (Esdorn 1940), Flück (1963) reports the total alkaloid production decreasing effect of Nitrogen (Flück 1963). According to this author, potassium (K) fertilization exerted an increasing effect on plant yield, while the alkaloid production was reduced.

In the case of phosphorus controversial findings were reported in alkaloid producing species. It appears that certain amount must be available but that increases over such amounts can either increase or decrease the production of alkaloids (Flück 1963) although in the case of *L. inflata* only increases were detected (Esdorn 1940).

More recently, within the framework of a GVOP R+D Project (Máthé et al. 2006) the effect of N- and Mg fertilization on the biomass and alkaloid production, was studied in open field trials by Máthé and his research team, at the University of West Hungary, Faculty of Agriculture and Food Sciences, Mosonmagyaróvár. In these experiments, the establishment of plants, propagated by *in vitro* and *in vivo* methods, was also studied.

The experiments established the favorable effect of fertilization on the alkaloid production (Fig. 7.5) and similarly also on the dry-mass production.

7.5.2   In Vitro Nutrients

The *in vitro* nutrient requirements of *L. inflata* have been studied primarily in relation to callus (Szöke 1994) and hairy root cultures (Yonemitsu et al. 1990) (Bálványos 2002) (Bálványos et al. 2004) (Bányai et al. 2003).

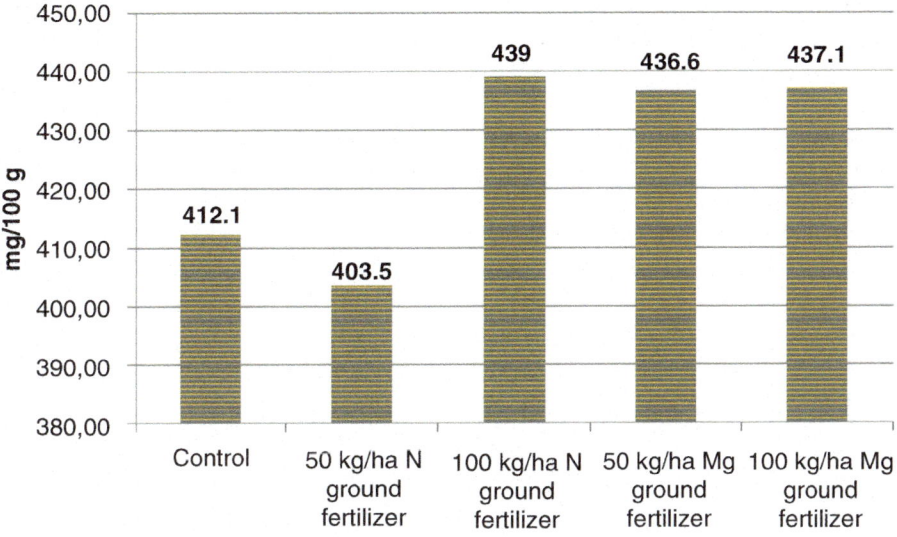

Fig. 7.5 Effect of N- and Mg-fertilizers on the total alkaloid production of *Lobelia inflata* above-ground plant parts (Mosonmagyaróvár 2011)

7.5.3   Water Requirements

Esdorn (Esdorn 1940) refers to the observation that the biological activity of *L. inflata* plants was variable in relation to the water content of soils stating that it was superior in humid soils. This fact has been verified under experimental conditions also by Krochmal et al. (1972) and more recently by Geller (2013) according to whom plant individuals watered more frequently attained greater size and flowered earlier. She has also stated that apparently, for this summer annual, reproductive phenological events are phenotypically correlated, and timing of reproduction is resource and size dependent.

Canadian researchers (Simons et al. 2010) studied the effect of snow-cover on the overwintering of *L. inflata* in the rosette stage of development. They have stated that without a consistent snowpack, survival was low (11%); with an insulation effect of R3 or greater, survival increased dramatically (81%).

7.5.4   Light Requirements

Seeds exhibit physiological dormancy, so that pre-planting treatments, such as cold stratification is recommended for the seeds to germinate at 20/10 C°, in the light (Baskin and Baskin 1992). Remarkably, *Lobelia inflata* does not require vernalization for bolting. According to Simons and Johnston (2000) seeds exhibit non-deep physiological dormancy, have a strict light requirement for germination, and may germinate throughout the growing season.

According to Muenscher (1936) light is necessary for the germination of *L. inflata* seeds. Under experimental conditions, seeds were sown on the soil surface in a garden: good germination was obtained, but practically no germination was obtained when the seeds were covered with 1 cm of soil. Seed viability was retained for at least 5 years, in dry storage under laboratory conditions.

In a technical paper on *L. inflata* cultivation (Krochmal and Wilken 1970) stated that light (in the red end of the spectrum) is required for germination. This can be provided by the use of incandescent household light bulbs.

7.5.5 Flowering and Reproductive Biology

Recently, the reproductive biology of *L. inflata* has been the object of numerous studies (Simons and Johnston 2000), (Hughes and Simons 2014, 2015).

A study into the reproductive phenology of *L. inflata* by Kelly (1992) has indicated that larger plants flowered earlier and produced more flowers than smaller plants. The onset of flowering was positively correlated with the onset of fruit maturation. It was also observed that at the flower level, the earlier a flower bloomed, the longer the resulting fruit took to develop. Also, fruit development times varied significantly among individual plants.

7.6 Traditional Uses of *Lobelia inflata* L.

The most commonly cited traditional use of *Lobelia inflata* L. (known as "Indian Tobacco") is smoking cessation and for the treatment of respiratory diseases such as asthma and bronchitis (Folquitto et al. 2019).

In the United States, Lobelia has always held a strong place in North American herbal history (Stanbury et al. 2013). According to Krochmal (1968) plant materials of *Lobelia inflata* have been used to manufacture drugs, since the early colonial times. For generations, the collection of wild growing MAPs has been an important income source for the native residents of Appalachia.

Indigenous peoples used Lobelia for wheezing, respiratory problems, and to relax muscle spasms. Generally it was referred to as "Indian Tobacco" and was used in smoking mixtures in traditional medicine. The dried leaves were also sprinkled over burning coals to inhale the vapors in cases of acute wheezing. Lobelia was used in teas and tinctures for asthma, cough, broncho-constriction—as well as for nervous tension and muscle spasm. Lobelia was also used to prepare lung plasters and compresses placed directly over the chest and lungs.

Farther ethnobotanical uses include the use of the roots by the Iroquois to treat venereal diseases, ulcers, and legs sores. The leaves were smashed and applied as a poultice to treat an abscess at the side of the neck. The plant was used to counteract sickness produced by witchcraft. The Cherokee mashed the roots of Indian tobacco and used them as a poultice for body aches. The leaves were rubbed on sores, aches, stiff necks, and chapped places. The Crow used the plant in religious ceremonies (Anderson 2019).

The herb lobelia was popularized by Dr. Samuel Thomson, who – as early as 1773 - became aware of its power to procure vomiting (Felpin and Lebreton 2004a). He was the founder of an ideosyncratic form of medicine called Thomsonianism. The enduring popularity of lobelia is one of the legacies of this nineteenth century enthusiasm. (Goldenseal is another herb popularized by Thomson). The so called "Thomsonians" additionally claimed that lobelia could relax muscles and nerves: therefore, they used it for anxiety, epilepsy, kidney stones, insomnia, menstrual cramps, muscle spasms, spastic colon, and tetanus.

In "A Modern Herbal" published in the 1930s, Maude Grieve (cited by Craker et al. 2003) described a poultice formula for respiratory problems using *Lobelia* and *Ulmus* bark powders combined with a weak lye solution. Historical medical journals from the 1800s and early 1900s describe the use of Lobelia for treatment of diphtheria and angina. Herbalists still commonly use Lobelia for respiratory problems.

According to Weil (2019), in 1993, the FDA banned their sale anti-smoking purpose, on the grounds that they are ineffective in helping people reduce or quit smoking. Although some experts still think that lobeline may reduce the effects of nicotine and that it might help in treating other drug addictions, in a controversial way, so far, no studies have confirmed this activity (McVaugh 2006).

7.7 Modern Medicine Based on its Uses in Traditional Medicine

The new, up-to-date analytical methods have opened up a possibility for the nearly complete exploration of active principles from *Lobelia inflata*, including the pharmacodynamics / synergistic effects (Nagananda et al. 2012).

The major active ingredient of *L. inflata* is lobeline has been reported to attenuate the behavioral effects of psychostimulants in rodents and to inhibit the function of nicotinic receptors (nAChRs) (Miller et al. 2004). Although, chemically it is, in fact, not similar to nicotine, still it has been a popular component of anti-smoking preparations.

Lobeline has been also found to diminish certain effects of nicotine in the body, specifically nicotine-induced release of dopamine. Since dopamine is believed to play a significant role in both drug and cigarette addiction, these findings seem to imply that lobeline might be useful for treating drug addiction (Miller et al. 2004). Potential benefits have been found for addiction to amphetamines. However, despite its widespread marketing for this purpose, there has never been any meaningful evidence that it works.

Lobeline has been described as a high-affinity nicotinic ligand that can cause nicotine like effects but is generally less toxic than nicotine.

Remarkably, however, several proposed uses of lobelia seems to lack supporting evidence. For example, while studies in horses have found that injected lobeline causes the animals to breath more deeply, this is a long way from a finding to support the widespread claims that lobelia is helpful for asthma (Meyler 1952). Similarly, animal studies hint that lobeline might enhance memory (Martin et al. 2018) and reduce pain. Beta-amyrin palmitate, another constituent of lobelia, might have antidepressant (Roni and Rahman 2014) and sedative properties (Bahramsoltani et al. 2015). However, there have not yet been sufficient human studies on these potential benefits of the *L. inflata*.

According to St-Pierre et al. (2017), Indian tobacco and its derived products are used to treat asthma and bronchitis, to repel bugs, and to relieve respiratory and muscular disorders. Lobelia plants produce bioactive piperidine alkaloids, including lobeline, which is used in the treatment of addictions. Experimental and clinical research studies on lobelia piperidine alkaloids have determined that they have many pharmacological properties including stimulant, diuretic, expectorant, antimicrobial and anti-cancer. Also it was reported that piperidine alkaloids act on the nicotinic acetylcholine receptors and interact with neurotransmitter transporters such as transporters for dopamine, serotonin and monoamine vesicle transporter. This type of interaction in the central nervous system makes this plant interesting not only for its beneficial effect in the treatment of addiction but also for the treatment of diseases such as Alzheimer's and Parkinson's diseases (Felpin and Lebreton 2004b).

7.8   Veterinary Uses of *Lobelia inflata* L.

Behavior problems in small animals have always been major problems in animal keeping. Thus the volume by Schwartz (2005) „Psychoactive Herbs in Veterinary Behavior Medicine" has been a most welcome publication on psychoactive herbal remedies for the possible treatment of these issues.

Regarding the veterinary use of *Lobelia inflata*, a Summary Report by the European Agency for the Evaluation of Medicinal Products Veterinary Medicines Evaluation Unit (EMEA/MRL/070/96-FINALm March 1996) clearly states that limited information is available on the pharmacokinetics, metabolism, and excretion of lobeline in animals. Lobeline seems to disappear from blood after intravenous administration (European Medicines Agency 2018), therefore the likelihood of consumer exposure to lobeline residues is rather limited.

7.9   Collection vs. Cultivation

The domestication of wild-growing species offers several known advantages (Máthé and Máthé 2008) (Máthé 2011): despite this, the first systematic experiments to transfer *Lobelia inflata* into a so called "row-crop" were started only in the 1960ies (Krochmal and Wilken 1970). It was shown that *Lobelia inflata* L. --a prime source of the alkaloid lobeline used in anti-smoking preparations - can be successfully grown as a row crop.

7.9.1   Collection

There is rather scarce information in the special literature that would indicate the sourcing of *Lobelia inflata* from open field production. According to unverified information by Richters Herbs Ltd. (2019), the world market size for *L. inflata* production amounts to 9000 hectares, from which 1700 ha is located in North America.

In view of the sound and possibly increasing demand on this crude drug, it seems necessary to pay more attention to the Good Collection and/or sustainable collection practices in wildcrafting. United Plant Savers (a non-profit education corporation dedicated to preserving native medicinal plants) emphasize that in its undisturbed, natural populations, under natural conditions *L. inflata* behaves like an annual species that self-sows a multitude of its small seeds. It should fare well in culture (provided common sense wildcrafting procedures are followed). However, if harvested before the lower seedpods have been able to mature and drop their seeds, the wildcrafted populations may already be at risk. Sustainably minded wildcrafters leave some healthy plant individuals to re-sow and scatter their mature seeds, thus ensuring the emergence of new plants for the next vegetation season. The Unites Plant Savers' (UpS) recommendations state that "Very limited wild harvest is permissible when no other alternative will do" (Geller 2013).

Due to the frequent overharvest of natural populations of wild growing medicinal plants in North America, according to USDA, it is advisable - before engaging in wild-crafting of this species – to consult the PLANTS Web site (https://plants.sc.egov.usda.gov/java/) of the United States Department of Agriculture, as well as the State Departments of Natural Resources for the current status of this species (e.g.: state noxious status and wetland indicator values, etc.). BONAP's North American Plant Atlas (NAPA) prepared in The Biota of North America Program with its US County-Level Species Maps can be of farther help, in this respect (Fig. 7.6).

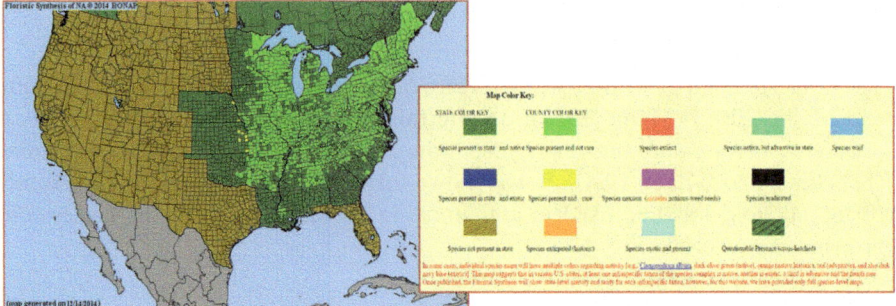

Fig. 7.6 Occurrence of *Lobelia inflata* in BONAP's North American Plant Atlas

7.9.2 Cultivation

According to (Anderson 2019), in the US, cultivars, improved and selected materials, seeds and plants are available from many nurseries. The origin and occasionally the identity of these propagating materials remains, however, questionable.

Based on the available sporadic information (Krochmal and Wilken 1970), Scott Peterson (n.d.), Bown and Herb Society of America (1995), Richters Herbs and (Máthé et al. 2006), and others, the following production protocol can be recommended for *Lobelia inflata* Table 7.3.

Early cultivation trials were conducted by Krochmal and Wilken (1970) and Krochmal et al. (1972). In these experiments wild growing and cultivated plants were compared for their biomass and lobeline production. It was established that both plant weight and lobeline yields were markedly greater in cultivated plants. The lobeline content of cultivated plants, on a dry-weight basis ranged from 0.36 to 2.25%, with an average of 1.05% (14 plants). In 1969, the lobeline content of wild growing plants varied between 0.19 and 2.38%, with an average of 0.65% (Krochmal et al. 1972). The lobeline content of wild plants was established to be also more variable.

This species seems to be only occasionally cultivated commercially as a medicinal plant. There have also been reports of field-cultivation trials in Poland, though no public reports – in English - are available. In Hungary, as reported by Máthé et al. (2006) the field production of *Lobelia inflata*, as a source of piperidine alkaloids has been successful.

These experiments have verified earlier observations according to which field production of *L. inflata* propagated by transplant raising (Photo 7.1) can be feasible.

As a novel result of above Hungarian experiments, the successful application of *in vitro* micro propagation technology should be mentioned.

Briefly described: the experiments were started from *in vitro* cultured surface sterilized seeds, germinated under *in vitro* conditions and on hormone free ½ Murashige Skoog (MS) medium, subsequently raised in ½ MS and B5 solid medium, containing 3% sacharose. 2.5 week old, rooted plants were washed and transferred to pots filled with soil/peat/Perlite-mixture under shaded glasshouse

Table 7.3 A simplified production protocol for *Lobelia inflata* L.

Propagation:	(a) from seeds by in vivo and/or *in vitro* transplant raising, (b) *in vitro* micro-propagation /cloning) by adventitious shoot induction. Small seeds germinate in light, therefore open field sowing is not preferred.
Transplant raising	In glasshouse, in propagating trays/boxes
Date of sowing:	First – Second decade of February
Propagation medium:	Tray-substrate (small granular, white sphagnum peat moss and black (frozen) peat mixture, with addition of 1.3 kg/m³ 14:16:18 NPK starter fertilizer, pH: 6.5
Mode of sowing	Seeds are mixed with sand and spread on the surface of sufficiently settled finely sifted soil. No covering is required in order to secure light for germination. Cover seeding with greenhouse plastic film (foil) until emergence.
Plant protection:	To prevent damping off disease of seedlings, watering in with Fundazol 50 WP
Glasshouse temperature	23–24 °C/day, 19–20 °C/night
Time of emergence	Relatively sluggish, 4–6 weeks
Pricking:	Into any commercially available peat and garden-soil mixture 4,5 × 5 cm TEKU trays after transplanting the emerging plantlets are watered with 0,1% WUXAL nutrient solution.
Transplanting into the field:	
Planting date::	Second – Third decade of may
Plant spacing:	30–45 cm × 25 cm
Soil preparation:	Subsoiling by combined disc cultivators
Mode of planting	Manual planting, due to the relatively small size of transplants
Irrigation:	Min 2 × 25 mm water
Plant care	In the open field, no diseases or pest caused any damage to Lobelia, therefore plant care in April focused on weeding, watering and topdressing application of between row cultivator 2 times, and manual weed-killing on 2–3 occasions
	Important. The formation of topsoil crusting should be avoided by soil loosening.
	Complementary irrigation to natural precipitation, in June (under continental conditions), with 4–6-times (5–10 mm irrigation water)
Harvesting:	
Date:	First harvest: At the time of full flowering phenophase of plants from middle of June when plants have reached the height of 35–40 cm. Second harvest (flowering): Under favorable conditions, the irrigated plants are capable of a second/third flowering. These plants, however, do not constitute rosettes. They consist mainly of flowering stalks.
Mode of harvest	Cutting above ground plant parts by trimming secators
Cutting height	ca. 5 cm above the root crown
Yield:	ca. 1100 kg/ha dry herb

(continued)

Table 7.3 (continued)

Primary processing:	Harvested herb should dried a.s.a.p., preferably in a drying facility (either natural drying or artificial driers at 35 co can be applied).
Remark:	Transplants started in March – April and transplanted into the open field in the third decade of august, under favorable conditions, are capable of overwintering. In spring they make a head start and produce higher yields. This method, however, seems to be rather weather dependent and need farther investigation.

Photo 7.1 Generatively propagated *Lobelia inflata* transplants raised in glasshouse (Mosonmagyaróvár 2007)

Photo 7.2 Rooting of *in vitro* micropropagated *Lobelia inflata* transplants (Mosonmagyaróvár 2007)

conditions. Until transplanting into the experimental fields, the plantlets were cultured for 5 weeks under mist irrigated conditions (Bálványos 2002).

In vitro micro-propagation (cloning) from seedlings by adventitious shoot induction was also possible on solid MS medium, containing 2% sucrose (pH 5.7) and subsequent rooting. Culture media were solidified with 7 g L-1 agar and autoclaved at 121 °C for 20 min. Cultures were incubated in air conditioned growth room at 22 ± 2 °C, under a 12:12 h photoperiod regime, at 1400 lx light (cool and warm white fluorescent lamps. Photos 7.1 and 7.2 illustrate the different stages of the experiment. An important outcome of *in vitro* micro-propagation experiments

Photo 7.3 *Lobelia inflata* in field cultivation, in Tordas (Hungary), 2007

was that although the propagated plantlets attained a relatively quick growth, still the offsprings turned out to be inhomogenous, with several plantlets developing flowering stems (Photo 7.2). *Lobelia inflata* in open field experiment Photo 7.3).

7.10   Footprint of *Lobelia inflata* L. in the Scientific Literature

In view of briefly surveyed history of this species in this chapter, it is interesting to take a look at the footprints of *Lobelia inflata* in the scientific special literature. A search in the scientific database SCOPUS, over a time range of nearly two hundred years (1833–2019) has resulted 129 entries. Their distribution according to subject areas is comprised by - Distribution of documents on *Lobelia inflata* in SCOPUS according to subject areas Fig. 7.7 and Table 7.4.

These figures well reflect the history of the species: 39.5% of the documents deal with health related aspects which seems to be quite in line with the both traditional and evolving modern uses of *L. inflata*. In view of the nearly perfect lack of publications on the field production and related aspects (e.g.: agrotechnology, plant protection, etc.) the rate of agriculture and biotechnology related publications (20.5%). seems to be far-fetched. This statistical distortion can be explained by the fact that frequently *in vitro* culturing (tissue culture, hairy roots cultures) are erroneously

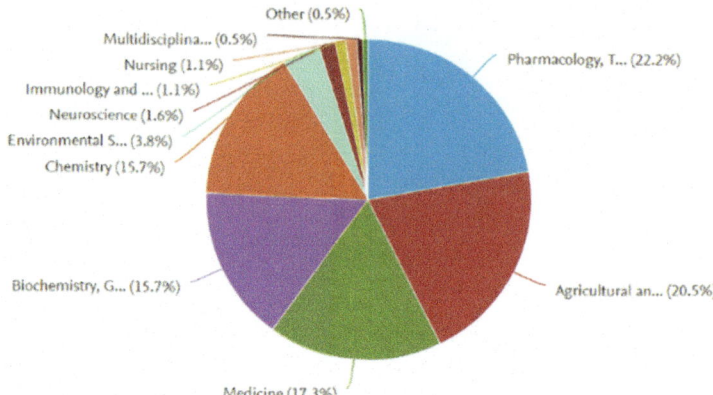

Fig. 7.7 Distribution of documents on *Lobelia inflata* in SCOPUS according to subject areas

Table 7.4 Distribution of documents on *Lobelia inflata* in SCOPUS according to subject areas

Subject area	Number of documents
Pharmacology	41
Agricultural and biological sciences	38
Medicine	32
Biochemistry	29
Chemistry	29
Environmental	7
Neuroscience	3
Immunology	2
Nursing	2
Multidisciplinary	1
Veterinary	1

called cultivation: as a consequence these publications are ranked among the topic agriculture.

The time-line of publications (Fig. 7.8), however, is in perfect harmony with the past and emerging trends of interest in this plants species that seems to be a token for its bright future.

7.11 Conclusions

In 1990s, several research laboratories started to study the production of secondary metabolites by hairy root cultures (Yonemitsu et al. 1990). As a natural consequence, investigations into the scientific and technological aspects of *in vitro* production of pharmacologically important Lobelia secondary metabolites were also started.

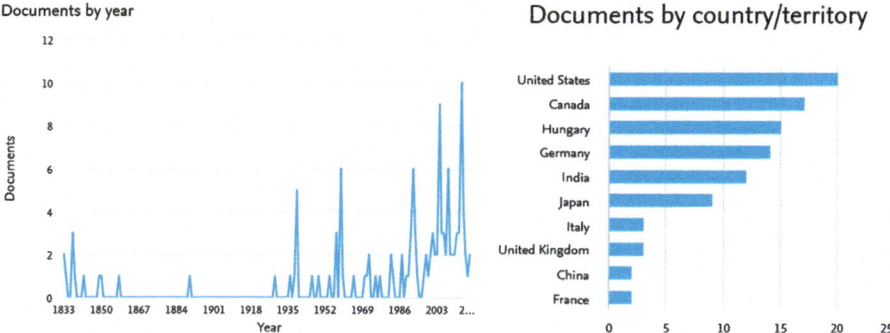

Fig. 7.8 Timeline (**a**) and Countries of origin (**b**) of *Lobelia inflata* publications in SCOPUS database (1833–2018)

To date, lobelin, the main alkaloid of *Lobelia inflata* is recognized as the source of important respiratory stimulants, that enhance and accelerate the respiratory movements. It has been used in cases if asthma, gas poisoning, and narcotic poisoning. Lobeline is currently also the subject of renewed interest for the treatment of drug abuse and neurological disorders, like Alzheimer's or Parkinson's disease.

Being a competitive nicotinic receptor antagonist, lobelin is still commercialized in anti-smoking preparations. Interest in Lobelia alkaloids, and in particular in the pharmacologically most active (−)-lobeline, has increased due to their effect on the central nervous system (CNS) and the multidrug-resistance (MDR) (Szőke et al. 2013).

Parallel with the quasi unceasing demand on this crude drug, the natural resources of *Lobelia inflata* have already become limited. In certain areas of US *Lobelia inflata* is already listed among Endangered Medicinal Plants (e.g.: in the Driftless Region, in Wisconsin) (Krall 2016).

Modern science, especially, biotechnology, also seems to remain interested in the already known and not yet discovered active principles of *Lobelia inflata*. Here, according to our present knowledge, especially hairy root cultures with applications like large-scale production of secondary metabolites and recombinant proteins, upscaling in bioreactors, "phyto-mining" and phytoremediation seem to be promising. Hairy roots cultures have also industrial applications and are used as important research tool for the elucidation of both secondary metabolite biosynthetic pathways and the expression, as well as function of key genes and regulatory elements. Hairy roots cultures are known to grow faster than the corresponding untransformed roots or cell suspension cultures, and are also potentially able to produce the root derived products of original plants (Hamill and Chandler 1994) (Aird et al. 1988). Relevant research conducted on the hairy root cultures of medicinally important plants, including *Lobelia inflata*, indicate that their secondary metabolites production, elicitation, recombinant proteins, genetic manipulation, phytoremediation and phyto-mining are promising.

Being an obligately self-fertilizing plant species, *Lobelia inflata* seems to be a good target for use in the study of temporal fluctuation in allele frequencies and of the genetic structures within and among populations (Hughes et al. 2014).

The old plant of the American Indians, seems to have not only a rich past but also a promising future for pharmaceutical and biological research.

References

Aird ELH, Hamill JD, Rhodes MJC (1988) Cytogenetic analysis of hairy root cultures from a number of plant species transformed by *Agrobacterium rhizogenes*. Plant Cell Tissue Org. Cult. 15(1):47–57. https://doi.org/10.1007/BF00039888

Anderson MK (2019) Plant guide: Indian Tobacco *Lobelia inflata* L. PLANTS Database, Davis. Available at: https://plants.sc.egov.usda.gov/plantguide/pdf/cs_loin.pdf. Accessed 9 Apr 2019

Bae GS et al (2010) Inhibition of lipopolysaccharide-induced inflammatory responses by piperine. Eur J Pharmacol 642(1–3):154–162. https://doi.org/10.1016/j.ejphar.2010.05.026

Bahramsoltani R et al (2015) Phytochemical constituents as future antidepressants: a comprehensive review. Rev Neurosci 26(6):699–719. https://doi.org/10.1515/revneuro-2015-0009

Bálványos I (2002) Studies on the growth and secondary metabolite production of *Lobelia inflata* L. hairy root cultures. PhD theses, Semmelweis University. Available at: http://semmelweis.hu/wp-content/phd/phd_live/vedes/export/balvanyosisvan-m.pdf

Bálványos I, Szoke É, Kursinszki L (2004) Effect of macroelements on the growth and lobeline production of *Lobelia inflata* L. hairy root cultures. Acta Hortic 597:245–251. https://doi.org/10.17660/ActaHortic.2003.597.35

Bányai P et al (2003) Cultivation of *Lobelia inflata* L. hairy root culture in bioreactor. Acta Hortic:253–256. https://doi.org/10.17660/ActaHortic.2003.597.36

Barner J (1941) Alkaloidbestimmungen von Herba Lobeliae. (Alkaloid determinations of Lobelia.) Lavd. Jahrb. 90:234

Baskin JM, Baskin CC (1992) Role of temperature and light in the germination ecology of buried seeds of weedy species of disturbed forests. I. *Lobelia inflata*. Can J Bot 70(3):589–592. https://doi.org/10.1139/b92-075

Bigelow J (1817) ITIS standard report page: *Lobelia inflata*. In: American medical botany :being a collection of the native medicinal plants of the United States, containing their botanical history and chemical analysis, and properties and uses in medicine, diet and the arts, vol 1. Cummings and Hilliard, Boston, pp 177–186. Available at: https://www.itis.gov/servlet/SingleRpt/SingleRpt?search_topic=TSN&search_value=34524#null. Accessed: 4 February 2019

Bowden WM (1959) Phylogenetic relationships of twenty-one species of Lobelia L. section Lobelia. Bull Torrey Bot Club 86(2):94–108. Available at: https://www.jstor.org/stable/2482990?readnow=1&refreqid=excelsior%3A5a2d374d99119112bd4cda0593a3fb646&seq=6#page_scan_tab_contents

Bown D, Herb Society of America (1995) Encyclopedia of herbs & their uses. Dorling Kindersley

Craker LE, Gardner Z, Etter SC (2003) Herbs in American fields: a horticultural perspective of herb and medicinal plant production in the United States, 1903 to 2003. HortScience. Available at: https://journals.ashs.org/hortsci/view/journals/hortsci/38/5/article-p977.pdf. Accessed: 15 Mar 2019

Duke J (1992) Handbook of phytochemical constituents of GRAS herbs and other economic plants. CRC Press, Boca Raton. https://doi.org/10.1201/9780203752623-1

Dwoskin LP, Crooks PA (2002) A novel mechanism of action and potential use for lobeline as a treatment for psychostimulant abuse. Biochem Pharmacol 63(2):89–98. https://doi.org/10.1016/S0006-2952(01)00899-1

Esdorn J (1940) Untersuchungen über den Alkaloidgehalt der *Lobelia inflata* L. in Abhängigkeit von äussern und innern Faktoren. Heil und Gewürzpflanzen 19(1):9

European Medicines Agency (2018) Guideline on specifications: test procedures and 6 acceptance criteria for herbal substances 2, herbal 7 preparations 3 and herbal medicinal products 4 /traditional 8 herbal medicinal products. Available at: www.ema.europa.eu. Accessed: 25 Nov 2018

Evans WC (2009) In: Evans WC (ed) Trease and evans pharmacognosy, international edition E-book, 16th edn. Elsevier Health Sciences

Felpin F-X, Lebreton J (2004a) History, chemistry and biology of alkaloids from *Lobelia inflata*. Tetrahedron 60(Report 694):10127–10153. https://doi.org/10.1016/j.tet.2004.08.010

Felpin F-X, Lebreton J (2004b) Tetrahedron report number 694: history, chemistry and biology of alkaloids from *Lobelia inflata*. Tetrahedron 60(45):10127–10153. https://doi.org/10.1016/j.tet.2004.08.010

Fisher A et al (1998) Progress in Alzheimer's and Parkinson's diseases. Springer. Available at: https://books.google.hu/books?id=GLTSBwAAQBAJ&pg=PA474&lpg=PA474&dq=lobeline+Alzheimer&source=bl&ots=UPhDdnA3PZ&sig=ACfU3U0AJXF1P58gxJC2HPvUbBr33zIrHA&hl=en&sa=X&ved=2ahUKEwjGyZHNofziAhXPCuwKHQl7CzQQ6AEwBXoECAkQAQ#v=onepage&q=lobelineAlzheimer&f=false. Accessed: 22 June 2019

Flück H (1963) Intrinsic and extrinsic factors affecting the production of secondary plant products. In: Swain T (ed) Chemical plant taxonomy. Academic Press, Inc., pp 167–186. https://doi.org/10.1016/b978-0-12-395540-1.50011-7

Folquitto D et al (2019) Biological activity, phytochemistry and traditional uses of genus Lobelia (Campanulaceae): a systematic review. Fitoterapia. 134:23–38. https://doi.org/10.1016/J.FITOTE.2018.12.021

Geller CA (2013) Lobelia (*Lobelia inflata*) – united plant savers, planting the future. Available at: https://unitedplantsavers.org/lobelia-lobelia-spp/. Accessed: 16 June 2019

Hamill JD, Chandler SF (1994) Use of transformed roots for root development and metabolism studies and Progress in characterizing root-specific gene expression. In: Biology of adventitious root formation. Springer US, Boston, pp 163–179. https://doi.org/10.1007/978-1-4757-9492-2_13

Hecht W (1931) Versuche zur Erforschung der Ursachen der Gehaltschwankungen bei Arzneipflanzen. Heil und Gewürzpflanzen 14(1):15

Hoffmann FH (1949) Über den Einfluss einiger Bodenarten auf Wachstum und Gehalt von Arzneipflanzen. Eidgenössische Technische Hochschule in Zürich. https://doi.org/10.3929/ethz-a-000096713

Huber A (2012) The benefits of the use of lobelia in herbal preparations. Available at: www.drweil.com/drw/u/QAA400815/Is-Lobelia. Accessed 4 Mar 2019

Hughes PW, Simons AM (2014) Secondary reproduction in the herbaceous monocarp *Lobelia inflata:* time-constrained primary reproduction does not result in increased deferral of reproductive effort. BMC Ecol 14(1):15. https://doi.org/10.1186/1472-6785-14-15

Hughes PW, Simons AM (2015) Microsatellite evidence for obligate autogamy, but abundant genetic variation in the herbaceous monocarp *Lobelia inflata* (Campanulaceae). J Evol Biol 28(11):2068–2077. https://doi.org/10.1111/jeb.12734

Hughes PW et al (2014) Development of polymorphic microsatellite markers for Indian tobacco, *Lobelia inflata* (Campanulaceae)(1.). Appl Plant Sci 2(4). https://doi.org/10.3732/apps.1300096

Ishimaru K, Yonemitsu H, Shimomura K (1991) Lobetyolin and lobetyol from hairy root culture of *Lobelia inflata*. Phytochemistry 30:2255–2257. https://doi.org/10.1016/0031-9422(91)83624-T

Kelly CA (1992) Reproductive phenologies in *Lobelia inflata* (Lobeliaceae) and their environmental control. Am J Bot 79(10):1126–1133. https://doi.org/10.2307/2445212

Knox EB et al (2008) The predominantly South American Clade of Lobeliaceae. Syst Bot 33(2):462–468(7). https://doi.org/10.1600/036364408784571590

Krajewska A et al (1987) Effect of new synthetic regulations on biomass and alkaloid production by callus tissues of *Lobelia inflata* L. Acata Botanica Hungarica 33(3–4):407–411. Available at: https://hungary.pure.elsevier.com/en/publications/effect-of-new-synthetic-regulations-on-biomass-and-alkaloid-produ. Accessed 3 Feb 2019

Krall L (2016) Endangered medicinal plants of the driftless region. J Med Plant Conserv. Edited by S. Leopold, (Spring), p 8. Available at: https://unitedplantsavers.org/wp-content/uploads/2016/05/UpS-Journal.Final_.pdf. Accessed: 10 May 2019

Krochmal A (1968) Medicinal plants and appalachia. Econ Bot. 22(4):332–337. https://doi.org/10.1007/BF02908128

Krochmal A, Wilken L (1970) The culture of indian tobacco (Lobelia inflata L.). Res. Pap. NE-181. Upper Darby, PA: U.S. Department of Agriculture, Forest Servcie, Northeastern Forest Experiment Station. 9 p. (USDA Res. Pap. NE-181.), 181, p 9. Available at: https://www.fs.usda.gov/treesearch/pubs/23790. Accessed 10 Feb 2019

Krochmal A, Wilken L, Chien M (1970) Lobeline content of Lobelia inflata: structural, environmental and developmental effects. Upper Darby, PA Available at: https://www.fs.fed.us/ne/newtown_square/publications/research_papers/pdfs/scanned/OCR/ne_rp178.pdf. Accessed: 3 Feb 2019

Krochmal A, Wilken L, Chien M (1972) Plant and lobeline harvest of Lobelia inflata L. Econ Bot 26(3):216–220. https://doi.org/10.1007/BF02861034

Lammers TG (1990) Campanulaceae. In: Wagner WL, Herbst DR, Sohmer DR (eds) Manual of the flowering plants of Hawai. University of Hawaii Press, Honolulu

Lammers TG (2007) World checklist and bibliography of Campanulaceae.. Royal Botanic Gardens

Lammers TG (2011) Revision of the infrageneric classification of Lobelia L. (Campanulaceae: Lobelioideae). Ann Mo Bot Gard 98(1):37–62. https://doi.org/10.3417/2007150

Ma Y, Wink M (2008) Lobeline, a piperidine alkaloid from Lobelia can reverse P-gp dependent multidrug resistance in tumor cells. Phytomedicine 15(9):754–758. https://doi.org/10.1016/j.phymed.2007.11.028

Marion L (1950) Chapter V: The pyridine alkaloids. In: Manske RHF, Holmes HL (eds) Volume 1, vol 1. Elsevier Inc., pp 165–269. https://doi.org/10.1016/S1876-0813(08)60188-8

Martin CA et al (2018) Lobeline effects on cognitive performance in adult ADHD. J Atten Disord 22:1361–1366. https://doi.org/10.1177/1087054713497791

Mascré M, Génot H (1933) Nouvelles expériences sur la culture de la lobélie (Lobelia inflata). Bull Sci Pharmacol 4(40):453

Máthé, Á. (2011) A new look at medicinal and aromatic plants, Acta Hortic. DOI: https://doi.org/10.17660/ActaHortic.2011.925.1 Ákos MáthéÁkos Máthé

Máthé Á, Máthé I (2008) Quality assurance of cultivated and gathered medicinal plants. Acta Hortic:67–76. https://doi.org/10.17660/ActaHortic.2008.765.8

Máthé Á et al (2006) Introduction of Lobelia inflata L. to hungary: performance of in vitro and generatively propagated plants. In: 2006 Abstracts 27th international horticultural congress & exhibition. August 13–19, 2006. International Society for Horticultural Science – Korean Society for Horticultural Science, Seoul, pp 88–88

McCurdy C et al (1999) Lobeline: a natural product with high affinity for neuronal nicotinic receptors and a vast potential for use in neurological disorders. In: Cutler S, Cutler H (eds) Biologically active natural products: pharmaceuticals. CRC Press. https://doi.org/10.1201/9781420048650

McVaugh R (2006) A key to the North American species of Lobelia (Sect. Hemipogon). Am Midl Nat 24(3):681. https://doi.org/10.2307/2420867

Meyler L (1952) Increased sensitivity for opiates and barbiturates in anoxia. Acta Med Scand 142:256–258. https://doi.org/10.1111/j.0954-6820.1952.tb13863.x

Miller DK et al (2004) Lobeline analogs with enhanced affinity and selectivity for plasmalemma and vesicular monoamine transporters. J Pharmacol Exp Ther 310(3):1035–1045. https://doi.org/10.1124/jpet.104.068098

Muenscher WC (1936) Seed germination in Lobelia, with special reference to the influence of light on Lobelia inflata. J Agric Res 52(8):627–631. Available at: https://naldc.nal.usda.gov/download/IND43968767/PDF. Accessed: 10 February 2019

Nagananda G et al (2012) Phytochemical screening and evaluation of antimicrobial activities of in vitro and in vivo grown plant extracts of Lobelia inflata L. Int J Pharm Bio Sci 3(3 (B)):433–442. Available at: www.ijpbs.net. Accessed: 10 June 2019

Paz C et al (2018) Assessment of insecticidal responses of extracts and compounds of *Drimys winteri*, *Lobelia tupa*, *Viola portalesia* and *Vestia foetida* against the granary weevil *Sitophilus granarius*. Ind. Crop. Prod. 122:232–238. https://doi.org/10.1016/J.INDCROP.2018.06.009

Procter W Jr (1837) ART. XVIII.--On *Lobelia inflata:* botanical history. Properties. Medical history. Chemical history. Am J Pharm 3(2):98. Available at: http://search.proquest.com/openview/7f697901abf4f3aec8bfd2f85fa3be80/1?pq-origsite=gscholar&cbl=41445. Accessed: 10 May 2019

Richters ProGrowers Info (2019) Richters ProGrowers Info. Available at: https://www.richters.com/progrow.cgi?search. Accessed: 22 June 2019

Roni M, Rahman S (2014) The effects of lobeline on nicotine withdrawal-induced depression-like behavior in mice. Psychopharmacology 231(15):2989–2998. https://doi.org/10.1007/s00213-014-3472-y

Schöpf C et al (1965) Absolute configuration of (−)-lobeline and of its reduction products. Ann Cehm 63:18154

Schwartz S (2005) Psychoactive herbs in veterinary behavior medicine. Wiley. Available at: https://books.google.hu/books?hl=hu&lr=&id=ZP6QVep-x24C&oi=fnd&pg=PR8&dq=veterinary+uses+of+lobelia+inflata&ots=qP3wPggiWF&sig=HCSeJj3LIg-Yh-pKhIRVxoWZsBs&redir_esc=y#v=onepage&q=veterinaryusesofLobelia inflata&f=false. Accessed: 18 June 2019

Scott Peterson J (n.d.) Plant guide INDIAN TOBACCO *Lobelia inflata* L. Plant Symbol = LOIN. Available at: http://npdc.usda.gov. Accessed: 15 Mar 2019

Sievers AFA (1930) American medicinal plants of commercial importance. US Dept. of Agriculture, Washington, DC, p 40. Available at: https://books.google.com/books?hl=hu&lr=&id=5R4FAQAAIAAJ&oi=fnd&pg=PA4&dq=sievers+lobelia+1930&ots=6sQeuyPJTV&sig=Of1j4Z0DbNflYXfNw9_sQjwJdDo. Accessed: 15 Mar 2019

Simons AM, Johnston MO (2000) Variation in seed traits of *Lobelia inflata* (Campanulaceae): sources and fitness consequences. Am J Bot 87(1):124–132. https://doi.org/10.2307/2656690

Simons AM, Goulet JM, Bellehumeur KF (2010) The effect of snow depth on overwinter survival in *Lobelia inflata*. Oikos 119(10):1685–1689. https://doi.org/10.1111/j.1600-0706.2010.18515.x

Stanbury J, Saunders PR, Zampieron ER (2013) The use of Lobelia in the treatment of asthma and respiratory illness. J Restorat Med (2):94–100. https://doi.org/10.14200/jrm.2013.2.0109

St-Pierre A, Lajeunesse A, Desgagné-Penix I (2017) Determination of piperidine alkaloids from indian tobacco (*Lobelia inflata*) plants and plant-derived products. Austin Biochem 2(2):2014. Available at: file:///F:/Users/Mathe/Downloads/fulltext_biochemistry-v2-id1014.pdf

Subarnas A et al (1993a) A possible mechanism of antidepressant activity of beta-amyrin palmitate isolated from *Lobelia inflata* leaves in the forced swimming test. Life Sci 52(3):289–296. https://doi.org/10.1016/0024-3205(93)90220-w

Subarnas ANAS et al (1993b) Pharmacological properties of β-Amyrin palmitate, a novel centrally acting compound, isolated from *Lobelia inflata* leaves. J Pharm Pharmacol 45(6):545–550. https://doi.org/10.1111/j.2042-7158.1993.tb05596.x

Szőke, E. (1994) *Lobelia inflata* L. (Lobelia): *In vitro* culture and the production of lobeline and other related secondary metabolites, in Bajaj, Y. P. S. (ed.) Medicinal and aromatic plants VII. Biotechnology in agriculture and forestry, vol 28. . Berlin/Heidelberg: Springer, pp. 289–327. doi: https://doi.org/10.1007/978-3-662-30369-6_19

Szőke E et al (2001) Studies on the alkaloid production of genetically transformed and non-transformed cultures of *Lobelia inflata* L. Int J Hortic Sci 7(2):65–71. https://doi.org/10.31421/IJHS/7/2/269

Szőke É, Lemberkovics É, Kursinszki L (2013) Alkaloids derived from lysine: piperidine alkaloids. In: Natural products. Springer, Berlin/Heidelberg, pp 303–341. https://doi.org/10.1007/978-3-642-22144-6_10

USDA Agricultural Research Service (2019) *Lobelia inflata*, Germplasm Resources Information Network (GRIN). Available at: https://npgsweb.ars-grin.gov/gringlobal/taxonomydetail.aspx?id=22466. Accessed: 11 Feb 2019

Walsh N et al. (2010) Notes and new taxa in Lobelia sect. Holopogon (Campanulaceae: Lobelioideae), Muelleria. Available at: https://www.rbg.vic.gov.au/documents/Muelleria_28(2),_Walsh.pdf. Accessed: 16 March 2019

Wieland H (1921) Alkaloids of the lobelia plant (preliminary communication). Chem Ber 54B:1784. Available at: https://www.drugs.com/npp/lobelia.html. Accessed: 5 June 2019

Wysokinska H, Chmiel A (1997) Transformed root cultures for biotechnology. Acta Biotechnol 17(2):131–159. https://doi.org/10.1002/abio.370170205

Yonemitsu H et al (1990) Lobeline production by hairy root culture of *Lobelia inflata* L. Plant Cell Rep 9(6). https://doi.org/10.1007/BF00232857

Chapter 8
Mentha L. and *Pycnanthemum* L. Germplasm at the US National Clonal Germplasm Repository in Corvallis, Oregon

Kim Hummer, Nahla Bassil, and Kelly J. Vining

Abstract About 450 accessions representing 34 taxa and hybrid species of *Mentha* from around the world are preserved in the USDA ARS National Clonal Germplasm repository (NCGR) genebank in Corvallis, Oregon. This collection includes advanced breeder selections and F_1 hybrids. The initial collection was donated in 1984, the majority originating from the M.J. Murray collection of the A.M. Todd Company, Kalamazoo, Michigan. Other representatives of diverse mint taxa were received from collaborators in Australia, New Zealand, Europe, and Vietnam. Members of this collection have been evaluated for oil components, *Verticillium* resistance, cytology, and other key morphological characters. Voucher specimens have been prepared. An initial set of microsatellite markers have been developed to determine clonal identity and assess genetic diversity. In addition, the mountain mint, *Pycnanthemum*, a North American *Lamiaceae* relative, is conserved at this genebank. The primary donor for this collection was Dr. Henrietta Chambers. For this genus, 94 accessions representing 17 of the 19 described taxa from North America are preserved. Plants of *Mentha* and *Pycnanthem* are available for distribution for research as stem or rhizome cuttings or seed requested through the GRIN-Global database of the National Plant Germplasm System. Evaluation and characterization of these genera continue to focus on genetic determinants of essential oil content and disease resistance for potential use in breeding programs.

Keywords Mint · Mountain mint · Genetic resources · Crop wild relatives · Disease resistance · Cytology · Essential oils

K. Hummer (✉) · N. Bassil
USDA-ARS-National Clonal Germplasm Repository, Corvallis, OR, USA
e-mail: kim.hummer@usda.gov

K. J. Vining
Department of Horticulture, Oregon State University, Corvallis, OR, USA

© This is a U.S. government work and not under copyright protection in the U.S.; foreign copyright protection may apply 2020
Á. Máthé (ed.), *Medicinal and Aromatic Plants of North America*, Medicinal and Aromatic Plants of the World 6,
https://doi.org/10.1007/978-3-030-44930-8_8

187

8.1 Introduction

Mentha L., mint, (*Lamiaceae* formerly *Labiatae*) includes about 20 species world-wide (Tucker and Naczi 2007). The mint family includes many other genera, such as mountain mint (*Pycnanthemum*), lavender (*Lavendula* L.), sage (*Salvia* L.), rose-mary (*Rosmarinus* L.), and oregano (*Origanum* L.). Many of these plants can be field grown commercially for the essential oils distilled from leaves and stems. Mint shoots and leaves are used for medicinal and aromatic purposes, including extracts, distillation compounds, condiments, and food flavorings. *Mentha* comprises many economically valuable species, including peppermint (*Mentha* ×*piperita*), Scotch spearmint (*M.* ×*gracilis*), native spearmint (*M. spicata*), corn mint (*M. canadensis*) and field mint (*M. arvensis*), which are cultivated for their medicinal properties.

In the US, the main mint crops produced are peppermint and spearmint. Spearmint includes Scotch and native spearmints. Menthol is the primary constituent of pep-permint oil, and carvone is in spearmint oil. The US annual farm gate value of pep-permint production has been $100 to $150 million, and that for spearmint $45 to $50 million, during the past decade (NASS 2018). The United National Food and Agriculture Statistical Databases (FAOSTAT) reports on peppermint, but not on the production of other mint species. According to FAOSTAT (2018), Morocco, Argentina, Mexico, Bulgaria, and Spain countries have the largest peppermint pro-duction. In China and India, two other world producers, menthol is extracted from corn and field mint, crops which are not reported in FAOSTAT.

In 2011, the US produced about seven million pounds (3175 MT) of peppermint oil and 2.2 million pounds (998 MT) of spearmint oil. Since then, US production dropped to under six million pounds (2721 MT), because of menthol production from mostly *M. arvensis* and *M. canadensis* in China and India. Although *M. arven-sis* cultivars have a higher menthol content than *M.* ×*piperita* accessions, their oil has lower overall quality (http://nda.agric.za/docs/Brochures/ProGuiPeppermint. pdf). The demand for mint oil by Southeast Asian countries is driving the market for menthol production and is expected to increase by about 5% per year.

The 19 species of mountain mint, *Pycnanthemum*, are endemic to North America (Table 8.1). The native range for these species occurs from Ontario, Canada, in the north to Florida in the south. Most species for this genus are native to the central and eastern US though one disjunct, *P. californicum*, grows along the Pacific Coast of California.

Mountain mint has ethnobotanical uses, e.g., Mesquaki Indians used the plant for everything from animal attractants used for traps to leaf infusions in tea for human consumption to cure chills and fever (Pellett 1947). In 1945, Professor Arthur Schwarting, College of Pharmacy, Nebraska University, reported distilling fresh leaves of mountain mint to produce essential oils with a high yield of menthol, important at that time when the supply from Japan was unavailable due to war (Pellett 1947). Schwarting observed that the leaves of *Pycnanthemum pillosum* con-tained 80% pulegone, 10% menthone, 3–5% limonene, and 2% menthol. Pulegone and menthone can be converted to the higher value product, menthol (Pellett 1950).

Table 8.1 *Mentha* and *Pycnanthemum* species, ranges and number of accessions in the US Department of Agriculture, National Clonal Germplasm Repository, Corvallis Oregon

Species	Range of the species	Accessions
Mentha aquatica	Europe except extreme North	15
Mentha aquatica var. citrata	Cultivated form	14
Mentha arvensis	Europe	3
Mentha australis	Northern Territory, Queensland, New South Wales, Victoria, Tasmania, South Australia	2
Mentha canadensis	North America, Japan, Korea, Sakhalin, East Asia	43
Mentha cervina	Southwest Europe	3
Mentha diemenica	Australia	1
Mentha gattefossei	High Atlas Mountains, Morocco	5
Mentha hybr.		100
Mentha japonica	Hokkaido, Honshu	2
Mentha longifolia	Most of Europe from Sweden south	18
Mentha longifolia subsp. capensis	South Africa, Namibia, Zimbabwe, Lesotho	1
Mentha longifolia subsp. hymalaiensis	Afghanistan, Himalayas	1
Mentha longifolia subsp. polyadenia	South Africa, Lesotho	1
Mentha longifolia subsp. typhoides	Aegean Region, Iran, Northern Iraq, Turkey, Syria, Lebanon, Israel, Egypt	1
Mentha longifolia var. asiatica	Afghanistan, Russian Federation, Kazakhstan, Kyrgystan, Tajikistan, Tirkmenistan, Uzbekistan	2
Mentha pulegium	Southwest and Central Europe	9
Mentha requienii	Corsica, Sardinia, Montecristo Island	1
Mentha spicata	Europe from cultivation	95
Mentha spicata subsp. condensata	Italy, Sicily, Balkan Peninsula, Aegean Region	2
Mentha spicata subsp. spicata	Cultivated form	1
Mentha spp.		6
Mentha suaveolens	Southwest and Western Europe as far North as the Netherlands	6
Mentha suaveolens subsp. insularis	Mediterranean Region	1
Mentha suaveolens subsp. suaveolens	Europe	22
Mentha × *dalmatica*	Hybrid	5
Mentha × *dumetorum*	Hybrid	2
Mentha × *gracilis*	Hybrid	22
Mentha × *piperita*	Hybrid	48
Mentha × *smithiana*	Hybrid	3
Mentha × *suavis*	Hybrid	1
Mentha × *verticillata*	Hybrid	3

(continued)

Table 8.1 (continued)

Species	Range of the species	Accessions
Mentha × *villosa*	Hybrid	6
Mentha × *villosa nothovar. alopecuroides*	Hybrid	10
Total		**455**
Pycnanthemum		
Pycnanthemum beadlei	North Carolina	11
Pycnanthemum californicum	California	3
Pycnanthemum curvipes	Alabama, Georgia, North Carolina, Tennessee	5
Pycnanthemum flexuosum	Virginia	1
Pycnanthemum floridanum	Florida	3
Pycnanthemum hybr.		6
Pycnanthemum incanum	Connecticut	4
Pycnanthemum loomisii	Kentucky	3
Pycnanthemum montanum	North Carolina	1
Pycnanthemum muticum	Connecticut	16
Pycnanthemum pilosum	Indiana	4
Pycnanthemum pycnanthemoides	Alabama, West Virginia	4
Pycnanthemum setosum	New Jersey	2
Pycnanthemum tenuifolium	Connecticut	17
Pycnanthemum torreyi	Virginia	2
Pycnanthemum verticillatum	Connecticut	1
Pycnanthemum virginianum	Indiana	11
Total		**94**

The objective of this chapter is to describe the activities of the US mint genebank at the National Clonal Germplasm Repository (NCGR) in Corvallis, Oregon. The diversity of mint species and cultivar collections of *Mentha* and *Pycnanthemum* are described. Genebank management techniques and distribution activities along with the ongoing evaluation efforts of descriptive morphology, cytology, molecular genetic diversity, and identity determination through molecular analysis are discussed. The chapter will conclude with specific examples of uses of the collection in breeding efforts.

8.2   Diversity of Mint Species and Cultivars

More than 3000 specific epithets have been described for *Mentha* (Tucker and Naczi 2007). This plethora of names has created confusion in the literature. Wild species of mint hybridize over evolutionary time and native hybrid-species swarms occur in conspecific regions. Tucker and others (Tucker and Chambers 2002; Tucker and Naczi 2007) redefined the genus into four sections and reduced approved taxa to 18 species with 11 grex, i.e., hybrid-species.

The mint collection of the US Department of Agriculture (USDA), NCGR in Corvallis, Oregon, includes representatives of approximately 450 accessions of 20 taxa and 10 hybrid-species (Table 8.1). In addition, the NCGR houses mountain mint, *Pycnanthemum* L., a *Mentha* relative also in the *Lamiaceae*. The NCGR mountain mint collection has 95 accessions of 17 taxa (GRIN-Global 2018). The goals of the genebank are to collect, maintain, distribute, evaluate, and document information on the assigned genera. This active genebank maintains the primary collections as plants in containers in a protected environment.

8.3 History of the NCGR Collections

The initial clonal material of the NCGR mint collection originated from Dr. Merritt J. "Bill" Murray (MJM), a mint breeder who worked for the A.M. Todd Company of Kalamazoo, Michigan (Chambers and Hummer 1992). He was very active in mint research from 1950 to the early 1970s.

Dr. Murray obtained mints from North and South America, Europe, the Middle East, and the Russian Federation. He assigned each of the clones a shorthand code which was a brief summary of information on ploidy, species, and sources. Dr. Murray's notes described the morphology, disease resistance, vigor, odor, fertility, genetic data, and oil composition, as well as pedigrees for about 102 clones. In 1972, upon his retirement, he donated his mint germplasm to Dr. C.E. Horner, a USDA plant pathologist who worked in Corvallis, Oregon. Dr. Horner was a specialist on *Verticillium* wilt, one of the diseases plaguing mint production in the Pacific Northwestern United States and elsewhere.

When Dr. Horner resigned in 1978, Dr. Al Haunold, a USDA hop breeder, assumed the management of the collection. From 1979 to 1980, essential oil analyses and morphological evaluations were performed on the mint by Dr. Don Roberts, a USDA Corvallis mint breeder. In 1983, Dr. Haunold arranged for the donation of this invaluable germplasm to the NCGR. The original MJM notes and essential oil analyses were provided to the NCGR along with the plant material.

In 1988, Dr. Henrietta Chambers assumed the curation of the mint collection at the NCGR. Her doctoral thesis investigated the cytotaxonomy of *Pycnanthemum* (Chambers 1961a). During her curatorial assignment at the NCGR, Dr. Chambers performed karyotype analysis from dividing pollen mother cells in flowering buds or meristematic root-tip cells in the NCGR *Mentha* (Chambers and Hummer 1994). Her evaluation complemented previously published chromosome counts, providing information on about 73 accessions.

Dr. Chambers collaborated with Dr. Arthur Tucker, a *Mentha* taxonomy specialist at Delaware State College. They reviewed taxonomy of the collection and reclassified and updated species nomenclature (Tucker and Chambers 2002).

Recently, the NCGR has received new accessions from Vietnam, Canada, Romania, Hungary, and from domestic US sources. Dr. Kelly Vining, geneticist at Oregon State University, has been analyzing the *Mentha* collection at the NCGR, building genome-based resources for mint, and breeding new cultivars with improved essential oil composition and disease resistance.

In the early 1990s, Dr. Chambers donated her thesis study collection of *Pycnanthemum* to the NCGR. The collection includes 17 of the 19 described North American *Pycnanthemum* species. *Pycnanthemum albescens* T. & G., a diploid from Florida and *P. clinopodioides* T. & G., a tetraploid from Connecticut, are the two species not currently represented in this collection.

8.4 Greenhouse/Screenhouse Plantings

Mint and mountain mint accessions are propagated by stem cuttings or rhizomes. Propagules are placed in propagation media under mist. Rooted plants are usually ready to up-pot within 3 weeks. Though *Mentha* cuttings propagate readily from stems or rhizomes, *Pycnanthemum* cuttings may take 4–6 weeks to root well. The plants are transplanted to standard potting mix once established.

Plants are initially labeled using 1″ × 4″ plastic labels with #2 soft pencil or with a computer generated label. When large enough, permanent plants are placed in 2 gallon pots with a computer generated label attached to the pot. These labels are printed in photo-resistant ink on a laser printer. The temporary plastic propagation label is kept on the inside of the pot for the life of the plant. Great care is taken when labeling or replacing labels to prevent mistakes.

The mint plants are mowed every 4–6 weeks during the growing season for flower removal. Flowers are removed from species because seeds can disperse and result in mixed genotypes and contaminate other nearby clones. Rhizomes that penetrate the drainage holes in the bottom of the pots are trimmed. Loose moss and debris is removed from the pots to reduce slugs and snails (phylum *Mollusca*), and thrip (*Thrips*) habitat.

Early in March and again in June the collection is given slow release fertilizer which is scattered over the pot surface so that drip emitters will water it in. They are also given micronutrients early in the spring and supplemental soluble fertilizer applications. Each fall and spring the mints are checked for weak plants which require repotting and incubation in a heated greenhouse until regrown. Plants are potted into short two gallon containers in standard potting mix.

The mountain mint is not pruned until the end of the season when new growth is seen at the base of the plants. These plants do not form rhizomes like *Mentha* so do not easily tolerate the loss of foliage until after bloom and formation of new vegetative growth. These plant crowns are staked to keep them from spreading and to aid

---

a) Medium: MS salts and vitamins. Per liter: 30 g sucrose, 0.5 mg BA (N-benzyladenine), 0.1 mg IBA (Indole 3 Butyric acid), 3.0 g agar (Difco Labs, Detroit, USA), 1.25 g agar at pH 5.7 before autoclaving.

Growth room conditions are 16 h photoperiod ($25 \cdot mol \cdot s^{-1} \cdot m^{-2}$) at 25 °C.

b) Initiation/surface sterilization: Collect 7 shoot tips (5-10 mm each) and surface sterilize in a 10% bleach (bleach is 5.25% sodium hypochlorite) solution with 0.1 ml/L of Tween 20 on a rotary shaker for 10 min. Rinse twice for 5 min in sterile water.

Fig. 8.1 *Mentha* tissue culture media recipe. (Reed and DeNoma 2019)

in pest monitoring and control. A single stake is placed in the center of the pot with the stems fastened to it using stretch tape.

Both mint and mountain mint are prone to whitefly (*Bemisia*) and fungus gnat (superfamily *Sciaroidea*) infestations. Thrips can be a problem for this collection, and must be monitored. The regular removal of the mint foliage usually keeps spider mite populations under control. The mint can be prone to powdery mildew infections during humid spring and fall weather. When the greenhouse heat is shut off in the spring, fans are kept running 24 h a day to improve air circulation and reduce mildew.

8.5 Tissue Culture

Mint can be cultured *in vitro* by isolating meristems of specific clones and growing them on media in containers in the laboratory. Tissue cultured plantlets are used as (1) secondary security backup for the primary collections which are grown in containers as whole plants, (2) as part of pathogen elimination protocols, and (3) for international plant distribution.

The procedural protocol and the recipe for mint culture medium is based on standard micropropagation methods, generally starting with Murashige and Skoog (MS) (1962) and presented in Fig. 8.1. Careful aseptic technique is required for this work. For initiation into culture, cuttings are collected from greenhouse or screenhouse plants when available. This plant material harbors fewer bacterial and fungal contaminants than does material collected from the field. Meristems, growing apices or axes at nodes, are isolated from the cuttings. Plantlets are grown first with hormones, then transferred to hormone-free medium in culture bags and stored under refrigerated conditions (4 °C) for several years without active re-culturing (Reed 1991; Reed et al. 1995). The multiplication medium for *Pycnanthemum* is prepared similarly to that of the mint medium, but no hormones are used and only one-half of the nitrate concentration is used (Jenderek et al. 2013).

8.6   Plant Distribution

NCGR mint and mountain mint germplasm is available to researchers as stem cuttings or rhizomes throughout the year. Some species representatives are available as seeds. During the past 37 years, the NCGR has distributed >8654 *Mentha* and about 900 *Pycnanthemum* inventory items. The most requested mint cultivars include 'Chocolate mint', 'Pineapple mint', 'Black Mitcham', 'White Mitcham', 'Eau de Cologne', 'Todd Mitcham', 'Kentucky Colonel', 'Scotch spearmint', 'Murray Mitcham', and 'Orange mint'. Mint has been distributed to researchers throughout the world and particularly to industry looking to develop improved cultivars for industrial uses.

8.7   Germplasm Characterization

8.7.1   Essential Oils

Mints with unique essential oil profiles are valued by the herb, nursery, and pharmaceutical trades. The chemical constituents are important for commercial interests and end products. More than 230 accessions of the mint collection of the NCGR were initially screened by gas chromatography for unique essential oil profiles (Roberts et al. 1980). Vining et al. (2005) examined 14 accessions of *M. longifolia* from the collection for essential oil content. While oil analysis results can change depending on the age of the plant prior to leaf harvest (Burgsten, personal communication), these data were taken from as consistently similar plant material as possible. These essential oils can be searched online as descriptors through the GRIN-Global database (2019).

The chemical composition of essential oil of *Pycnanthemum floridum* was determined by Shu et al. (1994). They observed that the leaves contained 39.83% pulegone, 32.15% menthone, 10.43% piperone, and lower quantities of 37 other essential oils. According to Dr. Jim Duke, economic botanist retired from the USDA, *Pycnanthemum* is the best source of pulegone, the major ingredient in pennyroyal oil. This highly pungent compound repels fleas, birds, and is under examination as a tick repellant (Kaplan 1990). Other botanists have collected lemon-scented *Pycnanthemum* types (Sorensen and Matekaitis 1981).

8.7.2   Disease Resistance

Disease resistance is another character that is important to the commercial mint production industry. *M. longifolia*, *M. aquatica*, and *M. suaveolens* accessions from the NCGR collections have been screened for *Verticillium* resistance

(Vining et al. 2005; Vining et al. 2019). *M. suaveolens* accessions displayed relatively greater Verticillium resistance, while the other two species showed a range of resistance to susceptibility.

8.7.3   Cytology

Harley and Brighton (1977), defined five sections within the genus *Mentha* (*Mentha* sect. *Audibertia*, sect. *Eriodontes*, sect. *Mentha*, sect. *Preslia*, and sect. *Pulegium*) including 19 species and 13 named hybrids. They also listed chromosome counts for sect. *Mentha* including most of the mint taxa recognized today.

Chambers and Hummer (1994) summarized other documentation of chromosome counts and surveyed 73 *Mentha* accessions from the NCGR collection. This survey included chromosome counts for *M. australis* (two accessions), *M. japonica*, *M. diemenica*, and *M. cunninghamii*. In this study, ploidy determinations were obtained by cytological counts of dividing pollen mother cells in flower-buds, or meristematic cells in root tips.

Vining et al. (2019) recently reported several unusual ploidy counts, finding diploid and triploid clones in *M. suaveolens*, and eneuploid clones of *M. aquatica* in the NCGR mint collection.

Over the years, Dr. Chambers performed cyto-taxonomic studies on *Pycnanthemum* (1961a, b). She also surveyed chromosomes and analyzed natural and artificial hybrids (Chambers 1993; Chambers and Chambers 1971). In addition, Chambers and Chambers (2008) examined infrageneric classification in *Pycnanthemum*.

8.8   Molecular Analysis

At the NCGR, one objective in the molecular laboratory is to develop a fingerprinting set for use in establishing a genetic profile for each *Mentha* accession. So far, the focus is to develop and apply simple sequence repeat (SSR) markers with long core repeats ($\geq$3). These SSR markers can be easily scored and lack polymerase chain reaction (PCR) artifacts.

Few SSR markers are available for mint. Kumar et al. (2015) designed primers targeting 68 of 110 SSRs identified in publicly available expressed sequence tag (EST-SSR) sequences of *M. ×piperita*. Out of 54 SSRs that generated clear amplicons in 13 accessions of *M. ×piperita*, three SSRs with long core repeats amplified in representative samples from four species including *M. arvensis*, *M. citrata*, *M. longifolia*, and *M. spicata* indicating transferability among species. We recently screened these three SSRs in addition to 48 other primer pairs designed from a draft genome assembly of *Mentha longifolia* CMEN 585, from South Africa (Vining et al. 2017). The testing panel consisted of three accessions each of *M. suaveolens* and *M. aquatica* in addition to two accessions of *M. longifolia* from South Africa, CMEN 584 and CMEN 585, (Vining et al. 2019).

Screening the testing panel with these 51 SSRs identified nine primer pairs designed from the *Mentha longifolia* CMEN 585 assembly (Vining et al. 2017) that were polymorphic and appeared easy to score. These nine SSRs were thus used in two multiplexes to genotype 49 accessions propagated from the NCGR collection that included 24 accessions of *M. aquatica*, 23 accessions of *M. suaveolens* and the two *M. longifolia* plants in the testing panel. The nine SSRs developed in this study separated the accessions mostly according to species and identified each accession as unique. Three sets of accessions of *M. aquatica* were closely related and were distinguished by a single allele each: CMEN 116, CMEN 117; CMEN 110, CMEN 111; and CMEN 121, CMEN 122. The consecutive local numbers of each pair of closely related accessions indicate they are in neighboring pots and may suggest plant contamination and will be investigated further.

The number of SSRs in *Mentha* remains low and more SSRs need to be identified and evaluated across *Mentha* species. Development of a robust fingerprinting set will be able to distinguish separate genotypes within the genebank collection. *Pycnanthemum* has not yet been evaluated with DNA markers.

8.9 Breeding Efforts

The USDA NCGR *Mentha* collection is valuable as a source of genetic diversity for mint breeding. The origin of 'Black Mitcham' peppermint, the most widely grown American cultivar in the US and Europe, is unknown, because it resulted from natural hybridization events. However, phylogenetic studies (Tucker et al. 1980; Tucker and Naczi 2007) suggest that ancestors of *M. longifolia* and *M. suaveolens* are the diploid ancestors of *M. spicata*, which then further hybridized with *M. aquatica* to give rise to *M. ×piperita*. Therefore, breeding over the past 50 years has focused on evaluation and crossing of accessions of *M. longifolia*, *M. suaveolens*, and *M. aquatica*.

From the mint industry viewpoint, improved oil quality and disease resistance are two priority traits for new cultivar development. Both traits are multigenic with complex mechanisms. *Mentha* germplasm evaluations have therefore focused on those traits and efforts are ongoing to understand the underlying systems.

Verticillium wilt is a vascular wilt disease caused by the soil-borne fungus *Verticillium dahliae*. 'Black Mitcham' peppermint is highly wilt-susceptible. It is also a sterile hexaploid, which hinders traditional breeding techniques. Other genetic recombination techniques have been performed over the years. An irradiation program, working in collaboration with nuclear conductors in the 1960s, resulted in the release of two new cultivars with relatively higher wilt resistance, 'Todd's Mitcham' (Todd et al. 1977) and 'Murray Mitcham' (Murray and Todd 1972). In the 1990s–2000s, the Mint Industry Research Council initiated a genetic engineering project with the objective of genetically transforming peppermint cultivars with specific genes to increase oil yield and quality and confer wilt resistance.

However, this research was discontinued due to concerns about market acceptance of genetically modified organisms.

Today, breeding efforts are focused on traditional crossing techniques of the fertile ancestral species. To that end, NCGR accessions of the crop wild relatives of *M. longifolia*, *M. suaveolens*, and *M. aquatica* were screened for relative levels of *Verticillium* wilt disease resistance. Both *M. longifolia* and *M. aquatica* showed a range of disease resistance to susceptibility, with some highly susceptible accessions, some highly resistance accessions, and most displaying moderate susceptibility levels (Vining et al. 2005, 2019). *M. suaveolens* accessions, in contrast, were highly susceptible, with the exception of one moderately susceptible accession (Vining et al. 2019).

Oil yield and quality are of equally high priority to growers, oil buyers, and flavor companies where oils are mixed to produce proprietary blends for consumer products. Mint oils consist of a complex mixture of monoterpene molecules that are synthesized in specialized leaf hairs termed glandular trichomes. Carvone is the primary constituent of spearmint oil; menthol is the primary constituent of peppermint oil. Menthol is by far more commercially valuable because of its broader use. *M. aquatica* accessions are characterized by high levels of (+)-menthofuran, and lower levels of (−) limonene and 1,8-cineole. *M. suaveolens* accessions are high in either (−)-carvone, piperitenone oxide, or trans-piperitone oxide (Vining et al. 2019). *M. longifolia* accessions are more diverse, with some high-pulegone accessions, and a range of representation of other constituents (Vining et al. 2005).

Transgressive segregation, a phenomenon in which progeny traits differ from those of either parent, is well documented in oil types resulting from *Mentha* crosses (Tucker 2012). With the advent of a *Mentha* reference genome, the genetic underpinnings of this phenomenon are just beginning to be explored. The complex genetic regulatory mechanisms governing oil type will be a challenge to breeding going forward.

8.10 Conclusions

The *Mentha* and *Pycnanthemum* collections at the USDA NCGR-Corvallis genebank represent a significant portion of diverse global species for these genera. Several gaps in the collection remain including *M. pulegium* L.var. *micrantha* (Fisch.) Benth. from Russia and Kazakhstan and *M. repens* (J.D.Hook.) Briq. from Tasmania, Australia. Further examination and evaluation for key economic traits will be pursued. In addition, further development of a robust SSR set for clonal identification is being developed. Two species of *Pycnanthemum, P. albescens* and *P. clinopodioides*, need to be represented in the US collection, and will be collected when possible. This genus has potential for industrial development.

Cuttings and seeds can be distributed upon request for research purposes from the curator at the NCGR-Corvallis, through GRIN-Global database:

Search at https://npgsweb.arsgrin.gov/gringlobal/search.aspx

References

Chambers HL (1961a) A cytotaxonomic study of the genus *Pycnanthemum* (Labiatae). Ph. D. the-sis Yale University, New Haven, CT, 233 pp

Chambers HL (1961b) Chromosome numbers and breeding systems in *Pycnanthemum* (Labiatae). Brittonia 13:116–128

Chambers HL (1993) Chromosome survey and analysis of artificial hybrids in *Pycnanthemum*. Castanea 58:197–208

Chambers HL, Chambers KL (1971) Artificial and natural hybrids in *Pycnanthemum* (Labiatae). Brittonia 23:71–88

Chambers HL, Chambers KL (2008) Infrageneric classification and nomenclatural notes for *Pycnanthemum* (Lamiaceae). J Bot Res Inst Texas 2(11):193–199

Chambers HL, Hummer KE (1992) Clonal repository houses valuable mint collection in Corvallis, Oregon. Diversity 8(4):31–32

Chambers HL, Hummer KE (1994) Chromosome counts in the *Mentha* collection at the USDA-ARS National Clonal Germplasm Repository. Taxon 43(3):423–432

FAOSTAT (2018) Food and Agriculture Organization of the United Nations, Statistical database. http://www.fao.org/faostat/en/#data/QC. Accessed 10/02/2018

GRIN-Global (2018) Germplasm Resources Information Network-Global, US National Plant Germplasm System. https://npgsweb.ars-grin.gov/gringlobal/search.aspx. Accessed 10/02/2018

Harley RM, Brighton CA (1977) Chromosome numbers in the genus *Mentha* L. Bot J Linn Soc 74:71–96

Jenderek M, Holman GH, DeNoma J, Reed BM (2013) Medium and long-term storage of the *Pycnanthemum* (mountain mint) germplasm collection. Cryo Letters 34(5):490–496

Kaplan K (1990) For alternative crops – he's the Duke. Agric Res:18–20

Kumar B, Kumar U, Yadav HK (2015) Identification of EST–SSRs and molecular diversity analy-sis in *Mentha piperita*. Crop J 3:335–342

Murashige T, Skoog F (1962) A revised medium for rapid growth and bioassays with tobacco tis-sue cultures. Physiol Plant 15:474–494

Murray MJ, Todd WA (1972) Registration of Todd's Mitcham peppermint. Crop Sci 12:128

NASS (2018) US National Agricultural Statistical Service. https://www.nass.usda.gov/Statistics_by_Subject/index.php?sector=CROPS. Accessed 10/02/2018

Reed BM (1991) Application of gas-permeable bags for *in vitro* cold storage of strawberry germ-plasm. Plant Cell Rep 10:431–434

Reed BM, Buckley PM, DeWilde TN (1995) Detection and eradication of endophytic bacteria from micropropagated mint plants. In Vitro Cell Dev Biol 31P:53–57

Reed BM, DeNoma J (2019) *Mentha* and *Pycnanthemum* tissue culture media sheets. Station Publication. USDA ARS NCGR, Corvallis, 3 pp.

Roberts D, Burgsten D, Chambers H (1980) Gas Chromatograms of the USDA *Mentha* Germplasm, vol 1–5. Station Publication, USDA ARS NCGR, Corvallis

Shu CK, Lawrence BM, Miller KI (1994) Chemical composition of the essential oil of *Pycnanthemum floridanum* E. Grant and Epling. J Essent Oil Res 6:529–531

Sorensen PD, Matekaitis PA (1981) A lemon-scented *Pycnanthemum* (Lamiaceae). Rhodora 83:145–146

Todd WA, Green RJ, Horner CE (1977) Registration of Murray Mitcham Peppermint1 (Reg. No. 2). Crop Sci 17:188–188

Tucker AO (2012) Genetics and breeding of the genus *Mentha*: a model for other polyploid species with secondary constituents. J Medicinally Active Plants:1–7

Tucker AO, Harley RM, Fairbrothers DE (1980) The Linnean types of *Mentha* (Lamiaceae). Taxon 29:233–255

Tucker AO, Naczi RFC (2007) *Mentha*: an overview of its classification and relationships. Chapter 1. In: Lawrence B (ed) Mint: the genus *Mentha*. CRC Press, Boco Raton, pp 1–40

Tucker AO, Chambers HL (2002) *Mentha canadensis* L. (Lamiaceae) a relict amphidiploid from the lower Tertiary. Taxon 51:703–718

Pellett FC (1947) Mountain mint. Am Bee J 87:172–173

Pellett FC (1950) Mountain mint: a new source of essential oil. Am Bee J 90:66–67

Vining KJ, Zhang Q, Tucker AO, Smith C, Davis T (2005) *Mentha longifolia* (L.) L.: a model species for mint genetic research. HortScience 40:1225–1229

Vining KJ, Johnson SR, Ahkami A, Lange I, Parrish AN, Trapp SC, Croteau RB, Straub SC, Pandelova I, Lange BM (2017) Draft genome sequence of *Mentha longifolia* and development of resources for mint cultivar improvement. Mol Plant 10:323–339. https://doi.org/10.1016/j.molp.2016.10.018

Vining KJ, Pandelova I, Hummer KE, Bassil NV, Contreras R, Neill K, Chen H, Parrich AN, Lange BM (2019) Genetic diversity survey of *Mentha aquatica* L. and *Mentha suaveolens* Ehrh., mint crop ancestors. Genet Resour Crop Evol 66:825. https://doi.org/10.1007/s10722-019-00750-4

Chapter 9
Taxus brevifolia a High-Value Medicinal Plant, as a Source of Taxol

N. Z. Mamadalieva and N. A. Mamedov

Abstract The genus Taxus has created considerable interest due to the presence of taxol. Taxol is one of the best-selling medicaments used in the treatment of ovarian, lung and breast cancers and Kaposi's sarcoma. In this chapter, we briefly describe the current possibilities to produce the taxol from natural sources, synthetic approaches for taxol and its analogues, anticancer activity of taxol and secondary metabolites of T. brevifolia.

Keywords *Taxus brevifolia* · Taxol · Anticancer · Semi-synthesis · Cell culture · Endophytic fungus

9.1 Introduction

Anticancer drug taxol (paclitaxel) is a highly functionalized taxane diterpene amide was first isolated from the bark of the *Taxus brevifolia* Nutt. (Moss 1998; Wani et al. 1971). *T. brevifolia* (Pacific yew) is grows along the Pacific Coast of southeastern Alaska southward through western British Columbia to central California. In the Rocky Mountain region, it occurs from southeastern British Columbia through northwestern Montana and northern Idaho into eastern Washington and Oregon (Little 1979).

Related species are found in Canada, the Himalayas, Europe, and Russia. *T. brevifolia* is a slow growing and long-living needle-leaved tree. Traditionally the wood of this species has been used by Native Americans for making weapons, canoes, bows, tool handles and poles (Hosie 1969; Preston 1948).

N. Z. Mamadalieva
Institute of the Chemistry of Plant Substances, Uzbekistan Academy of Sciences,
Tashkent, Uzbekistan

N. A. Mamedov (✉)
Medicinal Plants Program, University of Massachusetts, Stockbridge School of Agriculture,
Amherst, MA, USA
e-mail: mamedov@cas.umass.edu

© Springer Nature Switzerland AG 2020
Á. Máthé (ed.), *Medicinal and Aromatic Plants of North America*, Medicinal and Aromatic Plants of the World 6,
https://doi.org/10.1007/978-3-030-44930-8_9

The genus *Taxus* has created considerable interest due to the presence of taxol. Taxol is one of the best-selling medicaments used in the treatment of ovarian, lung and breast cancers and Kaposi's sarcoma (Cragg et al. 1993). Because of the special anticancer mechanics and its special curative effect for some kinds of tumors, taxol has attracted much recent attention. The scarcity of taxol and the ecological impact of harvesting it have prompted extension searches for alternative sources including total synthesis, cellular culture production, semi-synthesis and endophytic fungus production (Nicolaou et al. 1994).

In this chapter, we describe briefly the current possibilities to produce the taxol from natural sources, synthetic approaches for taxol and its analogues, anticancer activity of taxol and secondary metabolites of *T. brevifolia*.

9.2   Plant Resources of Taxol

There are at least ten species in *Taxus*. It has been established that not only *T. brevifolia* but other *Taxus* species such as *T. baccata* (Senilh et al. 1984), *T. cuspidata* (Neto and Di Cosmo 1992), *T. wallichiana* (Velde et al. 1994), and *T. xmedia* cv. Hicksii (Barboni et al. 1994; Appendino et al. 1994) also contain taxol. The content of taxol in *Taxus* spp. generally is very low, only 0.001% - 0.06% w/w of dry bark (Hoke et al. 1994; Park et al. 1999; Ketchum et al. 1999). In a low yield, it may be isolated from the bark of the *T. brevifolia*, however, its industrial production is largely dependent on the precursor. In *T. brevifolia*, taxol is found most abundantly in the following order: bark > needles > roots > branches > seeds > wood (Theodoridis and Verpoorte 1996). For treatment of one patient about 2 g of paclitaxel is required which can be obtained from the bark of 3–10 trees (Kikuchi and Yatagai 2003). As an average yield of taxol is approximately 0.015% 2000–2500 yew trees can produce only 1 kg taxol (Vidensek et al. 1990). The tree grows very slowly, and if harvested for taxol, the natural stands of the tree would be depleted. To save this valuable population of needles of *Taxus* spp. has been replaced with the bark by many researchers to obtain docetaxel instead of paclitaxel (Fumoleau et al. 1997; Vaishampayan et al. 1999; Theodoridis and Verpoorte 1996).

9.3   Total Synthesis of Taxol

In 1994, a fully chemical synthesis of taxol was achieved (Holton et al. 1994; Nicolaou et al. 1994). The total organic synthesis of taxol presents a considerable challenge because of its high fictionalization, and its complex and unusual ring structure (Hamburger and Hostettmann 1991). Besides total chemical synthesis (Holton et al. 1994; Nicolaou et al. 1994; Morihira et al. 1998) and isolation from plant, there are several other possible routes to industrialize taxol production:

semi-synthesis (Denis et al. 1988), cell culture (Christen et al. 1989; Hu et al. 2003), fungal fermentation (Stierle et al. 1993; Strobel et al. 1996; Li et al. 1996; Wang et al. 2000).

9.4 Cell Cultures Production of Taxol

Taxol can be produced in plant cell culture of *Taxus* spp. Particular modifications of culture conditions have been discovered to enhance the yield of various taxanes from cell culture of all species of *Taxus* (Bringi et al. 2007). A method proposed by Christen et al. (1991) that used yew cell-suspension culture instead of tissue culture increased the yield of taxol. In 1992, ESC Agenetics Co. reported that the yield of taxol with this method was 2–5 times higher than that in the yew bark. Gan (1997) obtained a yield of 0.119% in medium with the cell suspension of *T. yunnanensis*, approximately 12 times of that in the bark of adult trees. So far, although many *Taxus* species have been explored for production of taxol using plant cell cultures and gained considerable success, it is still limited for large-scale commercial use because of the low and unstable product yield, as well as high production cost (Ji et al. 2006).

9.5 Semi-Synthesis of Taxol

Taxol may be derived by semi-synthesis of taxol analogues; of which, the most important are 10-deacetylbaccatin III, baccatin III and 10-deacetyltaxol (Denis et al. 1988). Semi-synthesis starting with the precursor of taxol, 10-deacetylbaccatin III, which can be extracted from leaves of yew trees, proved to be an appropriate method with commercial feasibility. In addition, other related taxanes such as bac-catin III, cephalomannine, 10-deacetyl-7-xylosylpaclitaxel, 10-deacetyl-7-epipaclitaxel, 10-deacetyl-paclitaxel etc. can also be used for semisynthesis of taxol. It can be produced on a large scale by semisynthesis from baccatin III in the needles of *Taxus* trees. A unique taxane, brevifoliol, which lacks the oxetane ring, was isolated from *T. brevifolia* leaves (Balza et al. 1991). Although the productivity has been improved with semi-synthesis techniques, virtually, it is quite similar regarding the consumption of yew trees.

9.6 Taxol Producing Endophytic Fungus

Some taxoids are also synthesized by various endophytic fungi, which often live in association with *Taxus* trees. In 1993, Stierle et al. were the first to report that an endophytic fungus *Taxomyces andreanae* isolated from *T. brevifolia* could independently synthesize taxol. Although the yield of taxol was only as low as 24–50 ng/L, this finding stirred great interest in biotechnologists. Ever since, there have been a few reports on the isolation of taxol-producing endophytic fungi (Strobel et al. 1996; Li et al. 1996; Wang et al. 2000), demonstrating that organisms other than *Taxus* species could produce taxol. Till date, highest level of taxol has been synthesized after 7 days of incubation period from *Cladosporium* sp. F3 isolated from *T. baccata* producing taxol at 139.2 mg/kg (Chowdhary et al. 2017). Meanwhile, the biggest problem of using fungi fermentation to produce taxol is its very poor yield and unstable production. To solve such a problem, current studies mainly focuses on the tedious work of finding and isolating fungi with high, stable yield of taxol, as well as optimization of fermenting conditions like that of *Taxus* cell cultures (Fu et al. 2009).

9.7 Mechanism of Action of Taxol

Initial studies by Fuchs and Johnson (1978) indicated that taxol inhibited proliferation at the G2-M phase in the cell cycle and blocked mitosis. In 1979 by Horwitz et al. its unique mechanism of action was revealed. They reported that taxol accelerated polymerization of tubulin and acted as a microtubule-stabilizing agent and as colchicine, vinblastine, vincristin, podophyllotoxin and others induced microtubule destabilization (Horwitz et al. 1979; Horwitz 2004). Taxol does not interact directly with nuclear components (DNA and RNA), unlike some other chemotherapeutic drugs. The main site of action taxol is the microtubules, however, unlike the vinca alkaloids or colchicine derivatives, which induce depolymerization of the microtubules (Dumontet and Sikic 1999). Taxol acts during the mitotic phase of cell division and promotes the polymerization of the tubulin proteins and their assembly, resulting in the stabilization of microtubules and block the cell cycle, with consequent cell death. In the presence of taxol, tubulin can be assembled *in vitro* into microtubules without the normal conditions including microtubule-associated proteins (MAPs), guanosine triphosphate (GTP) and ethylene glycol-bis(β-aminoethyl ether)-N,N´-tetraacetic acid (EGTA) (Horwitz et al. 1979; Horwitz 2004).

Since 1992, taxol has been well established and approved by FDA (Food and Drug Administration) as a very important effective chemotherapeutic agent against a wide range of tumours such as ovarian, breast, and lung carcinoma. In addition, taxol has demonstrated promising activity against Kaposi sarcoma, bladder, prostate, esophageal, head and neck, cervical, and endometrial cancers (Suffness 1995; Rowinsky et al. 1992; Rowinsky 1997; Miele et al. 2012; Venook et al. 1998). Taxol quickly became a huge commercial success-with annual sales peaking at nearly $1.6 billion in 2000.

At present taxol routinely used in combination with other drugs such as anthracyclines (anti-tumor antibiotics) and trastuzumab (Henderson et al. 2003;

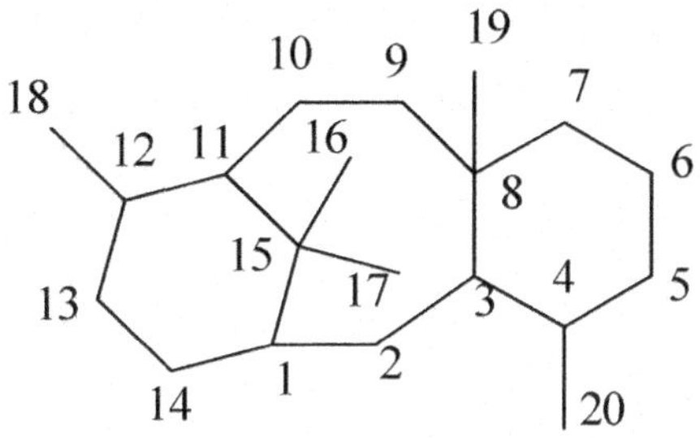

Fig. 9.1 Taxoid ring
system and numbering

Tubiana-Hulin 2005). Another analog of taxol - docetaxel is structurally differs from taxol in the substituents at the C10- and C30-positions and exhibits different clinical efficacies than those of taxol (Verweij et al. 1994). Docetaxel has exhibited improved potency for microtubule stabilization and better antimitotic activity than taxol (Gueritte-Voegelein et al. 1991) and has been found to be effective in malignancies that are resistant to taxol (Valero et al. 1998; Verschraegen et al. 2000).

9.8 Phytochemical aspects of taxol

The chemical structure of taxol was published in 1971 (Wani et al. 1971) and it was identified as a highly functionalized taxane diterpene amide -[(1S,2S,3R,4S,7R,9S,10S,12R,15S)-4,12-diacetyloxy-15-[(2R,3S)-3-benzamido-2-hydroxy-3-phenylpropanoyl]oxy-1,9-dihydroxy-10,14,17,17-tetramethyl-11-oxo-6-oxatetracyclo[11.3.1.03,10.04,7]heptadec-13-en-2-yl] benzoate with molecular formula $C_{47}H_{51}NO_{14}$, corresponding to molecular weight of 853.9 g/mol. Taxol consists of a taxane nucleus to which an uncommon four-membered oxetane ring was linked to C-4 and C-5, and an ester was attached at the C-13 position which is essential for its antitumor action (Table 9.1). At present, more than 550 taxanes have been isolated from leaves, stems, roots, bark, seeds, twigs, and branches of various *Taxus* spp. (Miele et al. 2012). Taxoides are secondary metabolites mainly found in *T. brevifolia*. They constitute a group of diterpene compounds characterized by a typical tricyclic skeleton. These taxane possess the same skeleton structure of three rings (Fig. 9.1) but carry different substituents.

T. brevifolia contains taxanes, coumarinolignans, flavanolignans, terpenolignan, xanthanolignans, sterols etc. (Arslanian et al. 1995) (Table 9.1).

9.9 Conclusions

Taxus brevifolia is a high value medicinal plant as a source of taxol. Taxol is still considered as one of the most used anticancer plant agents emerged from massive drug screening program of the National Cancer Institute (NCI). At this time NCI

Table 9.1 Chemical structures of secondary metabolites isolated from *T. brevifolia*

| Taxol | | Wani et al. (1971) |
| Brevifoliol | | Balza et al. (1991) and Chu et al. (1993a) |

7,13-Dideacetyl-9,1O-debenzoyltaxchinin C		Chen and Kingston (1994)
9-Deacetyl-9-benzoyl-1O-debenzoylbrevifoliol		Chen and Kingston (1994)

(continued)

Table 9.1 (continued)

5β,20-Epoxy-1β-hydroxy-4α,7β,13α-triacetoxy-2α,9α,10β-tribenzoxy-tax-11-ene	R_1 = Ac, R_2=R_3 = COPh	Chu et al. (1992)
2α,10β-Dibenzoxy-5β,20-epoxy-1β-hydroxy-4α,7β,9α,13α-tetraacetoxy-tax-11-ene	R_1 = R_2 = Ac, R_3 = COPh	Chu et al. (1992)
2α,7β-Dibenzoxy-5β,20-epoxy-1β-hydroxy-4α,9α,10β,13α-tetraacetoxy-tax-11-ene	R_1 = COPh, R_2 = R_3 = Ac	Chu et al. (1992)
1β,9α-Dihydroxy-4β,20-epoxy-2α,5α,7β,10β,13α-pentaacetoxy-tax-11-ene	R_1 = H, R_2 = Ac	Chu et al. (1993b)
1β,7β-Dihydroxy-4β,20-epoxy-2α,5α,9α,10β,13α-pentaacetoxy-tax-11-ene	R_1 = Ac, R_2 = H	Chu et al. (1993b)

Compound	Structure	Reference
10β-Benzoxy-5α-(3'-dimethylamino-3'-phenyl)-propanoxy-1β-hydroxy-7β,9α,13α-triacetoxy-11(15→1)-abeotaxa-4(20),11-diene	R =	Chu et al. (1994)
10β-Benzoxy-1β-hydroxy-5α-(3'-methylamino-3'-phenyl)-propanoxy)-7β,9α,13α-triacetoxy-11(15→1)-abeotaxa-4(20),11-diene	R =	Chu et al. (1994)
10β-Benzoxy-5α-cinnamoxy-1β-hydroxy-7β,9α,13α-triacetoxy-11(15→1)-abeotaxa-4(20),11-diene	R =	Chu et al. (1994)
10β-Benzoxy-1β,5α-dihydroxy-7β,9α,13α-triacetoxy-11(15→1)-abeotaxa-4(20),11-diene	R = H	Chu et al. (1994)
Taxinine M		Beutler et al. (1991)

(continued)

Table 9.1 (continued)

Baccatin III		Hoke et al. (1994)
Cephalomannine		Hoke et al. (1994)
7-epi-Taxol		Huang et al. (1986)

10-Deacetyl-10-oxo-7-epi-taxol		Huang et al. (1986)
Taxa-4(20),11-diene-2α,5α,7β,9α,10β,13α-hexaol 2,7,9,10,13-pentaacetate (decinnamoyltaxinine J)		Kingston et al. (1982)
10-Deacetylbaccatin III		Kingston et al. (1982)

(continued)

Table 9.1 (continued)

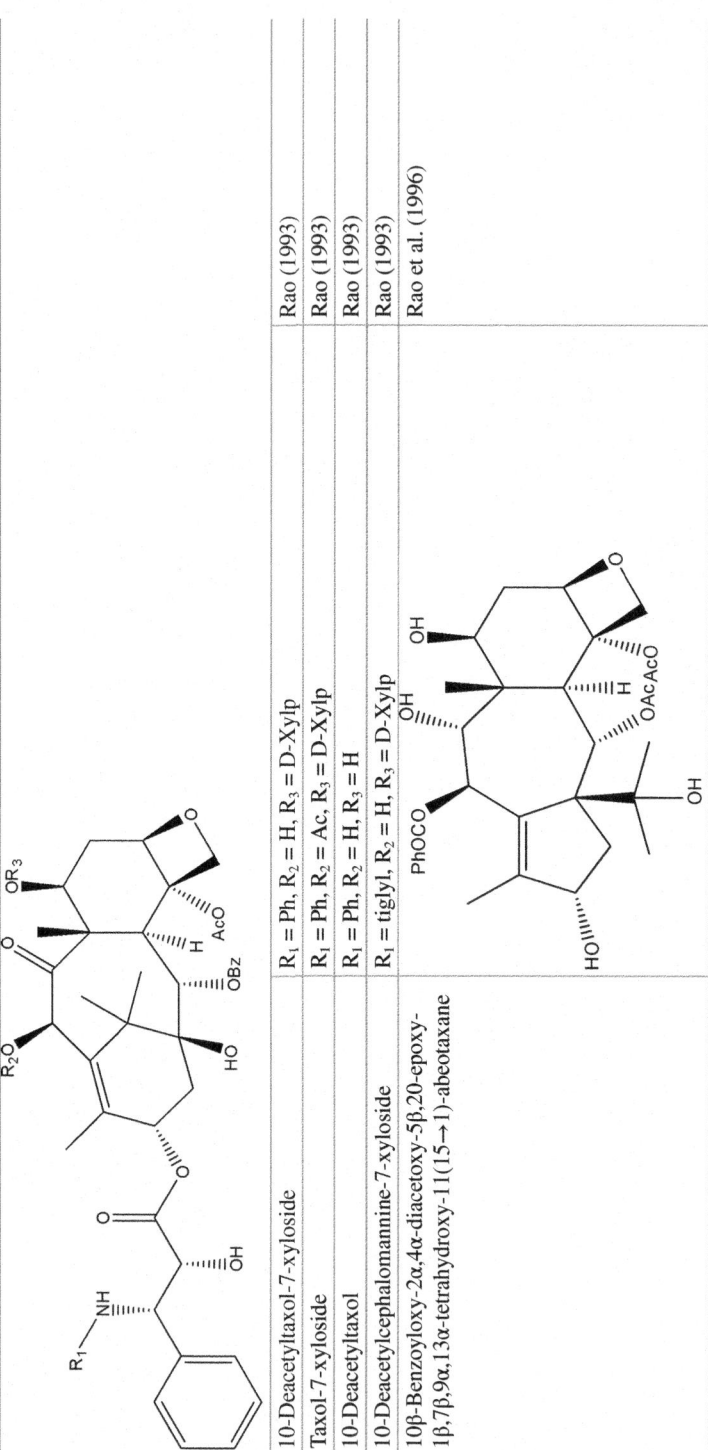

10-Deacetyltaxol-7-xyloside	R_1 = Ph, R_2 = H, R_3 = D-Xylp	Rao (1993)
Taxol-7-xyloside	R_1 = Ph, R_2 = Ac, R_3 = D-Xylp	Rao (1993)
10-Deacetyltaxol	R_1 = Ph, R_2 = H, R_3 = H	Rao (1993)
10-Deacetylcephalomannine-7-xyloside	R_1 = tiglyl, R_2 = H, R_3 = D-Xylp	Rao (1993)
10β-Benzoyloxy-2α,4α-diacetoxy-5β,20-epoxy-1β,7β,9α,13α-tetrahydroxy-11(15→1)-abeotaxane		Rao et al. (1996)

Compound		Reference
1β,7β-Dihydroxy-4β,20-epoxy-2α,5α,9α,10β,13α-penta-acetoxy-tax-11-ene		Rao et al. (1996)
7,9-Deacetyl baccatin IV	$R_1 = OAc$, $R_2 = H$, $R_3 = H$, $R_4 = OAc$	Rao et al. (1996)
Baccatin IV	$R_1 = OAc$, $R_2 = OAc$, $R_3 = OAc$, $R_4 = OAc$	Rao et al. (1996)
9-Dihydro-13-acetyl baccatin III	$R_1 = OAc$, $R_2 = H$, $R_3 = H$, $R_4 = OBz$	Rao et al. (1996)

(continued)

Table 9.1 (continued)

| Ponasterone A | | Rao et al. (1996) |
| 2(R)-Hydroxy-4-(4'-hydroxyphenyl)-butane | | Chu et al. (1994) |

Brevitaxin		Arslanian et al. (1995)
β-Sitosterol	R = H	Rao et al. (1995)
β-Sitosterol-3-O-β-D-glucoside	R = β-D-Glcp	Rao et al. (1995)

(continued)

continues to support basic research that explores how Taxol works in different cell types and how it can be used to treat other types of cancer. Taxol has shown success in treating cancers alone and in combination with other drugs. Taxol is obtained not only from the bark of Pacific yew tree, but from the needles of some other yew species, as well. The discovery of this tree's anticancer activity and the isolation and identification of its main bioactive compound, taxol, is among the most successful breakthroughs in the history of medicinal drug development.

References

Appendino G, Cravotto G, Enriu R, Gariboldi E, Barbo L, Torregiani E, Gabetta B, Zini G, Bombardelli E (1994) Taxoids from the Roots of *Taxus* × *media* cv. Hicksii. J Nat Prod 57:607–613

Arslanian RL, Bailey DT, Kent MC, Richheimer SL, Thornburg KR, Timmons DW, Zheng QY (1995) Brevitaxin, a new diterpenolignan from the bark of *Taxus brevifolia*. J Nat Prod 58:583–585

Balza F, Tachibana S, Barrios H, Towers GHN (1991) Brevifoliol, a taxane from *Taxus brevifolia*. Phytochemistry 30:1613–1416

Barboni L, Gariboldi P, Torregiani E, Appendino G, Gabetta B, Bombardelli E (1994) Taxol analogues from the roots of *Taxus x media*. Phytochemistry 36:987–990

Beutler JA, Chmurny GM, Look SA, Witherup KM (1991) Taxinine M, a new tetracyclic taxane from *Taxus brevifolia*. J Nat Prod 54:893–897

Bringi V, Kadkade PG, Prince CL, Roach BL (2007) Enhanced production of taxol and taxanes by cell cultures of *Taxus* species. US 7264951 B1 20070904

Chen R, Kingston DGI (1994) Isolation and structure elucidation of new taxoids from *Taxus brevifolia*. J Nat Prod 57:1017–1021

Chowdhary K, Jangir M, Sharma S (2017) Endosymbionts: a continuing source of cytotoxic metabolite taxol. Int J Adv Res 5:1–8

Christen AA, Bland J, Gibson DM (1989) Cell cultures as a means to produce Taxol. Proc Am Assoc Cancer Res 30:566

Christen AA, Gibson DM, Bland J (1991) Production of Taxol or taxol-like compounds with *Taxus brevifolia* callus cell culture. US Patent No 5019504

Chu A, Zajicek J, Davin LB, Lewis NG, Croteau RB (1992) Mixed acetoxy-benzoxy taxane esters from *Taxus brevifolia*. Phytochemistry 31:4249–4252

Chu A, Zajicek J, Towers GHN, Soucy-Breau CM, Lewis NG, Croteau R (1993a) Brevifoliol: A structure revision. Phytochemistry 34:269–271

Chu A, Davin LB, Zajicek J, Lewis NG, Croteau R (1993b) Intramolecular acyl migrations in taxanes from *Taxus brevifolia*. Phytochemistry 34:473–476

Chu A, Furlan M, Davin LB, Zajicek J, Towers GHN, Soucy-Breau CM, Rettig SJ, Croteau R, Lewis NG (1994) Phenylbutanoid and taxane-like metabolites from needles of *Taxus brevifolia*. Phytochemistry 36:975–985

Cragg GM, Schepartz SA, Suffness M, Grever MR (1993) The taxol supply crisis. New NCI policies for handling the large-scale production of novel natural product anticancer and anti-HIV agents. J Nat Prod 56:1657–1668

Denis JN, Greene AE, Guenard D, Gueritte-Voegelein F, Mangatal L, Potier P (1988) A highly efficient, practical approach to natural taxol. J Am Chem Soc 110:5917–5919

Dumontet C, Sikic BI (1999) Mechanisms of action of and resistance to antitubulin agents: Microtubule dynamics, drug transport, and cell death. J Clin Oncol 17:1061–1061

Fu Y, Li S, Zu Y, Yang G, Yang Z, Luo M, Jiang S, Wink M, Efferth T (2009) Medicinal chemistry of paclitaxel and its analogues. Curr Med Chem 16:3966–3985

Fuchs DA, Johnson RK (1978) Cytologic evidence that taxol, an antineoplastic agent from *Taxus brevifolia*, acts as a mitotic spindle poison. Cancer Treat Rep 62:1219–1222

Fumoleau P, Seidman AD, Trudeau ME, Chevallier B, Ten Bokkel Huininik WW (1997) Docetaxel: a new active agent in the therapy of metastatic breast cancer. Expert Opin Investig Drugs 6:1853–1865

Gan F, Zheng G, Peng L, Luo J (1997) Study on cell suspension culture of *Taxus yunnanensis*. Acta Phytophysiol Sinica 23:43–46

Gueritte-Voegelein F, Guenard D, Lavelle F, LeGoff M-T, Mangatal L, Potier P (1991) Relationships between the structure of taxol analogues and their antimitotic activity. Jour. Med Chem 34:992–998

Hamburger M, Hostettmann K (1991) Bioactivity in plants: the link between phytochemistry and medicine. Phytochemistry 30:3864–3874

Henderson IC, Berry DA, Demetri GD, Cirrincione CT, Goldstein LJ, Martino S, Ingle JN, Cooper MR, Hayes DF, Tkaczuk KH, Fleming G, Holland JF, Duggan DB, Carpenter JT, Frei E, Schilsky RL, Wood WC, Muss HB, Norton L (2003) Improved outcomes from adding sequential paclitaxel but not from escalating doxorubicin dose in an adjuvant chemotherapy regimen for patients with node-positive primary breast cancer. J Clin Oncol 21:976–983

Hoke SH, Cooks RG, Chang CJ, Kelly RC, Qualls SJ, Alvarado B, McGuire MT, Snader KM (1994) Determination of taxanes in Taxus bervifolia extracts by MS/MS and HPLC. J Nat Prod 57:277–286

Holton RA, Somoza C, Kim HB, Liang F, Biediger RJ, Boatman PD, Shindo M, Smith CC, Kim S, Nadizadeh H, Suzuki Y, Tao C, Yu P, Tang S, Zhang P, Murthi KK, Gentile LN, Liu JH (1994) First total synthesis of Taxol. J Am Chem Soc 116:1597–1600

Horwitz SB (2004) Personal recollections on the early development of taxol. J Nat Prod 67:136–138

Horwitz SB, Fant J, Schiff PB (1979) Promotion of microtubule assembly *in vitro* by taxol. Nature 277:665–667

Hosie RC (1969) Native trees of Canada, 7th edn. Canadian Forestry Service, Department of Fisheries and Forestry, Ottawa, ON, 380 p

Hu YM, Gan FY, Lu CH, Ding HS, Shen YM (2003) Production of taxol and related taxanes by cell suspension cultures of *Taxus yunnanensis*. Acta Bot Sin 45:373–378

Huang CHO, Kingston DGI, Magri NF, Samaranayake G, Boettner FE (1986) New taxanes from *Taxus brevifolia*, 2. J Nat Prod 49:665–669

Ji Y, Bi J-N, Yan B, Zhu X-D (2006) Taxol-producing fungi: A new approach to industrial production of taxol. Chin J Biotech 22:1–6

Ketchum REB, Luong JV, Gibson DM (1999) Efficient extraction of paclitaxel and related taxoids from leaf tissue of *Taxus* using a potable solvent system. J Liq Chromatog Related Tech 22:1715–1732

Kikuchi Y, Yatagai M (2003) The commercial cultivation of *Taxus* species and production of Taxoids. In: Itokawa H, Lee KH (eds) Taxus: the Genus *Taxus*. CRC Press, pp 151–178

Kingston DGI, Hawkins DR, Ovington L (1982) New taxanes from *Taxus brevifolia*. J Nat Prod 45:466–470

Li JY, Stroble G, Sidhu R, Hess WM, Ford EJ (1996) Endophytic taxol producing fungi from bald cypress, *Taxodium distichum*. Microbiology 142:2223–2226

Little EL Jr (1979) Checklist of United States trees (native and naturalized). Agric. Handb. 541. U.S. Department of Agriculture, Forest Service, Washington, DC, 375 p

Miele M, Anna Mumot AM, Zappa A, Romano P, Ottaggio L (2012) Hazel and other sources of paclitaxel and related compounds. Phytochem Rev 11:211–225

Morihira K, Hara R, Kawahara S, Nishimori T, Nakamura N, Kusama H, Kuwajima I (1998) Enantio-selective total synthesis of Taxol. J Am Chem Soc 120:12980–12981

Moss RW (1998) Herbs against Cancer: history and controversy Equinox Press Inc, 300 p

Neto AGF, Di Cosmo F (1992) Distribution and amounts of taxol in different shoot parts of *Taxus cuspidata*. Planta Med 58:464–466

Nicolaou KC, Yang Z, Liu JJ, Ueno H, Nantermet PG, Guy RK, Claiborne CF, Renaud J, Couladouros EA, Paulvannan K (1994) Total synthesis of taxol. Nature 367:630–634

Park YK, Chung ST, Row KH (1999) Preparative chromatographic separation of Taxol from yew trees. J Liq Chromatogr Related Technol 22:2577–2761

Preston RJ Jr (1948) North American trees. The Iowa State College Press, Amess, IA, 371 p

Rao KV (1993) Taxol and related taxanes. I Taxanes of *Taxus brevifolia* bark. Pharm Res 10:521–524

Rao K, Hanuman JB, Alvarez C, Stoy M, Juchum J, Davies RM, Baxley R (1995) A new large-scale process for taxol and related taxanes from *Taxus brevifolia*. Pharm Res 12:1003–1010

Rao KV, Bhakuni RS, Hanuman JB, Davies R, Johnson J (1996) Taxanes from the bark of *Taxus brevifolia*. Phytochemistry 41:863–866

Rowinsky EK (1997) The development and clinical utility of the taxane class of anti-microtubule chemotherapy agents. Annu Rev Med 48:353–374

Rowinsky EK, Onetto N, Canetta RM, Arbuck SG (1992) Taxol: the first of the taxanes, an important new class of anticancer agents. Semin Oncol 19:646–662

Senilh V, Blechert S, Colin M, Guenard D, Picot F, Potier P (1984) Mise en evidence de Nouveaux analogues du taxol extraits de *Taxus baccata*. J Nat Prod 47:131–137

Stierle A, Strobel G, Stierle D (1993) Taxol and taxane production by *Taxomyces andreanae*, an endophytic fungus of pacific yew. Science 260:214–216

Strobel G, Yang XS, Sears J, Kramer R, Sidhu RS, Hess WM (1996) Taxol from *Pestalotiopsis microspora*, an endophytic fungus of *Taxus wallachiana*. Microbiology 142:435–440

Suffness M (1995) Taxol: Science and Applications. CRC Press, New York

Theodoridis G, Verpoorte R (1996) Taxol analysis by high performance liquid chromatography: a review. Phytochem Anal 7:169–184

Tubiana-Hulin M (2005) How to maximize the efficacy of taxanes in breast cancer. Cancer Treat Rev 31:S3–S9

Vaishampayan U, Parchment RE, Jasti BR, Hussain M (1999) Taxanes: an overview of the pharmacokinetics and pharmacodynamics. Urology 54:22–29

Valero V, Jones SE, von Hoff DD, Booser DJ, Mennel RG, Ravdin PM, Holmes FA, Rahman Z, Schottstaedt MW, Erban JK, Esparza-Guerra L, Earhart RH, Hortobagyi GN, Burris HA 3rd. (1998) A phase II study of docetaxel in patients with paclitaxelresistant metastatic breast cancer. Jour Clin Oncol 16:3362–3368

Velde DGV, Georg GI, Gollapudi SR, Jampani HB, Liang X-Z, Mitscher LA, Ye Q-M (1994) Wallifoliol, a taxol congener with a novel carbon skeleton, from Himalayan *Taxus wallichiana*. J Nat Prod 37:862–867

Venook AP, Egorin MJ, Rosner GL, Brown TD, Jahan TM, Batist G, Hohl R, Budman D, Ratain MJ, Kearns CM, Schilsky RL (1998) Phase I and pharmacokinetic trial of paclitaxel in patients with hepatic dysfunction: Cancer and Leukemia Group B 9264. J Clin Oncol 16:1811–1819

Verschraegen C, Sittisomwong T, Kudelka A, Guedes E, Steger M, Nelson-Taylor T, Vincent M, Rogers R, Atkinson E, Kavanagh J (2000) Docetaxel for patients with paclitaxel resistant Mullerian carcinoma. J Clin Oncol 18:2733–2739

Verweij J, Clavel M, Chevalier B (1994) Paclitaxel (taxol) and docetaxel (Taxotere): not simply two of a kind. Ann Oncol 5:495–505

Vidensek N, Lim P, Campbell A, Carlson C (1990) Taxol content in bark, wood, root, leaf, twig, and seedling from several *Taxus* species. J Nat Prod 53:1609–1615

Wang JF, Li GL, Lu HY, Zheng ZH, Huang YJ, Su WJ (2000) Taxol from *Tubercularia sp.* strain TF5, an endophytic fungus of *Taxus mairei*. FEMS Microbiol Lett 193:249–253

Wani MC, Taylor HL, Wall ME, Coggon P, McPhail AT (1971) Plant antitumor agents. VI. The isolation and structure of taxol, a novel antileukemic and antitumor agent from *Taxus brevifolia*. J Am Chem Soc 93:2325–2327

Chapter 10
Dr. Duke's Phytochemical and Ethnobotanical Databases, a Cornerstone in the Validation of Ethnoveterinary Medicinal Plants, as Demonstrated by Data on Pets in British Columbia

Cheryl Lans and Tedje van Asseldonk

Abstract In the 1990s we published Ethnoveterinary Medicine (EVM) research in veterinary journals for the first time. Many of the plants used were either considered weeds or were not considered at all. The lack of scientific research on the plants being published resulted in Dr. Duke's Ethnobotanical database being utilised to fill in the scientific gaps. This chapter reports on the plants used for urinary and other problems in pets. Cushings /hyperadrenocorticism was treated with ginkgo leaf (*Ginkgo biloba*). Benign prostatic hyperplasia in dogs was treated with saw palmetto (*Serenoa repens*) fruit. *Aphanes arvensis*, *Hydrangea arborescens* and *Urtica dioica* were used as kidney tonics. The understudied plants parsley-piert (*Aphanes arvensis*) and *Hydrangea arborescens* were evaluated using Dr. Duke's ethnobotanical database.

Keywords Urinary problems · Pets · Ethnoveterinary medicine · British Columbia

10.1 Introduction Background

In the 1990s we had EVM (Ethnoveterinary Medicine) research published in established veterinary journals for the very first time. The journals were not very concerned with long Latin names of plants and which families the plants belonged to, they were more concerned about the safety and efficacy of the plant remedies.

C. Lans (deceased) · T. van Asseldonk (✉)
Institute for Ethnobotany & Zoopharmacognosy, Beek, The Netherlands
e-mail: info@ethnobotany.nl

© Springer Nature Switzerland AG 2020
Á. Máthé (ed.), *Medicinal and Aromatic Plants of North America*, Medicinal and Aromatic Plants of the World 6,
https://doi.org/10.1007/978-3-030-44930-8_10

Establishing the treatment value of the plant remedies in the 1990s was not always easy since many of the plants used were either considered weeds or had limited studies available such as brine shrimp lethality assays. In order to assess the usefulness of the EVM remedies Dr. Duke's Ethnobotanical database was utilized to fill in the scientific gaps. The database had been online since 1972. The database had several functions but the ones we used most commonly were:

1. obtaining a list of chemicals and activities for a specific plant, using its scientific name
2. producing a list of plants with chemicals known for a specific biological activity
3. producing a list of plants with the highest levels of a specific chemical

The EVM remedies were evaluated for safety and efficacy with a non-experimental method that included Dr. Duke's ethnobotanical databases as an alternative to the systematic reviews that were gaining popularity in the scientific literature. Many systematic reviews eliminate the poorly-funded studies conducted in developing countries on medicinal plants because they do not fit the criteria established in rich countries for being of high quality. However, these studies still contained information that could be useful to resource-poor farmers who need alternatives to First World drugs.

This chapter reports on the plants used for kidney, liver and related problems in cats and dogs, and demonstrates the use of Dr. Duke's databases in the discussion section. In 2003 participatory research on the ethnoveterinary remedies (medicinal plants used for animals) utilized in British Columbia, Canada (BC) was begun (Lans et al. 2006).

Benign prostatic hyperplasia was one of the conditions treated in dogs in BC. Steenkamp et al. (2006) reported that prostatitis (inflammation of the prostate, acute bacterial prostatitis, chronic bacterial prostatitis and chronic non-bacterial prostatitis), and benign prostatic hyperplasia (BPH), enlargement of the prostate due to non-cancerous growth within the gland, are common prostate disorders. Chronic non-bacterial prostatitis is common and is usually caused by fungi, mycoplasmas or viruses. Acute bacterial prostatitis is caused by a urinary tract infection, typically *Escherichia coli*, which has spread to the prostate, and chronic bacterial prostatitis is often the result of partial blockage of the male urinary tract, as occurs with BPH, with the resulting proliferation of bacteria. BPH has two phases, one with no clinical signs and the second showing as disorders of urination resulting from urinary tract obstruction by an enlarged prostate (Steenkamp et al. 2006). A hormone-induced chronic inflammation that results from the infiltration of inflammatory cells (and prostaglandins, leukotrienes and growth factors) into the prostate is often associated with BPH.

Studies have suggested that phytotherapy with agents such as a lipidosterolic extract of *Serenoa repens* can provide equivalent efficacy to tamsulosin in BPH treatment. In a comparison with finasteride, *Serenoa repens* extract had little effect on androgen-dependent parameters, but both treatments relieved the symptoms of BPH in two thirds of patients (Hutchison et al. 2007).

Addison's disease and Cushing's syndrome were treated by a veterinarian in this study. Essential hypertension with raised blood pressure is increasing worldwide. 'Secondary' hypertension with a known cause is linked to adrenal corticosteroids (Hammer and Stewart 2006). Hypoadrenocorticism (Addison's disease) is caused by deficient secretion of endogenous glucocorticoid and mineralocorticoids (aldosterone), with many resulting clinical signs that may be mistaken for renal failure and various gastrointestinal disorders (Meeking 2007; Greco 2007). Naturally occurring primary hypoadrenocorticism is usually caused by immune-mediated destruction of the adrenal cortex in both cats and dogs. Secondary hypoadrenocorticism, in which the pituitary gland produces inadequate amounts of adrenocorticotrophic hormone (ACTH), can be caused by chronic steroid therapy or less commonly by tumors, trauma, or congenital defects of the pituitary gland. Inadequate secretion of corticosteroids in Addison's disease causes life threatening hypotension, whereas mineralocorticoid and glucocorticoid excess in primary hyperaldosteronism and Cushing's syndrome, respectively, result in high blood pressure (Hammer and Stewart 2006). Cushing's disease results from glucocorticoid excess or is an indirect result of obesity (Meij et al. 1997). Dogs with the pituitary dependent Cushing's disease have benign tumours but half of the tumours in adrenal Cushing's Syndrome are malignant (Zeugswetter et al. 2007).

10.2   Materials and Methods

Ethnoveterinary data for British Columbia was collected over a 6-month period, in 2003. and analyzed in the subsequent months. The research area in British Columbia consisted of the most populated areas: the Lower Mainland, the Thompson/Okanagan region and south Vancouver Island (Lans et al. 2006).

A purposive sample of livestock farmers and pet owners (n = 60) was created in order to obtain the key informants with the ethnoveterinary knowledge sought. The sample was obtained from membership lists of organic farmers, other specialists in alternative medicine and holistic veterinarians. The information on pets came from two naturopaths, ten herbalists, five dog trainers, breeders and pet shop owners, nine holistic veterinarians and the majority of twenty-seven organic farmers who had pets.

Two visits were made to each respondent. All of the first interviews were open-ended and unstructured. A draft outline of the respondents' ethnoveterinary remedies was then prepared, delivered and subsequently discussed at the second visit in order to confirm the data provided at the first interview and to obtain more details on the treatments. Medicinal plant voucher specimens were collected whenever possible and were identified and deposited in the University of Victoria Herbarium.

The plant-based remedies were evaluated for safety and efficacy with a non-experimental method, prior to including them in the draft outline. This evaluation process then continued throughout the data collection until the publication of the results. Journal articles, books and databases on pharmacology, ethnoveterinary medicine and ethnomedicine available on the Internet (PubMed, Science Direct) were searched to identify the plants' chemical compounds and clinically tested physiological effects. This type of ethnopharmacological review and evaluation is based on previous work and the use of these methods in the same research study has been published (Lans et al. 2006). If a plant had been poorly studied then Dr. Duke's Phytochemical and Ethnobotanical Databases were consulted to evaluate their efficacy. For this chapter parsley-piert (*Aphanes arvensis*) and *Hydrangea arborescens* are considered poorly studied. The compounds found in *Aphanes arvensis* were not documented by Duke (2017), only its use as a tonic in Spain. In its original formulation the databases could be searched in several different ways. There was a plant search with all of the available research conducted on the plant and the chemical compounds found in the plant. Another search option was to search for plants containing a particular compound. This latter search would list the plants with the highest quantities of the relevant chemical first. This would be useful if a well-researched plant contained a chemical that was considered relevant for the EVM treatment under discussion. The understudied plant would then be searched to see if it had the relevant compound, and/or the compound would be searched to see if the understudied plant was in the list of plants containing the compound. The non-experimental validation of the plants is presented in the discussion section of this chapter.

10.3 Results

Several plants were used for urinary and other problems in pets, both as fresh preparations and as purchased products. Some of the plants, for example parsley piert (*Aphanes arvensis*), juniper berries (*Juniperus communis*), *Hydrangea arborescens* and goldenrod (*Solidago virgaurea*) have historic roots in European and North American indigenous knowledge. Table 10.1 documents the plants used for kidney, liver and other urinary problems in pets.

10.3.1 Treatments for Benign Hyperplasia and Prostate Support in General

Benign prostatic hyperplasia in dogs was treated with ¼ tsp. saw palmetto (*Serenoa repens*) fruit extract every day for 4 days (39 kg dog). The treatment was stopped for a few days and then restarted. A tea was made using 1 tbsp of berries in 0.46 L of water; 57 ml of the tea was put in the drinking water.

Table 10.1 Plants used for kidney, liver and urinary problems in pets in British Columbia

Scientific name	Common name	Plant part used	Uses
Aphanes arvensis (Rosaceae)	Parsley piert	Whole herb	Kidney tonic
Althaea officinalis L. (Malvaceae)	Marshmallow	Root	Ingredient in commercial kidney and blood pressure tea
Arctium lappa L. (Asteraceae)	Burdock	Roots	Liver tonic
Arctostaphylos uva ursi L. (Ericaceae)	Uva ursi	Berry, leaf	Ingredient in commercial kidney and blood pressure tea, spay related incontinence, cystitis, urinary problems
Berberis aquifolium Pursh. *Mahonia aquifolium* (Berberidaceae)	Oregon grape	Root	Spay related incontinence
Curcuma longa L. (Zingiberaceae)	Turmeric	Root	Ingredient in commercial liver products
Cynara scolymus L. (Asteraceae)	Artichoke	Leaf	Ingredient in purchased liver product A, tonic
Dioscorea villosa L. (Dioscoreaceae)	Wild yam	Root	Ingredient in commercial liver products, Addison's disease
Eleutherococcus senticosus (Rupr. & Maxim.) Maxim. (Araliaceae)	Siberian ginseng	Root	Addison's disease, ingredient in commercial liver products
Equisetum palustre L. (Equisetaceae)	Horsetail	Aerial parts	Ingredient in kidney and blood pressure tea
Eupatorium purpureum L. (Asteraceae)	Gravel root	Whole plant, roots	Ingredient in kidney and blood pressure tea
Ginkgo biloba L.	Ginkgo	Leaf	Cushing's disease, hyperadrenocorticism
Glycyrrhiza glabra L. (Fabaceae)	Licorice	Root	Addison's disease, ingredient in commercial liver product A
Hydrangea arborescens L. Hydrangeaceae	Hydrangea	Root	Kidney tonic
Hydrastis canadensis L. (Ranunculaceae)	Goldenseal	Root	Ingredient in commercial kidney and blood pressure tea
Juniperus communis L. (Cupressaceae)	Juniper	Berry	Ingredient in commercial kidney and blood pressure tea, urinary problems
Melissa officinalis L. (Lamiaceae)	Lemon balm	Aerial parts	Toner for urinary problems
Petroselinum crispum (P. Mill.) Nyman ex A.W. Hill(Apiaceae)	Parsley	Aerial parts	Urinary problems
Rehmannia glutinosa (Gaertn.) Steud. (Gesneriaceae)	Rehmannia	Root	Ingredient in purchased liver tonic A

(continued)

Table 10.1 (continued)

Scientific name	Common name	Plant part used	Uses
Rosmarinus officinalis L. (Lamiaceae)	Rosemary	Leaf	Ingredient in commercial liver tonic B
Rumex crispus L. (Polygonaceae)	Curly dock	Roots	Liver support, urinary problems
Schisandra chinensis (Turcz.) Baill. (Schisandraceae)	Schisandra	Fruit	Ingredient in commercial liver tonic B
Serenoa repens (Bartr.) Small (Arecaceae)	Saw palmetto	Fruit	Benign hyperplasia & prostate support
Silybum marianum (L.) Gaertn. (Asteraceae)	Milk thistle	Seed	Ingredient in commercial liver products A & B
Solidago virgaurea L., (Asteraceae)	Goldenrod	Aerial parts	Ingredient in commercial kidney and blood pressure tea
Taraxacum officinale (L.) Weber (Asteraceae)	Dandelion	Root	Ingredient in commercial liver product A
Urtica dioica L. (Urticaceae)	Nettles	Roots, leaves	Kidney tonic, malformed kidneys, ingredient in kidney and blood pressure tea
Vaccinium macrocarpon Aiton (Ericaceae)	Cranberry	Fruit juice	Urinary problems, cystitis
Vaccinium myrtillus L. (Ericaceae)	Bilberry	Leaves	Urinary problems
Zea mays L. (Poaceae)	Corn	Silk	Cat straining to urinate, kidney tonic
Zingiber officinale Roscoe (Zingiberaceae)	Ginger	Rhizome	Ingredient in commercial kidney and blood pressure tea

10.3.2 Plants Used as a Kidney Tonic

A tonic for kidney complaints consisted of parsley-piert (*Aphanes arvensis*), root of *Hydrangea arborescens*, cornsilk (*Zea mays*) and nettles (*Urtica dioica*). Chopped hydrangea root (28 gm) was soaked overnight. The next day a decoction was made with 1 L of water (in a glass pot, simmered for 15 min). The decoction was supplemented with 29 gm each of parsley piert, cornsilk and nettles added in the last 5 min of simmering. A dog was given 57 ml of the strained decoction twice daily (per 11 to 14 kg patient bodyweight) for 5 days.

10.3.3   Treatment for a Dog Born with a Tiny Malformed Kidney

This dog was given a tincture of nettles aerial parts (*Urtica dioica*) in 30% alcohol as a kidney tonic (10 drops twice daily per 23 kgs bodyweight for 12 months). This remedy and other homeopathic remedies (not presented) were claimed to have extended the life of the pet by 10 months.

10.3.4   Treatment for Cushings/Hyperadrenocorticism

Cushings/hyperadrenocorticism was diagnosed by a veterinarian (an excess of cortisol produces the clinical signs excessive urination and thirst, panting). Patient: 8 year old female Doberman X, 50kgs. The treatment below was used in one case. Capsules of ginkgo leaf (*Ginkgo biloba*) by mouth (purchased human brand name product – 500 mg capsules. The dose used was 2/3 human dose). After 3 weeks of treatment the veterinarian noticed less panting.

10.3.5   Treatments for Urinary Problems

A case study of incontinence linked to spaying with blood in the urine. A veterinarian diagnosed the dog's condition. The pet was given a purchased uva-ursi (*Arctostaphylos uva-ursi*) tincture (20 drops twice a day). The pet was also given an Oregon grape root tincture (*Berberis aquifolium*) 1:1 twice a day to clear up bacteria in the urine (the 16 kg animal was given 10 drops/day for 5 days, repeated after 5 days if necessary). The first 5 days was sufficient in this case.

10.3.6   Treatment for Pets with Liver Problems

Pets with liver problems were given 60–80% alcoholic tincture of milk thistle seed (*Silybum marianum*) (30 drops twice a day). Or the powdered seed was added to the pet's food. Or seeds were ground and placed in capsules before administration. A tincture was made with ¼ cup seeds in 284 ml of vodka, brandy or rum (46 kg dog got 2.5 ml four times daily). This treatment was considered an early response to chronic liver problems.

Another 27 kg dog was given 30 drops of each purchased herb tincture below per day (or 1 drop per 1 kg body weight per day by mouth): licorice root (*Glycyrrhiza glabra*), Siberian ginseng root (*Eleutherococcus senticosus*), and wild yam root (*Dioscorea villosa*). An adrenal glandular supplement was also used as a supportive treatment (given indefinitely). This pet was also given a purchased capsule that contained these ingredients: milk thistle seed (*Silybum marianum*), dandelion root (*Taraxacum officinale*), artichoke leaf (*Cynara scolymus*) and turmeric root (*Curcuma longa*).

10.3.7   Treatment for a Dog with Addison's Disease

Patient: A Bouvier male that was 6 years old and 34 kgs. Addison's disease (early hypoadrenocorticism) was diagnosed by a veterinarian. The symptoms were weakness, vomiting and diarrhoea at times of stress. Single tinctures were used. The dose was 30 drops tincture of each herb per day (ie: 1 drop per 1 kg body weight per day orally). The tinctures were made from licorice root (*Glycyrrhiza glabra*), Siberian ginseng root/eleuthero root (*Eleutherococcus senticosus*) and wild yam root (*Dioscorea villosa*). Episodes of weakness, vomiting/diarrhoea were resolved after 2 weeks using the herbs. These episodes were intermittent for the previous 3 years and did not recur in the 6 months after using the herbs. An adjunctive treatment was also used. This consisted of a dried purchased product which is an adrenal glandular supplement (the pet took the product for 3 years prior to this research).

10.3.8   Treatment for Urinary Problems

A purchased product was used for urinary problems that contained uva-ursi/bearberry (*Arctostaphylos uva-ursi*), bilberry (*Vaccinium myrtillus*) and juniper berries (*Juniperus communis*) (200 milligrams per 9 kgs. body weight). This combination of herbs was not given during pregnancy or for serious kidney problems and was not given for longer than 2 weeks. For uva-ursi's effectiveness the pet's urine was made alkaline by including the following in the pet's diet: vegetables, fruit, fruit juices and potatoes. Pets were given lots of water to ease elimination and prevent renal calculi.

One dog breeder used a purchased kidney and blood pressure tea prepared by a local herbalist. The ingredients are listed in Table 10.1.

Other treatments and dosages for kidney and liver problems are given in Table 10.2.

10.4   Discussion

The non-experimental validation of the plants is presented in Table 10.3.

It provides evidence for the ethnoveterinary remedies. Some medicinal plants used are discussed in more detail below.

Table 10.2 Treatments and dosages for kidney and liver problems in pets in British Columbia

Plant name	Condition	Tea preparation	Tincture preparation	Dosages
Arctium lappa	Liver tonic	Decoction of 1 tsp of thinly-sliced, fresh or dried burdock root in 224 ml water		250 ml per 46 kg patient bodyweight put in the pet's food until healthy
Arctostaphylos uva-ursi	Cystitis			
Melissa officinalis	Toner for urinary problems			2 tbsp of the herb in 114 ml of boiling water per 11–14 kg patient bodyweight). Given as the drinking water
Rumex crispus	Urinary problems	250 ml water and 1 cup chopped, fresh or dried yellow dock root		1/4 cup of the liquid a day per 11–14 kg patient bodyweight for 2 days
Rumex crispus	Liver problems		A 1:5 tincture was made in 50% alcohol. Or a 1:2 tincture of fresh root in 50% alcohol.	3 g of dried root. 1:5 tincture: 15 drops /day of the tincture per 46 kg patient bodyweight. 1:2 tincture-10 drops /day per 46 kg patient bodyweight
Silybum marianum	Liver problems			23 kg dog got ground milk thistle seed 50 mg per 4.5 kgs twice a day. Broccoli and brussel sprouts were fed.
Vaccinium macrocarpon	Urinary problems			100–200 mg powder for 4.5 kg cats and 300 mg for 14 kg dogs three to 4 days a month (10 mg/0.5 kg patient bodyweight), given with food.
Vaccinium macrocarpon	Cystitis in cats	Put in the water or was given as a liquid drench		2 drops put on their nose
Zea mays	Cats with bloody urine	¼ cup cut corn silk in 0.45 L of boiling water		
Zea mays	Cats with bloody urine	Chopped fresh corn silk was fed raw		

Table 10.3 Non-experimental validation of plants used for kidney and liver problems in pets in British Columbia

Medicinal plant	Validation information	References
Aphanes arvensis	A kidney tonic in our research, it showed no inhibitory effect in a prostaglandin test but did show PAF-inhibiting activity. An open-label study was conducted in 48 male and female patients aged 4–44 years found that a standard 3% extract of *Alchemilla vulgaris* in glycerine (Aphtarine) cured common minor oral ulcers. The active constituents in *A. vulgaris* include polyphenols, and its pharmacological activities include inhibition of intercellular matrix degrading proteases, antioxidant and angioprotective activities. The plant contains tannins (gallic and ellagic acid), flavonoids (quercetin, luteolin) and proanthocyanidins.	Tunón et al. (1995), Shrivastava and John (2006), and Shrivastava et al. (2007)
Althaea officinalis	An ingredient in a kidney/blood pressure tea given to pets. Primary chemical constituents of marshmallow include mucilage, polysaccharides, flavonoids (quercetin, kaaempferol), asparagine, tannins, lecithin, pectin and phenolic acids. Marshmallow contains large amounts of vitamin A, calcium, zinc and significant amounts of iron, sodium, iodine, and B-complex. *A. officinalis* had microbial activity against *E. coli*. Marshmallow may interfere with hypoglycaemic therapy and so should not be fed to diabetic animals.	De Smet and D'Arcy (1996) and Watt et al. (2007)
Arctium lappa	Used as a liver tonic *Arctium lappa* improved hepatic outcome in rats fed liquid ethanol and injected with carbon tetrachloride, both histopathologically and using biochemical parameters. It also has free radical scavenging activity in rats similarly treated with carbon tetrachloride. Arctigenin, a lignan in *Arctium lappa*, has antioxidant, antimicrobial and anti-inflammatory activities. *Arctium lappa* also has moderate antibacterial and anticandidal activity.	Jellinek and Maloney (2005), Cho et al. (2004), and Holetz et al. (2002)

(continued)

Table 10.3 (continued)

Medicinal plant	Validation information	References
Arctostaphylos uva-ursi	In our research *Arctostaphylos uva-ursi* was used for urinary problems, spay related incontinence and cystitis. *Arctostaphylos uva-ursi* is a well-known urinary antiseptic and anti-adhesion herb. Aqueous extracts of *Arctostaphylos uva-ursi* increased urine flow in rats. An extract of *Arctostaphylos uva-ursi* significantly reduced the MICs of beta-lactam antibiotics, such as oxacillin and cefmetazole, against methicillin-resistant *Staphylococcus aureus*. The active compound, corilagin, showed a synergistic bactericidal action when added to the growth medium in combination with oxacillin. This herb should not be administered with any substances that cause acidic urine, such as ascorbic acid and ammonium chloride, as it will reduce the antibacterial effect.	Yarnell (2002); Beaux et al. (1999), Shimizu et al. (2001), and Blumenthal (2000)
Berberis vulgaris	*Berberis aquifolium* was used for spay related incontinence. The pharmacologic actions of berberine include inhibition of intestinal fluid accumulation and ion secretion, inhibition of smooth muscle contraction and reduction of inflammation. *In vitro* studies showed that berberine inhibited activator protein 1 (AP-1), a key transcription factor in inflammation and carcinogenesis. In another study berberine had a direct affect on several aspects of the inflammatory process. It exhibited dose-dependent inhibition of arachidonic acid release from cell membrane phospholipids, inhibition of thromboxane A2 from platelets, and inhibition of thrombus formation.	Anon (2000)
Curcuma longa	Turmeric was an ingredient in commercial liver products. There is a strong correlation between antioxidant activity and antiinflammatory activities of curcuminoids. Curcumin can alleviate liver damage. Curcumin given intraperitoneally up-regulated HO-1 protein expression as well as activity. The induction of HO-1 expression by curcumin provides protection against hepatic damage associated with induced oxidative stress in rats. Curcumin can attenuate liver injury induced by diverse hepatotoxicants via multiple mechanisms. Curcumin administration increased several cytoprotective enzymes, especially in the liver.	Darshan and Doreswamy (2004), Vanisree and Sudha (2006), El-Ashmawy et al. (2006), and Farombi et al. (2007)

(continued)

Table 10.3 (continued)

Medicinal plant	Validation information	References
Cynara scolymus	Part of a purchased liver tonic in our study, artichoke leaf (*Cynara scolymus*) contains cynarin (choleretic activity), natural phenolic antioxidants such as hydroxycinnamic acids and flavones. A commerical extract was effective in MDA levels in liver as well as in serum enzymes levels; the extract with the highest content in phenolic derivatives (GAE) exerted the major effect on bile flow and liver protection. The different extracts had good antioxidant capacity. Consumption of wild artichoke (*C. cardunculus*) by healthy subjects may protect them from hyperglycemia and insulin resistance	Duke (1990), Speroni et al. (2003), Nomikos et al. (2007), and Goñi et al. (2005)
Dioscorea villosa	Wild yam root was used for liver problems and Addison's disease in our study. The mucilage of yam tubers contain soluble glycoprotein, fiber, and steroidal saponins (diosgenin) with an effect on lipid metabolism. In an animal study bitter yam sapogenin extract or commercial diosgenin did not significantly alter faecal magnesium, calcium, and zinc excretion but significantly decreased faecal sodium and potassium excretion. The absorption of iron was impaired by bitter yam sapogenin extract or commercial diosgenin during the first week of feeding. Bitter yam sapogenin extract or commercial diosgenin supplements significantly decreased intestinal lipids towards normal showing that it could improve insulin resistance in rats receiving fructose-rich chow.	Benghuzzi et al. (2003), Omoruyi et al. (2006), Chen et al. (2003), Hsu et al. (2007), and Ajdžanović et al. (2017)
Eleutherococcus senticosus	Pets were treated with *Eleutherococcus senticosus* for Addison's disease and it was an ingredient in commercial liver products. *E. senticosus* stems moderated fulminant hepatic failure in a mouse model and the active ingredients are anthocyanins, flavonoids, phenolic acids and phenols (eleutherosides).	Blumenthal (2000), Anon (2006), Park et al. (2004), and Zaluski et al. (2017)

(continued)

Table 10.3 (continued)

Medicinal plant	Validation information	References
Equisetum arvense	Kidney problems were treated with *Equisetum telmateia*. The petroleum extract had an MIC value of 39.1 µg/ml against *Staphylococcus epidermis*. *Equisetum arvense* has demonstrated hypoglycaemic and diuretic activity. The hydroalchoholic extract of stems of *Equisetum arvense* produced an antinociceptive effect and anti-inflammatory effects. The anti-inflammatory effects may be due to beta-sitosterol, campesterol and isofucosterol. The flavonoid isoquercitrin has anti-oxidant effects. A standardized extract from horsetail increased hippuric acid, the glycine conjugate of benzoic acid, in the urine of 11 volunteers.	Uzun et al. (2004), Gurbuz et al. (2002), Do Monte et al. (2004), and Graefe and Veit (1999)
Eupatorium purpureum	A kidney and blood pressure tea included *Eupatorium purpureum*. Cistifolin from the antirheumatic herb, gravel root (rhizome of *Eupatorium purpureum*), showed activity both in *in vitro* and *in vivo* models of inflammation.	Habtemariam (2001)
Gentiana lutea	Gentian roots were used for digestive problem in pets. The methanolic extracts of the aerial parts and the roots of *G. scabra* both have certain hepatoprotective effects on acute liver injury models.	Jiang and Xue (2005)
Ginkgo biloba	Ginkgo was used for Cushing's disease in our research. Ginkgolide B (normally obtained from the inner root bark) shortens the half-life of cortisol produced by the adrenals. It has been shown to have a regulatory effect on glucocorticoid levels. On rat tissue its inhibitory effect was limited to the adrenal cortex, corticosterone levels were decreased but aldosterone levels did not change. *Ginkgo biloba* extract also has immunostimulatory activity. *Ginkgo biloba* phytosomes (25 mg/kg & 50 mg/kg i.p.) have hepatoprotective activity similar to 200 mg/kg p.o. silymarin.	Papadopoulos et al. (1998), Wynn and Marsden (2003), Amri et al. (2003), Villaseñor-García et al. (2004), and Sultana et al. (2018)

(continued)

Table 10.3 (continued)

Medicinal plant	Validation information	References
Glycyrrhiza glabra	Licorice was used for Addison's disease and was present in commercial liver products such as Livatona. Stronger neominophagen C (SNMC) is a Japanese preparation that contains 0.2% glycyrrhizin, 0.1% cysteine, and 2% glyceine. SNMC has anti-inflammatory activity and is used as an cytoprotective drug. It suppressed hepatic steatosis in ovariectomized mice. It reduces mortality in patients with subacute liver failure and improves liver functions in patients with subacute hepatic failure, chronic hepatitis, and cirrhosis. Glycyrrhizin is the active compound of licorice (*Glycyrrhiza glabra*) and can be used long term for chronic hepatitis C patients and may alleviate alcohol-related liver disease.	Dhiman and Chawla (2005), Lin et al. (2008), Baselga-Escudero et al. (2017), Sultana et al. (2018), and Kleiner et al. (2016)
Hydrangea arborescens	Hydrangea formed one ingredient in a purchased kidney tonic used for pets. The fermented and dried leaves of *Hydrangea macrophylla* SER. var. *thunbergii* MAKINO, suppressed D-galactosamine-induced liver injury by 85.2% when added to the diet at 1% and fed to rats for 15 days. The hepatoprotective effect was more potent than that of a milk thistle extract and turmeric powder. The water-ethanol extract was active suggesting that the active compounds are lipophilic. Hydrangenol, phyllodulcin, thunberginol A, and hydrangeaic acid from the processed leaves of *H. macrophylla* var. *thunbergii* promoted adipogenesis in 3 T3-L1 cells. Hydrangenol significantly lowered blood glucose and free fatty acid levels 2 weeks after its administration at a dose of 200 mg/kg/d.	Nakagiri et al. (2003) and Zhang et al. (2007a)
Hydrastis canadensis	Goldenseal was used for urinary problems in pets. Berberine extracts and decoctions have demonstrated significant antimicrobial activity against a variety of organisms including bacteria, viruses, fungi, protozoans and chlamydia. The predominant clinical uses of berberine include bacterial diarrhea. Goldenseal can modulate the antigen-specific immune response, by enhancing the acute primary IgM response.	Anon (2000)
Juniperus communis	Juniper was used for kidney problems in pets. The aqueous ethanol extract of *Juniperus communis* (bark) showed pancreatic lipase inhibitory activity. The hydro-alcoholic extracts of branches of *Juniperus sabina* showed inhibitory activity against MDA-MB-468.	Kim and Kang (2005) and Madari and Jacobs (2004)

(continued)

Table 10.3 (continued)

Medicinal plant	Validation information	References
Melissa officinalis	*Melissa officinalis* was used for urinary problems in pets. Major bacteriostatic effects were exerted against Grampositive bacteria, *S. aureus* and *S. epidermidis*, by rosmarinic acid (MIC 0.12 mg/mL) and against *B. spizizenii* (MIC 0.5 mg/mL) by both of the dried leaf extracts EtOH-H2O (1:1), and *n*-BuOH. In the case of gram-negative bacteria, yeasts, and molds the MICs were greater than 2.0 mg/mL. No antimicrobial effect was observed for rosmarinic acid at lower concentrations (5–250 μg/mL) in previous work. Rosmarinic acid is the major bioactive compound present in *Melissa* extracts.	Uzun et al. (2004), Savino et al. (2005), and Mencherini et al. (2007)
Petroselinum crispum	Urinary problems in pets are treated with parsley. There is evidence for the diuretic effect of parsley (*Petroselinum crispum*). Rats offered an aqueous parsley seed extract to drink, eliminated a significantly larger volume of urine per 24 h as compared to when they were drinking water. Other experiments using an *in situ* kidney perfusion technique demonstrated a significant increase in urine flow rate with parsley seed extract. The mechanism of action of parsley seems to be mediated through an inhibition of the Na + -K+ pump that would lead to a reduction in Na + and K+ reabsorption leading thus to an osmotic water flow into the lumen, and diuresis.	Kreydiyyeh and Usta (2002)
Rehmannia glutinosa	*Rehmannia glutinosa* is found in purchased liver preparations. Alone and in combination with Astragalus it produced improved proteinuria, hematuria and renal function.	Geraghty (2000) and Kang et al. (2005)
Rosmarinus officinalis	*Rosmarinus officinalis* is an ingredient in commercial liver preparations used for pets. Rosemary (*Rosmarinus officinalis*) showed significant antithrombotic activity *in vitro* and *in vivo*. Both water and alcohol extracts of *Rosmarinus officinalis* have hepatoprotective effects against CCL4-induced liver damage. The liver protectant effects were attributed to antioxidant compounds in RO water extract (rosmarinic acid, diterpenoids like carnosic acid, carnosol, rosmanol and epirosmanol, carotenoids and alpha-tocopherol. A related plant *Rosmarinus tomentosus*, protected rat liver in an experimental model of cirrhosis.	Yamamoto et al. (2005), Amin and Hamza (2005), and Galisteo et al. (2006)

(continued)

Table 10.3 (continued)

Medicinal plant	Validation information	References
Rumex crispus	*Rumex crispus* was used for liver support and urinary problems in pets. Crude extracts of leaves of *Rumex nervosus* and the root of *Rumex abyssinicus* have antibacterial activity against *Streptococcus pyogenes* and *Staphylococcus aureus* and activity against Coxsackie virus B3. The methanol extract of the entire plant in Saudi Arabia possessed no activity against *E. coli, P. vulgaris, S. aureus* and *P. aeruginosa* and *C. albicans*. A similar extract of the leaves of the Mexican species showed weak activity against *E. coli, P. aeruginosa, S. aureus, Bacillus subtilis* and *C. albicans*. The 50% methanol of the flowers and leaves of the Nigerian species showed antibacterial activity against *B. subtilis, E. coli, Proteus* species, *P. aeruginosa* and *S. aureus*. Roots of *R. patientia* have significant antioxidant, hepatoprotective and antihyperproliferation properties.	Getie et al. (2003) and Lone et al. (2007)
Schisandra chinensis	An incredient in a commercial liver tonic used for pets, Schisandra (*Schisandra chinensis*) is a plant adaptogen or compound that increases the ability of an organism to adapt to environmental factors and to avoid damage from such factors. The SAS-mediated stimulating effects of single doses of adaptogens derived from *Rhodiola rosea, Schizandra chinensis* and *Eleutherococcus senticosus* were reviewed. The active principles of the three plants that exhibit single dose stimulating effects are glycosides of phenylpropane- and phenylethane-based phenolic compounds such as salidroside, rosavin, syringin and triandrin, the latter being the most active. Male Sprague-Dawley rats (n = 8–12 per group) were divided into the control, CCl4, CCl4 + silymarin (0.35%), CCl4 + low-dose herbal extract (0.24% of *Ginkgo biloba, Panax ginseng*, and *Schizandra chinensis* extract at 1:1:1; LE), and CCl4 + high-dose herbal extract (1.20% of the same herbal extract; HE) groups. Silymarin or herbal extract was orally given to rats a week before chronic intraperitoneal injection with CCl4 for 6 weeks. The high dose herbal extract improved hepatic antioxidant capacity through enhancing catalase activity and glutathione redox status, but the low-dose herbal extract inhibited liver fibrosis through decreasing hepatic TGF-β1 level in rats with CCl4-induced liver injury.	Panossian and Wagner (2005) and Chang et al. (2007)

(continued)

Table 10.3 (continued)

Medicinal plant	Validation information	References
Serenoa repens	Saw palmetto is traditionally used for benign hyperplasia and prostate support and has this use for pets. Extracts from the fruits of saw palmetto (*Sabal serrulata*, syn. *Serenoa repens*) and the roots of stinging nettle (*Urtica dioica*) are the most commonly used. Short-term randomised trials and some metaanalyses in the recent literature suggest clinical efficacy and good tolerability for extracts from *Serenoa repens* and also *Pygeum africanum*, products with high concentrations of beta-sitosterol, and pumpkin seeds. Studies have claimed that the efficacy of an extract from *S. repens* is comparable to that of finasteride and alpha-blockers.	Koch (2001) and Dreikorn (2002)
Silybum marianum	Milk thistle was found in commercial liver products given to pets. Legalon is the name of one commercial product. *Silybum marianum* and some of its derivatives promote liver cell regeneration. Silymarin (including silychristin, silybin), has shown hepatoprotective action in animals (dogs, rabbits, rats, mice) at a dose of 15 mg/kg of silymarin. Oral administration of silybin to dogs in a phosphatidylcholine complex (phytosome) produced greater bioavailability of silybin compared to a standard silymarin extract.	Boerth and Strong (2002), Lieber et al. (2003), Ball and Kowdley (2005), Desplaces et al. (1975), Dhiman and Chawla (2005), Filburn et al. (2007), Chassagne et al. (2018), Sultana et al. (2018), and Latief and Ahmad (2018)
Solidago canadensis	Goldenrod was used for urinary problems in pets. In European phytotherapy *Solidago canadensis* were used to treat chronic nephritis, cystitis, urolithiasis, rhematism and as an antiphlogistic. Compounds found in the plant include flavonoids, sesquiterpenes, diterpenes, triterpenes, saponins and a phenolic glycoside. Assessment of peroxyl radical scavenging activity showed *Solidago canadensis* extracts were greater than green tea, ascorbic acid and Trolox.	Zhang et al. (2007b) and McCune and Johns (2002)
Taraxacum officinale	Dandelions were given to pets for liver problems. The European Scientific Cooperative on Phytotherapy (ESCOP) recommends dandelion root (*Taraxacum officinale*) for kidney and liver problems, indigestion and loss of appetite. The German Commission E authorizes the use of combination products containing dandelion root and herb for biliary abnormalities, appetite loss, dyspepsia, and for stimulation of diuresis (urine flow). *Taraxacum officinalis* leaves contain potassium that replaces any lost from the body.	Petlevski et al. (2003), NSRC (2004), and Hoffman (2003)

(continued)

Table 10.3 (continued)

Medicinal plant	Validation information	References
Urtica dioica	The efficacy and tolerability of a fixed combination of 160 mg sabal fruit (burdock) extract WS 1473 and 120 mg urtica root extract WS 1031 per capsule (PRO 160/120) was investigated in elderly, male patients suffering from lower urinary tract symptoms (LUTS) caused by benign prostatic hyperplasia in a prospective multicenter trial in 257 patients (127 and 126 were evaluable for efficacy). Using the International Prostate Symptom Score (I-PSS), patients treated with PRO 160/120 had a higher total score reduction after 24 weeks of double-blind treatment than the placebo group. This applied to obstructive as well as to irritative symptoms, and to patients with moderate or severe symptoms at baseline. PRO 160/120 was clearly superior to the placebo for the amelioration of LUTS as measured by the I-PSS. *Urtica urens* can protect the body against environmental toxins.	Koch (2001), Petlevski et al. (2003), Lopatkin et al. (2005), Blumenthal (2000), and Ozkarsli et al. (2008)
Vaccinium macrocarpon s	Bilberry leaves were used to treat pets with urinary problems. Cranberry was used for urinary problems. Extracts of *Vaccinium angustifolium* possess insulin-like and glitazone-like activities and show anti-diabetic effects in pancreatic beta cells. Foliage of huckleberry (*Vaccinium* spp.) is fairly high in carotene, manganese, and energy. Bilberry extract protected against KBrO3-induced kidney damage, the active constituents may be anthocyanins.	Martineau et al. (2006), Petlevski et al. (2003), Acuña et al. (2002), and Bao et al. (2008)
Zea mays	As for humans, corn silk was used for urinary problems in pets. Studies have shown *in vivo* diuretic and hypotensive activity. Corn silk contains amines, fixed oils, saponins, tannins, bitter glycosides, allantoin, cryptoxanthin, flavone and phytosterols including beta-sitosterol and stigmasterol. The last two compounds are known to have antiinflammatory activity *in vivo* and may have a beneficial effect in treating prostate problems.	Habtemariam (1998), Maksimović et al. (2004), and Velazquez et al. (2005)

(continued)

Table 10.3 (continued)

Medicinal plant	Validation information	References
Zingiber officinale	Pets with urinary problems were treated with ginger. The rhizome of ginger contains curcumin in addition to a dozen phenolic compounds known as gingerols and diarylheptanoids. These compounds possess significant antioxidant activity. The ethanol extract of *Z. officinale* alone and in combination with vitamin E provided significant protection against cisplatin-induced nephrotoxicity in rats which was evident from the lowered serum urea and creatinine levels. However the concentration of urea and creatinine in the *Z. officinale* treated groups were not normalized even in the 500 mg/kg treated group. The nephroprotection exhibited by *Z. officinale* can be correlated with its GST stimulating activity in the kidneys (the glutathione-s-transferase (GST) group of enzymes plays a major role in the detoxification pathway). Ginger may also protect the liver of mice from damage.	Craig (1999), Bub et al. (2006), Ajith et al. (2007), and Verma and Asnani (2007)

Previous studies have shown that the polyphenol curcumin (in *Curcuma* spp) can decrease the degree of inflammation associated with experimental colitis (Camacho-Barquero et al. 2007).

Conflicting results have been found in studies dealing with the benefits of curcumin in different nephrotoxic models of kidney injury. ADR-induced kidney injury in rats was prevented by treatment with curcumin (Venkatesan et al. 2000). Curcumin prevented renal lesions in streptozotocin diabetic rats and reduces ischaemic renal injury. In cyclosporin-induced acute renal failure, curcumin suspended in low concentration of 5, 10 and 15 mg/kg for 21 days, attenuated oxidative stress as well as BUN and serum creatinine (Vlahović et al. 2007). Quercetin and its sugar conjugates are commonly found bioflavonoids (Chander et al. 2005). In a randomized, placebo-controlled trial, a combination of curcumin 480 mg and quercetin 20 mg was administered orally in capsule form to cadaveric kidney transplant recipients for 1 month, starting immediately after transplantation. The investigators observed better early graft function in treated patients than in controls (71% [lowdose] versus 93% [high-dose] versus 43% [controls]) and the high-dose regimen lowered the incidence of acute graft rejection at 6 months post-transplantation (0% versus 14.3%) (Goel et al. 2007).

Artichoke (*Cynara scolymus*) is traditionally used against hepato-biliary diseases (Speroni et al. 2003). This activity is attributed to the antioxidative activity of artichoke extracts or to active compounds such as flavones, flavanones, flavonols, coumarins and phenolic acids. Acute hepatotoxicity induces inflammation, necrosis and oxidative stress of hepatocytes. Scientists currently believe that hepatocellular injury is not caused by the damaging agent itself but by the inflammatory cells that have been attacked by the stressed hepatocytes (Muriel and Rivera-Espinoza 2007). Artichoke's choleretic activity is attributed to mono- and di-caffeoylquinic acids.

Goñi et al. (2005) attributes the plants beneficial effects on intestinal health to artichoke's indigestible fraction which may enhance the activity of beneficial species in the colonic microbiota. Artichoke (chlorogenic acid, cynaroise and luteolin) has been shown to improve the lipid profile of patients with liver disease by reducing the LDL, TC and TG serum levels (Santos et al. 2018).

Boldocynara is a commercial product composed of *Cynara scolymus*, *Silybum marianum*, *Taraxacum officinale* and *Peumus boldus* (Villiger et al. 2015). *Silybum marianum* had moderate pancreatic lipase (PL) inhibitory activities (30% at 100 μg/ml) and *Cynara scolymus* produced moderate ACE inhibitory activity (31% at 100 μg/ml). *Taraxacum officinale* had weak alpa-glucosidase activity (15% at 100 μg/ml) and weak XO inhibitory effects (Villiger et al. 2015).

Eleutherococcus senticosus, an adaptogen with immunomodulative activity attributed to polysaccharides, is widely used to treat hepatitis (Park et al. 2004). The hot water extract of *E. senticosus* reduced the levels of aspartate transaminase and alanine transaminase in serum and improved histological changes of liver in induced liver injury in rats. *E. senticosus* inhibits tumour necrosis factor and iplatelet aggregation. Hot water extracts are also reported to inhibit mast cell-mediated anaphylaxis, induce apoptosis in human stomach cancer KATO III cells, and protect against gastric ulcers in stressed rats. Park et al. (2004) concluded that water-soluble polysaccharides of *E. senticosus* stems have protective effects against induced fulminant hepatic failure in mice.

Hydrangea paniculata showed renal protection activity in an animal model (Zhang et al. 2017). Possible active compounds were esculentin, loganin, skimmin and umbelliferone. Dr. Duke's database shows loganin being present in *Hydrangea arborescens*.

Juniperus communis leaves and stems contain seven catechins, fourteen flavonoids, nine phenylpropanoids, six neolignans and five terpenoids (Iida et al. 2010). *Juniperus communis* berries contain (+)-isocupressic acid, (+)-junenol, (+)-sugiol, 1-(1,4-Dimethyl-3-cyclohexen-1-yl) ethanone, 6-hydroxyluteolin, 6-xyloside, 6-xyloside, afzelechin, bilobetin, vommunic acid, elliotinol, epiafzelechin, geijerone, junicedral, junionone, junipercomnoside A, junipercomnoside B, longifolene, quercetin, 3-O(−)rhamnoside, sciadopitysin and scutellarein (Li et al. 2018). Duke (2017) lists (+)-afzelichin, (+)-beta-elemen—alpha-ol, (+)-catechin (in fruit and leaves), and (+)-epigallocatechin. See the EMA assessment report on *Juniperus communis* for more information (EMA/HMPC/441930/ 2008).

Serenoa serrulata berries were used in the treatment of both BPH and prostatitis. The hydroxyl scavenging activity of this herbal remedy is a factor because free radicals and oxidative stress are associated with inflammation. Hydroxyl radical scavengers are reported to suppress upregulation of cyclooxygenase (COX) and subsequently reduce inflammation. Over-expression of COX-2 and a decrease in prostaglandin E-1 synthesis takes place in patients with BPH and prostatitis and the aqueous extracts of *Serenoa serrulata* showed a COX-2 inhibitory effect in the research of Steenkamp et al. (2006). The Gram negative bacteria *Escherichia coli*, is a pathogen in cases of BPH and prostatitis, but the crude extract of *Serenoa serrulata* had no effect against it. A large study with men found that *Serenoa repens* was not better than placebo (Bent et al. 2006). The study was criticized for the type of supplement used and for using patients with moderate-to-severe BPH symptoms, when *Serenoa serrulata* is recommended for mild-to-moderate symptoms (ABC 2006).

Herbal remedies based on goldenrod (*Solidago virgaurea* L.) have been used for centuries to treat urinary tract diseases, and it was included in a multi-plant kidney and blood pressure tea in our research. Diuretic activity was shown when the flavonoid fractions of *Solidago virgaurea* L. and *S. gigantea* Ait., were given to rats (Chodera et al. 1991). Extracts from *Solidago virgaurea* inhibited muscarinic receptor-mediated contraction of rat and human bladder strips (Borchert et al. 2004). Melzig (2004) claims that herbal preparations with a complex action spectrum (anti-inflammatory, antimicrobial, diuretic, antispasmodic, analgesic) should be used for treatment of infections and inflammations, to prevent formation of kidney stones and to help remove urinary gravel. Melzig (2004) concludes that the therapy is safe at a reasonable price and does not show drug-related side-effects. See the EMA Assessment report on *Solidago virgaurea* for more information (EMEA/HMPC/285759/ 2007). The European Medicines Agency report is listed under the title Solidaginis virgaureae herba.

Vaccinium myrtillus is rich in anthocyanins, tannis and phenypropanoids (chlorogenic acid) (Nohara et al. 2018). Fifteen anthocyanins are present, consisting of five glycoside classes of cyanidin, delphinidin, peonidin, petunidin and malvidin (Veljković et al. 2017). The anthocyanins and their antioxidant activity are responsible for the renoprotective effect of bilberry extracts (Veljković et al. 2017). Kurkin et al. (2016) found that the fresh fruit extract had a diuretic effect but the air-dried fruit extract and the shoot extract did not. The aqueous extract did not have a significantly different diuretic effect than the control. The fresh fruit extract had kaliuretic and natriuretic actions and increased renal creatinine excretion which is linked to an increase in glomerula filtration.

10.5   Conclusion

The plants used for urinary and kidney problems showed efficacy in the non-experimental validation process. This efficacy is plausible based on effects that were established *in vitro*, or in animal models, other than pets. Traditional satisfactory use as listed in Dr. Duke's database can indicate these remedies may also be used for pets. Of course definitive proof of safety and efficiency can be obtained with more clinical results and experimental tests on target animals.

Quercetin, an extract from the plant *Ginkgo biloba* ameliorates myoglobin attenuated renal functions due to their antioxidant and free radicals scavenging properties (Vlahović et al. 2007; Chander et al. 2005).

Curcumin has multiple therapeutic activities that are beneficial for kidney function (Venkatesan et al. 2000). Curcumin is safe even when consumed at a daily dose of 12 g for 3 months (Goel et al. 2007). Conventional medicine does not provide many therapeutic agents for hepato- biliary diseases (Aktay et al. 2000; Muriel and Rivera-Espinoza 2007); and holistic and other practitioners rely on *Silybum marianum* and *Cynara scolymus*, two of the plants used in this research. *Silybum marianum* is the most commonly used plant-based treatment for chronic or acute liver disease without major side-effects. Clinical trials have shown that silymarin, is effective in treating various forms of liver disease, including cirrhosis, hepatitis, necroses, and liver damage induced by drug and alcohol abuse (Lee et al. 2006). It has several activities (antioxidative, antilipid peroxidative, membrane stabilizing, immunomodulatory and liver regeneration) (Muriel and Rivera-Espinoza 2007). Silymarin is also useful for drug-induced nephrotoxicity that produces oxidative stress and leads to renal failure in dogs (Varzi et al. 2007).

It is also useful to have alternative treatments for Addison's disease and Cushing's syndrome. The literature review revealed that standardized *Ginkgo biloba* extract and its purified components gingkolide A or ginkgolide B can regulate glucocorticoid levels by controlling adrenal PBR expression at the transcriptional level. The extract and its purified component ginkgolide B decrease stress-induced elevations of serum corticosterone without affecting physiological basal levels and the extract contains an additional component that can regulate ACTH secretion (Amri et al. 2004).

Dr. Duke sent more than 1000 questionnaires asking for published and unpublished data on medicinal plants and received more than 500 replies. His inclusive and systematic approach helped provide necessary information to assess the safety and efficacy of EVM remedies in the early stages of EVM research.

References

American Botanical Council (2006. New clinical trial on saw palmetto inconsistent with positive results in previous studies. http://www.herbalgram.org/default.asp?c=sawpalmBPH

Acuña UM, Atha DE, Ma J, Nee MH, Kennelly EJ (2002) Antioxidant capacities of ten edible North American plants. Phytother Res 16:63–65

Ajith TA, Nivitha V, Usha S (2007) *Zingiber officinale* Roscoe alone and in combination with alpha-tocopherol protect the kidney against cisplatin-induced acute renal failure. Food Chem Toxicol 45:921–927

Aktay G, Deliorman D, Ergun E, Ergun F, Yeşilada E, Çevik C (2000) Hepatoprotective effects of Turkish folk remedies on experimental liver injury. J Ethnopharmacol 73:121–129

Anon (2000) Berberine monograph. Altern Med Rev 5:175–177

Anon (2006) Monograph. Eleutherococcus senticosus. Altern Med Rev 11:151–155

Ajdžanović V, Jarić I, Miler M, Filipović B, ŠoŠić-Jurjević B, Ristić N, Milenkovic D et al (2017) Diosgenin-caused changes of the adrenal gland histological parameters in a rat model of the menopause. Acta Histochem 119(1):48–56

Amin A, Hamza AA (2005) Hepatoprotective effects of *Hibiscus*, *Rosmarinus* and *Salvia* on azathioprine-induced toxicity in rats. Life Sci 77:266–278

Amri H, Drieu K, Papadopoulos V (2003) Transcriptional suppression of the adrenal cortical peripheral-type benzodiazepine receptor gene and inhibition of steroid synthesis by ginkgolide B. Biochem Pharmacol 65:717–729

Amri H, Li W, Drieu K, Papadopoulos V (2004) Identification of the adrenocorticotropin and ginkgolide B-regulated 90-kilodalton protein (p90) in adrenocortical cells as a serotransferrin precursor protein homolog (adrenotransferrin). Endocrinology 145:1802–1809

Ball KR, Kowdley KV (2005) A review of *Silybum marianum* (milk thistle) as a treatment for alcoholic liver disease. J Clin Gastroenterol 39:520–528

Bao L, Yao XS, Tsi D, Yau CC, Chia CS, Nagai H, Kurihara H (2008) Protective effects of bilberry (*Vaccinium myrtillus* L.) extract on KBrO3-induced kidney damage in mice. J Agric Food Chem 56:420–425

Baselga-Escudero L, Souza-Mello V, Rachid T, Pascual-Serrano A, Voci A, Demori I, Grasselli E et al (2017) Beneficial effects of the Mediterranean spices and aromas on non-alcoholic fatty liver disease. Trends Food Sci Technol 61:141–159

Beaux D, Fleurentin J, Mortier F (1999) Effect of extracts of *Orthosiphon stamineus* Benth, *Hieracium pilosella* L., *Sambucus nigra* L. and *Arctostaphylos uva-ursi* (L.) Spreng. In rats. Phytother Res 13:222–225

Benghuzzi H, Tucci M, Eckie R, Hughes J (2003) The effects of sustained delivery of diosgenin on the adrenal gland of female rats. Biomed Sci Instrum 39:335–340

Bent S, Kane C, Shinohara K, Neuhaus J, Hudes ES, Goldberg H, Avins AL (2006) Saw palmetto for benign prostatic hyperplasia. N Engl J Med 354:557–566

Blumenthal M (2000) Interactions between herbs and conventional drugs: introductory considerations. In: Herbs—everyday reference for health professionals. Ottawa, Canadian Pharmacists Association and Canadian Medical Association, pp 9–20

Boerth J, Strong KM (2002) The clinical utility of milk thistle (*Silybum marianum*) in cirrhosis of the liver. J Herbal Pharmacother 2:11–17

Borchert VE, Czyborra P, Fetscher C, Goepel M, Michel MC (2004) Extracts from *Rhois aromatica* and *Solidaginis virgaurea* inhibit rat and human bladder contraction. Naunyn Schmiedeburgs Arch Pharmacol 369:281–286

Bub S, Brinckmann J, Cicconetti G, Valentine B (2006) Efficacy of an herbal dietary supplement (smooth move) in the management of constipation in nursing home residents: a randomized, double-blind, placebo-controlled study. J Am Med Dir Assoc 7:556–561

Camacho-Barquero L, Villegas I, Sánchez-Calvo JM, Talero E, Sánchez-Fidalgo S, Motilva M, Alarcón de la Lastra C (2007) Curcumin, a *Curcuma longa* constituent, acts on MAPK p38 pathway modulating COX-2 and iNOS expression in chronic experimental colitis. Int Immunopharmacol 7:333–342

Chander V, Singh D, Chopra K (2005) Reversal of experimental myoglobinuric acute renal failure in rats by quercetin, a bioflavonoid. Pharmacology 73:49–56

Chang HF, Lin YH, Chu CC, Wu SJ, Tsai YH, Chao JC (2007) Protective effects of *Ginkgo biloba*, *Panax ginseng*, and *Schizandra chinensis* extract on liver injury in rats. Am J Chin Med 35:995–1009

Chassagne F, Haddad M, Amiel A, Phakeovilay C, Manithip C, Bourdy G, Deharo E et al (2018) A metabolomic approach to identify anti-hepatocarcinogenic compounds from plants used traditionally in the treatment of liver diseases. Fitoterapia: Revista Di Studi Ed Applicazioni Delle Piante Medicinali 127:226–236

Chen H, Wang C, Chang CT, Wang T (2003) Effects of Taiwanese yam (*Dioscorea japonica* Thunb var. *pseudojaponica* Yamamoto) on upper gut function and lipid metabolism in Balb/c mice. Nutrition 19:646–651

Cho MK, Jang YP, Kim YC, Kim SG (2004) Arctigenin, a phenylpropanoid dibenzylbutyrolactone lignan, inhibits MAP kinases and AP-1 activation via potent MKK inhibition: the role in TNF-alpha inhibition. Int J Immunopharmacol 4:1419–1429

Chodera A, Dabrowska K, Sloderbach A, Skrzypczak L, Budzianowski J (1991) Wpływ frakcji flawonoidowych gatunków rodzaju *Solidago* L. na diurezę i stężenie elektrolitów. Acta Pol Pharm 48:5–6

Craig WJ (1999) Health-promoting properties of common herbs. Am J Clin Nutr 70(suppl):491S–499S

Darshan S, Doreswamy R (2004) Patented anti-inflammatory plant drug development from traditional medicine. Phytother Res 18(5):343–357

De Smet PAGM, D'Arcy PF (1996) Drug interactions with herbal and other non-toxic remedies. Springer, Berlin

Desplaces A, Choppin J, Vogel G, Trost W (1975) The effects of silymarin on experimental phalloidine poisoning. Arzneimittelforschung 25:89–96

Dhiman RK, Chawla YK (2005) Herbal medicines for liver diseases. Dig Dis Sci 50:1807–1812

Do Monte FH, dos Santos JG Jr, Russi M, Lanziotti VM, Leal LK, Cunha GM (2004) Antinociceptive and anti-inflammatory properties of the hydroalcoholic extract of stems from *Equisetum arvense* L. in mice. Pharmacol Res (Lon) 49:239–243

Dreikorn K (2002) The role of phytotherapy in treating lower urinary tract symptoms and benign prostatic hyperplasia. World J Urol 19:426–435

Duke JA (1990) Promising phytomedicinals. In: Janick J, Simon JE (eds) Advances in new crops. Timber Press, Portland, pp 491–498

Duke JA (2017) Handbook of phytochemical constituent GRAS herbs and other economic plants: herbal reference library, 2017. Internet resource. https://www.bol.com/nl/p/handbook-of-phytochemical-constituent-grass-herbs-and-other-economic-plants/9200000086854118

El-Ashmawy IM, Ashry KM, El-Nahas AF, Salama OM (2006) Protection by turmeric and myrrh against liver oxidative damage and genotoxicity induced by lead acetate in mice. Basic Clin Pharmacol Toxicol 98:32–37

EMA/HMPC/441930/2008. https://www.ema.europa.eu/documents/herbal-report/final-assessment-report-juniperi-pseudo-fructus_en.pdf

EMEA/HMPC/285759/2007. https://www.ema.europa.eu/documents/herbal-report/assessment-report-solidago-virgaurea-l-herba_en.pdf

Farombi EO, Shrotriya S, Na HK, Kim SH, Surh YJ (2007) Curcumin attenuates dimethylnitrosamine-induced liver injury in rats through Nrf2-mediated induction of heme oxygenase-1. Food Chem Toxicol 2007 Sep 26 [Epub ahead of print]

Filburn CR, Kettenacker R, Griffin DW (2007) Bioavailability of a silybin-phosphatidylcholine complex in dogs. J Vet Pharmacol Ther 30:132–138

Galisteo M, Suárez A, Montilla MP, Fernandez MI, Gil A, Navarro MC (2006) Protective effects of *Rosmarinus tomentosus* ethanol extract on thioacetamide-induced liver cirrhosis in rats. Phytomedicine 13:101–108

Geraghty M (2000) Herbal supplements for renal patients: What do we know? March 2000 Nephrology News & Issues

Getie M, Gebre-Mariam T, Rietz R, Hohne C, Huschka C, Schmidtke M, Abate A, Neubert RH (2003) Evaluation of the anti-microbial and anti-inflammatory activities of the medicinal plants *Dodonaea viscosa*, *Rumex nervosus* and *Rumex abyssinicus*. Fitoterapia 74:139–143

Goel A, Kunnumakkara AB, Aggarwal BB (2007) Curcumin as "Curecumin": from kitchen to clinic. Biochem Pharmacol 2007 Aug 19 [Epub ahead of print]

Goñi I, Jiménez-Escrig A, Gudiel M, Saura-Calixto F (2005) Artichoke (L) modifies bacterial enzymatic activities and antioxidant status in rat cecum. Nutr Res 25:607–615 I

Graefe EU, Veit M (1999) Urinary metabolites of flavonoids and hydroxycinnamic acids in humans after application of a crude extract from *Equisetum arvense*. Phytomedicine 6:239–246

Greco DS (2007) Hypoadrenocorticism in small animals. Clin Tech Small Anim Pract 22:32–35

Gurbuz I et al (2002) Invivo gastroprotective effects of five Turkish folk remediesagainst ethanol-induced lesions. J Ethnopharmacol 83(3):241–244

Habtemariam S (1998) Extract of corn silk (stigma of *Zea mays*) inhibits the tumour necrosis factor-α- and bacterial lipopolysaccharide-induced cell adhesion and ICAM-1 expression. Planta Med 64:314–318

Habtemariam S (2001) Antiinflammatory activity of the antirheumatic herbal drug, gravel root (*Eupatorium purpureum*): further biological activities and constituents. Phytother Res 15:687–690

Hammer F, Stewart PM (2006) Cortisol metabolism in hypertension. Best Pract Res Clin Endocrinol Metab 20:337–353

Hoffman D (2003) Medical herbalism – the science and practice of herbal medicine. Healing Arts Press, Rochester

Holetz FB, Pessini GL, Sanches NR, Cortez DA, Nakamura CV, Filho BP (2002) Screening of some plants used in the Brazilian folk medicine for the treatment of infectious diseases. Memorias do Instituto Oswaldo Cruz 97:1027–1031

Hsu JH, Wu YC, Liu IM, Cheng JT (2007) Dioscorea as the principal herb of Die-Huang-Wan, a widely used herbal mixture in China, for improvement of insulin resistance in fructose-rich chow-fed rats. J Ethnopharmacol 112:577–584

Hutchison A, Farmer R, Verhamme K, Berges R, Navarrete RV (2007) The efficacy of drugs for the treatment of LUTS/BPH, a study in 6 European countries. Eur Urol 51:207–215

Iida N, Inatomi Y, Murata H, Nakanishi T, Inada A, Murata J, Lang FA et al (2010) New phenylpropanoid glycosides from *Juniperus communis* var. *depressa*. Chem Pharm Bull 58(5):742–746

Jellinek N, Maloney M (2005) Escharotic and other botanical agents for the treatment of skin cancer: a review. J Am Acad Dermatol 53:487–495

Jiang WX, Xue BY (2005) Hepatoprotective effects of *Gentiana scabra* on the acute liver injuries in mice. Zhongguo Zhong Yao Za Zhi 30:1105–1107. Article in Chinese

Kang DG, Sohn EJ, Moon MK, Lee YM, Lee HS (2005) *Rehmannia glutinose* ameliorates renal function in the ischemia/reperfusion-induced acute renal failure rats. Biol Pharm Bull 28:1662–1667

Kim HY, Kang MH (2005) Screening of Korean medicinal plants for lipase inhibitory activity. Phytother Res 19:359–361

Kleiner D, Hegyi G, Urbanics R, Dézsi L, Robotka H, Fehér E, Sárdi E et al (2016) Hepatoprotective liposomal glycyrrhizin in alcoholic liver injury. Eur J Integrative Med 8:23–28

Koch E (2001) Extracts from fruits of saw palmetto (*Sabal serrulata*) and roots of stinging nettle (*Urtica dioica*): viable alternatives in the medical treatment of benign prostatic hyperplasia and associated lower urinary tracts symptoms. Planta Med 67:489–500

Kreydiyyeh SI, Usta J (2002) Diuretic effect and mechanism of action of parsley. J Ethnopharmacol 79:353–357

Kurkin VA, Zaitseva EN, Ryazanova TK, Dubishchev AV (2016) Effects of Bilberry fruit and shoot extracts on renal excretory function. Pharm Chem J 50(4):239–243

Lans C, Turner N, Brauer G, Lourenco G, Georges K (2006) Ethnoveterinary medicines used for horses in Trinidad and in British Columbia, Canada. J Ethnobiol Ethnomed 2:31

Latief U, Ahmad R (2018) Herbal remedies for liver fibrosis: a review on the mode of action of fifty herbs. J Tradit Complement Med 8(3):352–360

Lee DYW, Zhang X, Ji XS (2006) Preparation of tritium-labeled Silybin-a protectant for common liver diseases. J Label Compd Radiopharm 49:1125–1130

Li S, Pasquin S, Eid HM, Gauchat J-F, Haddad PS, Li S, Eid HM et al (2018) Anti-apoptotic potential of several antidiabetic medicinal plants of the eastern James Bay Cree pharmacopeia in cultured kidney cells. BMC Complement Altern Med 18(1)

Lieber CS, Leo MA, Cao Q, Ren C, DeCarli LM (2003) Silymarin retards the progression of alcohol-induced hepatic fibrosis in baboons. J Clin Gastroenterol 37:336–339

Lin YL, Hsu YC, Chiu YT, Huang YT (2008) Antifibrotic effects of a herbal combination regimen on hepatic fibrotic rats. Phytother Res 22:69–76

Lone IA, Kaur G, Athar M, Alam MS (2007) Protective effect of *Rumex patientia* (English Spinach) roots on ferric nitrilotriacetate (Fe-NTA) induced hepatic oxidative stress and tumor promotion response. Food Chem Toxicol 45:1821–1829

Lopatkin N, Sivkov A, Walther C, Schlafke S, Medvedev A, Avdeichuk J, Golubev G, Melnik K, Elenberger N, Engelmann U (2005) Long-term efficacy and safety of a combination of sabal and urtica extract for lower urinary tract symptoms--a placebo-controlled, double-blind, multi-centre trial. World J Urol 23:139–146

Madari H, Jacobs RS (2004) An analysis of cytotoxic botanical formulations used in the traditional medicine of ancient Persia as abortifacients. J Nat Prod 267(8):1204–1210

Maksimović Z, Dobrić S, Kovacević N, Milovanović Z (2004) Diuretic activity of Maydis stigma extract in rats. Pharmazie 59:967–971

Martineau LC, Couture A, Spoor D, Benhaddou-Andaloussi A, Harris C, Meddah B, Leduc C, Burt A, Vuong T, Mai Le P, Prentki M, Bennett SA, Arnason JT, Haddad PS (2006) Anti-diabetic properties of the Canadian lowbush blueberry *Vaccinium angustifolium* Ait. Phytomedicine 13:612–623

McCune LM, Johns T (2002) Antioxidant activity in medicinal plants associated with the symptoms of diabetes mellitus used by the indigenous peoples of the North American boreal forest. J Ethnopharmacol 82:197–205

Meeking S (2007) Treatment of acute adrenal insufficiency. Clin Tech Small Anim Pract 22:36–39

Meij BP, Mol JA, Bevers MM, Rijnberk A (1997) Alterations in anterior pituitary function of dogs with pituitary-dependent hyperadrenocorticism. J Endocrinol 154:505–512

Melzig MF (2004) Goldenrod–a classical exponent in the urological phytotherapy. Wien Med Wochenschr 154:523–527. Article in German

Mencherini T, Picerno P, Scesa C, Aquino R (2007) Triterpene, antioxidant, and antimicrobial compounds from *Melissa officinalis*. J Nat Prod 70:1889–1894

Muriel P, Rivera-Espinoza Y (2007) Beneficial drugs for liver diseases. J Appl Toxicol 2007 Oct 26 [Epub ahead of print]

Nakagiri R, Hashizume E, Kayahashi S, Sakai Y, Kamiya T (2003) Suppression by Hydrangeae Dulcis Folium of D-galactosamine-induced liver injury *in vitro* and *in vivo*. Biosci Biotech Bioch 67:2641–2643

Nohara C, Yokoyama D, Tanaka W, Sogon T, Sakono M, Sakakibara H (2018) Daily consumption of bilberry (*Vaccinium myrtillus* L.) extracts increases the absorption rate of anthocyanins in rats. J Agric Food Chem 66(30):7958–7964

Nomikos T, Detopoulou P, Fragopoulou E, Pliakis E, Antonopoulou S (2007) Boiled wild artichoke reduces postprandial glycemic and insulinemic responses in normal subjects but has no effect on metabolic syndrome patients. Nutr Res 27:741–749

NSRC (2004) A study of the nutritional composition of the dandelion by part (Taraxacum officinale). Rural Resources Development Institute, NIAST, Korea

Omoruyi FO, McAnuff-Harding MA, Asemota HN (2006) Intestinal lipids and minerals in streptozotocin-induced diabetic rats fed bitter yam (*Dioscorea polygonoides*) sapogenin extract. Pak J Biol Sci 19:269–275

Ozkarsli M, Sevim H, Sen A (2008) *In vivo* effects of *Urtica urens* (dwarf nettle) on the expression of CYP1A in control and 3-methylcholanthrene-exposed rats. Xenobiotica 38:48–61

Panossian A, Wagner H (2005) Stimulating effect of adaptogens: an overview with particular reference to their efficacy following single dose administration. Phytother Res 19:819–838

Papadopoulos V, Widmaier EP, Amri H, Zilz A, Li H, Culty M, Castello R, Philip GH, Sridaran R, Drieu K (1998) *In vivo* studies on the role of the peripheral benzodiazepine receptor (PBR) in steroidogenesis. Endocr Res 24:479–487

Park EJ, Nan JX, Zhao YZ, Lee SH, Kim YH, Nam JB, Lee JJ, Sohn DH (2004) Water-soluble polysaccharide from *Eleutherococcus senticosus* stems attenuates fulminant hepatic failure induced by D-galactosamine and lipopolysaccharide in mice. Basic Clin Pharmacol Toxicol 94:298–304

Petlevski R, Hadzija M, Slijepcevic M, Juretic D, Petrik J (2003) Glutathione S-transferases and malondialdehyde in the liver of NOD mice on short-term treatment with plant mixture extract P-9801091. Phytother Res 17:311–314

Santos HO, Bueno AA, Mota JF (2018) The effect of artichoke on lipid profile: a review of possible mechanisms of action. Pharmacol Res 137:170–178

Savino F, Cresi F, Castagno E, Silvestro L, Oggero R (2005) A randomized double-blind placebo-controlled trial of a standardized extract of *Matricariae recutita*, *Foeniculum vulgare* and

Melissa officinalis (ColiMil) in the treatment of breastfed colicky infants. Phytother Res 19:335–340

Shimizu M, Shiota S, Mizushima T, Ito H, Hatano T, Yoshida T, Tsuchiya T (2001) Marked potentiation of activity of beta-lactams against methicillin-resistant *Staphylococcus aureus* by corilagin. Antimicrob Agents Chemother 45:3198–3201

Shrivastava R, John GW (2006) Treatment of aphthous stomatitis with topical *Alchemilla vulgaris* in glycerine. Clin Drug Investig 26:567–573

Shrivastava R, Cucuat N, John GW (2007) Effects of *Alchemilla vulgaris* and glycerine on epithelial and myofibroblast cell growth and cutaneous lesion healing in rats. Phytother Res 21:369–373

Speroni E, Cervellati R, Govoni P, Guizzardi S, Renzulli C, Guerra MC (2003) Efficacy of different *Cynara scolymus* preparations on liver complaints. J Ethnopharmacol 86:203–211

Steenkamp V, Gouws MC, Gulumian M, Elgorashi EE, van Staden J (2006) Studies on antibacterial, anti-inflammatory and antioxidant activity of herbal remedies used in the treatment of benign prostatic hyperplasia and prostatitis. J Ethnopharmacol 103:71–75

Sultana B, Yaqoob S, Zafar Z, Bhatti HN (2018) Escalation of liver malfunctioning: a step toward herbal awareness. J Ethnopharmacol 216:104–119

Tunón H, Olavsdotter C, Bohlin L (1995) Evaluation of anti-inflammatory activity of some Swedish medicinal plants. Inhibition of prostaglandin biosynthesis and PAF-induced exocytosis. J Ethnopharmacol 48:61–76

Uzun E, Sariyar G, Adsersen A, Karakoc B, Ötük G, Oktayoglu E, Pirildar S (2004) Traditional medicine in Sakarya province (Turkey) and antimicrobial activities of selected species. J Ethnopharmacol 95:287–296

Vanisree AJ, Sudha N (2006) Curcumin combats against cigarette smoke and ethanol-induced lipid alterations in rat lung and liver. Mol Cell Biochem 288:115–123

Varzi HN, Esmailzadeh S, Morovvati H, Avizeh R, Shahriari A, Givi ME (2007) Effect of silymarin and vitamin E on gentamicin-induced nephrotoxicity in dogs. J Vet Pharmacol Ther 30:477–481

Velazquez DV, Xavier HS, Batista JE, de Castro-Chaves C (2005) *Zea mays* L. extracts modify glomerular function and potassium urinary excretion in conscious rats. Phytomedicine 12:363–369

Veljković M, Pavlović DR, Stojiljković N, Ilić S, Jovanović I, Poklar UN, Rakić V et al (2017) Bilberry: chemical profiling, *in vitro* and *in vivo* antioxidant activity and nephroprotective effect against gentamicin toxicity in rats. Phytother Res 31(2017):115–123

Venkatesan N, Punithavathi D, Arumugam V (2000) Curcumin prevents adriamycin nephrotoxicity in rats. Br J Pharmacol 129:231–234

Verma RJ, Asnani V (2007) Ginger extract ameliorates paraben induced biochemical changes in liver and kidney of mice. Acta Pol Pharm 64:217–220

Villaseñor-García MM, Lozoya X, Osuna-Torres L, Viveros-Paredes JM, Sandoval-Ramírez L, Puebla-Pérez AM (2004) Effect of *Ginkgo biloba* extract EGb 761 on the nonspecific and humoral immune responses in a hypothalamic-pituitary-adrenal axis activation model. Int Immunopharmacol 4:1217–1222

Villiger A, Sala F, Suter A, Butterweck V (2015) *In vitro* inhibitory potential of *Cynara scolymus*, *Silybum marianum*, *Taraxacum officinale*, and *Peumus boldus* on key enzymes relevant to metabolic syndrome. Phytomedicine 22(1):138–144

Vlahović P, Cvetković T, Savić V, Stefanović V (2007) Dietary curcumin does not protect kidney in glycerol-induced acute renal failure. Food Chem Toxicol 45:1777–1782

Watt K, Christofi N, Young R (2007) The detection of antibacterial actions of whole herb tinctures using luminescent *Escherichia coli*. Phytother Res 21:1193–1199

Wynn SG, Marsden SA (2003) Manual of natural veterinary medicine: science and tradition. Mosby, St Louis

Yamamoto J, Yamada K, Naemura A, Yamashita T, Arai R (2005) Testing various herbs for antithrombotic effect. Nutrition 21:580–587

Zaluski D et al (2017) LC-ESI-MS/MS profiling of phenolics from Eleutherococcusspp. Inflorescences, structure-activity relationship as antioxidants, inhibitors of hyaluronidase and acetylcholinesterase. Saudi Pharm J 25(5):734–743

Yarnell E (2002) Botanical medicines for the urinary tract. World J Urol 20:285–293

Zeugswetter F, Hoyer MT, Pagitz M, Benesch T, Hittmair KM, Thalhammer JG (2007) The des-mopressin stimulation test in dogs with Cushing's syndrome. Domest Anim Endocrinol Aug 27 [Epub ahead of print]

Zhang H, Matsuda H, Kumahara A, Ito Y, Nakamura S, Yoshikawa M (2007a) New type of anti-diabetic compounds from the processed leaves of *Hydrangea macrophylla* var. *thunbergii* (Hydrangeae Dulcis Folium). Bioorg Med Chem 17:4972–4976

Zhang J, Zhang X, Lei G, Li B, Chen J, Zhou T (2007b) A new phenolic glycoside from the aerial parts of *Solidago canadensis*. Fitoterapia 78:69–71

Zhang S, Ma J, Sheng L, Zhang D, Chen X, Yang J, Wang D (2017) Total coumarins from *Hydrangea paniculata* show renal protective effects in lipopolysaccharide-induced acute kidney injury via anti-inflammatory and antioxidant activities. Front Pharmacol 8

Chapter 11
Catnip (*Nepeta cataria* L.): Recent Advances in Botany, Horticulture and Production

Erik N. Gomes, Kirsten Allen, Katharine Jaworski, Martin Zorde, Anthony Lockhart, Thierry Besancon, Theodore Brown, William Reichert, Qingli Wu, and James E. Simon

Abstract Catnip (*Nepeta cataria* L.), a popular aromatic herb used as a traditional medicine is more widely recognized for its use in the pet toy industry due to the behavioral effects it elicits on cats and other felids. A major interest in catnip is also due to its repellent activity against arthropods. Essential oil of catnip is an effective repellent against several species of mosquitoes, flies, ticks, mites, and other disease vectors, with results comparable to DEET. Both the repellency to arthropods and the characteristic effects on cats are mainly attributed to nepetalactone, a bicyclic oxygenated monoterpene in the essential oil of catnip. While catnip is grown as a garden herb and in the open field for dried biomass and essential oil, the lack of improved genetic

E. N. Gomes
New Use Agriculture and Natural Plant Products Program, Department of Plant Biology, Rutgers University, New Brunswick, NJ, USA

CAPES Foundation, Ministry of Education of Brazil, Brasília DF, Brazil

K. Allen · K. Jaworski · M. Zorde · T. Brown · W. Reichert
New Use Agriculture and Natural Plant Products Program, Department of Plant Biology, Rutgers University, New Brunswick, NJ, USA

A. Lockhart
Department of Medicinal Chemistry, Ernest Mario School of Pharmacy, Rutgers University, Piscataway, NJ, USA

T. Besancon
Department of Plant Biology, Rutgers University, Philip E. Marucci Center for Blueberry and Cranberry Research and Extension, Chatsworth, NJ, USA

Q. Wu · J. E. Simon (✉)
New Use Agriculture and Natural Plant Products Program, Department of Plant Biology, Rutgers University, New Brunswick, NJ, USA

Department of Medicinal Chemistry Ernest Mario School of Pharmacy, Rutgers University, Piscataway, NJ, USA
e-mail: jimsimon@rutgers.edu

© Springer Nature Switzerland AG 2020
Á. Máthé (ed.), *Medicinal and Aromatic Plants of North America*, Medicinal and Aromatic Plants of the World 6,
https://doi.org/10.1007/978-3-030-44930-8_11

materials makes it difficult for North American growers to expand production and ensure adequate product supply. The present chapter provides an overview of the recent advances in breeding, biochemistry, production systems, biological activities and potential new uses of *N. cataria* and other *Nepeta* species in North America.

Keywords Nepetalactones · Terpenes · Essential Oil · Insect Repellent · Arthropod Repellent

11.1 Introduction

Nepeta cataria L., commonly known as catnip or catmint, is a perennial, herbaceous, shrub species belonging to the family Lamiaceae (Duda et al. 2015a; Linnaeus 1800). The geographic origin of the species is unclear from scientific literature, with reports pointing to Eastern and Southern Europe, Southeast and Central Asia, North Africa and even North America (Duda et al. 2015a; Ibrahim et al. 2017; Reichert et al. 2016; Said-Al et al. 2018; United States Department of Agriculture (USDA) 2019; Zhu et al. 2009). The species has been naturalized in many regions around the globe, and is found in the wild, in gardens, and in small commercial fields in Great Britain, Southwestern Scandinavia, Eastern Spain, the Middle East, China and North America (Duda et al. 2015a; Ibrahim et al. 2017).

Catnip is a popular ornamental aromatic herb and recognized widely for its use in the pet toy industry due to the effects on members of the feline family, both domesticated and wild (Clapperton et al. 1994). The plant elicits active behaviors in felids, including domestic cats, such as rolling over, chin and cheek rubbing, head shaking, pawing, floor scratching, persistent sniffing, vocalizations and licking and chewing of the catnip source (Espín-Iturbe et al. 2017; Herron 2003; Waller et al. 1969). Such behavioral effects are attributed to the volatile compound nepetalactone, a bicyclic oxygenated monoterpene present in the essential oil of catnip (Bol et al. 2017; Clapperton et al. 1994; McElvain et al. 1941; Reichert et al. 2016). In addition to the effects on felids, catnip is consumed medicinally as an herbal tea in some traditional cultures for the treatment of fever, diarrhea, insomnia and adjustment of menstrual cycles (Herron 2003), inflammation (Reichert et al. 2018), toothaches, as well as a sedative and for alleviation of digestive ailments, including infantile colic, and other conditions (Gilani et al. 2009; Lewis and Elvin-Lewis 1982; Naghibi et al. 2010; Smitherman et al. 2005). Studies on different catnip extracts have confirmed antimicrobial (Bandh et al. 2011; Zomorodian et al. 2013), antioxidant (Duda et al. 2015b; Reichert et al. 2018), trypanocide (Saeidnia et al. 2008), spasmolytic and bronchodilator properties (Gilani et al. 2009), among others.

A major current interest in catnip (*N. cataria*) relative to commercialization opportunities is in the repellent activity of its essential oil against arthropods (Amer and Mehlhorn 2006; Marmy et al. 2018; Patience et al. 2018; Peterson and Coats 2001; Peterson et al. 2002; Reichert et al. 2019). The plant has been reported to be a viable tool for preventing the transmission of various arthropod-vectored diseases on human medicine and livestock production (Patience et al. 2018; Zhu et al. 2009). Laboratory entomological studies with catnip essential oil and isolated nepetalactones have demonstrated a repellent effect against the *Aedes aegypti* mosquito, vector of dengue, yellow fever and Zika viruses (Amer and Mehlhorn 2006; Polsomboon et al. 2008; Reichert et al. 2019; Simon et al. 2019; Zhu and Zeng 2006). Studies have also shown high efficiency of catnip essential oil and its major compound nepetalactone to repel malaria and lymphatic filariasis transmitting mosquitoes (*Anopheles gambiae* and *Culex quinquefasciatus*, respectively), houseflies (*Musca domestica*), stable flies (*Stomoxys calcitrans*), ticks (*Rhipicephalus appendiculatus*) and chicken mites (*Dermanyssus gallinae*), among many other species (Birkett et al. 2011; Polsomboon et al. 2008; Simon et al. 2019; Zhu et al. 2009, 2012; Zhu and Zeng 2006).

Catnip essential oil and isolated nepetalactones repellency against arthropods have been reported to be at least comparable to the industry standard repellent DEET (Bernier et al. 2005; Feaster et al. 2009; Petterson and Coats 2011; Reichert et al. 2019). Additional studies showed that *N. cataria* essential oils containing various nepetalactone stereoisomers where comparable to DEET at repelling mosquitoes, while offering better spatial repellency (Bernier et al. 2005; Peterson and Coats 2011; Schultz et al. 2004). In addition to better spatial repellency, catnip oil has also been shown to be relatively safer when compared to DEET, picaridin and *p*-menthane-3,8-diol, mosquito repellents approved by the U.S. Environmental Protection Agency (Peterson et al. 2002; Zhu et al. 2009, 2012).

The use of catnip essential oil as a preventive approach for vector-borne diseases can be a sustainable alternative not only because of its effectiveness, but also because it can be grown, distilled, locally produced and distributed by local communities, representing a source of economic growth for rural areas (Patience et al. 2018). Additionally to rural poor areas, it could be affordable to local communities that may otherwise not have access to nor could afford to purchase synthetic insecticides. Use of live plants as sentries against arthropod pests or the applications of volatile oils from locally produced insect repellents that are effective could require a smaller initial investment. When compared to other food and industrial crops, niche crops such as catnip have promise to be profitable as part of small farm diversification that could provide income generation opportunities while maintaining their main commercial crops (Kaiser and Ernst 2019).

Worldwide, the insect repellent market is expected to reach more than US$11 billion by 2019, while in the United States it is growing at a rate of 4–7% a year, reaching about US$600 million in the same year (Technavio 2019). In the United States, over the next 5 years, the insect repellent industry is expected to invest in research and development to produce more plant-based natural repellents, given the growing consumer preference for natural products (IBISWorld 2019; TMR 2019).

Despite the increasing commercial interest in catnip for these different applications, there are serious constraints in this crop's commercialization particularly in North America. Catnip is a low growing, low yielding (relative to biomass and total essential oils) aromatic herb, making it difficult to successfully mechanize for use in more traditional modern farming systems. Catnip is not a high essential oil yield producer making it difficult for essential oils farmers to easily incorporate catnip into their production systems (Park et al. 2007). Furthermore, there are few registered herbicides and pesticides approved for use on catnip, leading to serious constraints, added expenses and grower risk relative to weed and pest control. The introduction of higher yielding more aggressive and taller genetic materials that lend themselves to mechanical harvesting coupled with the development of improved production systems are needed to overcome the current constraints to large-scale commercial introduction and production to ensure reliable adequate product supplies in a manner that minimizes grower risk and increases crop profitability. The few genetic and breeding studies that have been conducted on catnip, only confirm that this species is a minor or orphan crop, with the commercially available varieties more reflective of landraces than domesticated and improved genetic varieties. The few improved lines for the essential oil market and the commercial industry for use as a culinary herb and source of essential oils and/or the bioactive nepetalactone have begun to address this major production constraint (Figs. 11.1 and 11.2). The development and introduction of improved germplasm and the selection of highly productive genotypes and genetic improvements are such initial fundamental steps for catnip cultivation (Park et al. 2007; Reichert et al. 2016). Considering the dynamics of secondary metabolite production, the study of environmental effects on the essential oil and nepetalactone biosynthesis is also fundamental for understanding the ecophysiology of the species and its implications

Fig. 11.1 Field photo of the Rutgers University catnip cultivars at full bloom growing at the Rutgers Clifford and Melda E. Snyder Research and Extension Farm in Pittstown, New Jersey, August, 2019

Fig. 11.2 Close-up of catnip in full flower, note some post-full bloom, others just flowering at the Rutgers Clifford and Melda E. Snyder Research and Extension Farm in Pittstown, New Jersey, August, 2019

on the productivity of different genetic materials yet lacking in catnip. In contrast, there has been a wide number of catmint hybrids developed by the nursery industry for their ornamental applications (e.g. the non-*N. cataria* spp.) which have become popular garden and landscape plants. In this paper, we review the current status of catnip as a commercial aromatic and medicinal plant.

11.2   Taxonomy and Genetics of *Nepeta cataria* L.

Catnip belongs to the Viridiplantae subkingdom, the Streptophyta infra-kingdom, the Embryophyta superdivision, the Tracheophyta division, the Spermatophytina subdivision, the Magnoliopsida class, the Asteranae superorder, the Lamiales order, the Lamiaceae family, the Nepetoidae subfamily, the tribe Menthae, and the genus *Nepeta* L. (ITIS 2019; Payandeh et al. 2015). The Missouri Botanical Garden lists approximately 795 taxa in *Nepeta* including different subspecies, varieties, and cultivars in addition to *Nepeta cataria* (Missouri Botanical Garden 2019; Setzer 2016). *Nepeta* is one of the largest genera within the mint family encompassing approximately 300 species (Jamzad et al. 2003). The genus name may honor the Italian city of Nepete (known as Nepi today) located north of Rome in Etruria which was the ancient country located between the Arno and Tiber Rivers and was recognized, prior to the rise of Rome, as the center of Etruscan civilization (Missouri Botanical Garden 2019).

Catnip is a perennial plant with the square stems and terminal flower spikes typical of other species in the mint family (Mountain Rose Herbs 2019). The plants generally grow 0.3–1 m tall with heart-shaped, toothed, opposite leaves where the inflorescence is terminal in verticillaster units (Simpson 2010; Mountain Rose Herbs 2019). Due to the morphological nature of the bilabiate bisexual flowers, the plant can both self-pollinate and outcross (Claßen-Bockhoff et al. 2003). Individual plants can accumulate up to several thousand flowers on loose terminal spikes on the primary axis and throughout most branches (Sih and Baltus 1987). Less than 10% of the flowers will tend to flower at any given time (Sih and Baltus 1987). The individual flowers are zygomorphic, tend to be 10–12 mm long and depending on the cultivar will be colored dull white, purple-pink or white with colored speckles (Sih and Baltus 1987). Each flower is staminate for 1–2 days on average (Sih and Baltus 1987). Flowers that have been pollinated produce 1–4 (usually 4) seeds per nutlet (Sih and Baltus 1987). In the absence of pollinators, the catnip flowers can self-pollinate; yet selfing seldom produces maximum seed-set (Sih and Baltus 1987).

Catnip is a diploid species and the results for a study conducted by researchers in Iran found a consistent chromosome number for four different *Nepeta* species to be $2n = 2x = 18$ (Payandeh et al. 2015). Additional reports have identified chromosome numbers $2n = 14, 16, 18, 34, 36$ and 54 within the *Nepeta* genus (Martin et al. 2013). In *Nepeta cataria*, four microsporangia differentiate into each anther with walls that form according to the dicotyledonous type (Daskalova 2004). The anther walls are 4-layered and built up of epidermis, endothecium, middle layer and tapetum of the secretory type (Daskalova 2004). The epidermal cells form 1–4 glandular trichomes of 6–8 cells and arrange on the connective side of the anther (Daskalova 2004). The cells reach the maximum of their development when flowers reach peak maturity after ripening of the embryo sacs (Daskalova 2004). Meiosis is simultaneous in type since the division of the four formed nuclei takes place only at the end of meiotic division (Daskalova 2004). After simultaneous meiosis in the anthers, only tetrahedral tetrads form (Daskalova 2004). Also, the mature pollen is 3-celled 6-colpate (Daskalova 2004). Most of these embryological features are characteristic of other *Lamiaceae* species as well (Daskalova 2004).

The genetic factors controlling the morphological and metabolic pathways in catnip still require a greater depth of exploration to more fully understand the mechanisms involved at a molecular and cellular level. While the genetic control of the pathways forming the carbon backbones of terpenoids have been widely studied, more work needs to be done regarding the species-specific biosynthesis of secondary metabolites (Croteau and Gershenzon 1994). The genes responsible for the expression of terpenoids in catnip need to be isolated within a single homogenous population in order to produce a desired population with target compounds (Reichert 2019).

The Mint Genome Project, a research consortium focused on uncovering the chemical pathways that synthesize secondary metabolites within *Lamiaceae*, provides a wealth of genetic information regarding genome and transcriptome sequencing data for all mint family species, including catnip (Mint Evolutionary Genomics Consortium 2018). One of the collective goals was to uncover the evolutionary mechanisms responsible for the extensive chemodiversity within the mint family (Mint Evolutionary Genomics Consortium 2018). In 2018, researchers used the transcriptome of *Nepeta* species including *Nepeta cataria* and *N. mussinii* to begin identifying candidate genes, like the iridoid synthase gene, involved in the biosynthetic pathways of catnip to establish a starting point for further investigation of nepetalactone biosynthesis (Sherden et al. 2018). The authors also reported the genome size of catnip to be 1.15 pg/2C (Mint Evolutionary Genomics Consortium 2018). This work will provide a critical foundation upon which further genetics and breeding studies can use to focus on both improving the plant architecture and custom designing the volatile secondary metabolites in the terpenoid pathway.

11.3 Biologically Active Substances of Catnip and *Nepeta* Species

Along with insect repellency, the essential oil of *Nepeta* species possess numerous other biological activities; specifically, antibacterial, antifungal (Adiguzel et al. 2009) and analgesic (Aydin et al. 1998) properties. Former studies have shown that the essential oil of *Nepeta cataria* (78.9% nepetalactones) was more effective as an antibacterial treatment compared to the methanol extract of the plant (Adiguzel et al. 2009). Adiguzel et al. (2009) tested both clinical and plant originated bacteria, as well as yeast and fungal isolates, showing that the nepetalactone-rich essential oil was effective against *Burkholderia cepacia* A225 and *Klebsiella pneumoniae* A137, two plant-based bacteria, and *Bacillus subtilis* ATCC-6633, *Bacillus subtilis* A57, *Brucella abortus* A77, *Bacillus* macerans A199, *Escherichia coli* A1, *Proteus vulgaris* KUKEM1329, *Staphylococcus aureus* A215, *Staphylococcus epidermis* A233, and *Streptococcus pyogenes* ATCC-176, which are of clinical origin. Furthermore, the essential oil showed anticandidal activity against *Candida albicans* A117 and was antifungal towards *Alternaria alternata*, *Aspergillus flavus*, *Aspergillus variecolor*, *Fusarium oxysporum*, *Fusarium solani*, *Fusarium tabacinum*, *Penicillum* spp., *Rhizopus* spp., *Rhizoctonia solani*, *Sclerotinia sclerotiorum*, *Trichophyton rubrum*, and *Trichophyton mentagrophytes*.

Aydin et al. (1998) showed that the essential oil of *Nepeta caesarea* (92–95% nepetalactones) could act as a pain-relieving agent. Intraperitoneal injection of the essential oil into albino mice showed increased latency in the tail-clip test, indicating an analgesic effect. The response was comparable to morphine, a well-known opioid medication. Interestingly, the essential oil treatment was inhibited by naloxone, an antagonist of opioid receptors, which implies nepetalactone may be an agonist for opioid receptors. The specific subtype of receptor it may have affinity to is unknown (Aydin et al. 1998).

In addition to nepetalactone, the ethanol extracts of *Nepeta* species also contain many flavonoids and phenolic acids which are known antioxidants (Reichert et al. 2018). Phenolic components like the flavanol epicatechin and the flavanone eriodictyol have been identified in *Nepeta cataria* (Proestos et al. 2006; Süntar et al. 2018). The presence of non-volatile antioxidant and anti-inflammatory agents in catnip is not surprising given such compounds are known in other Lamiaceae including *Mentha* spp. and *Origanum* spp. (Shen et al. 2010, 2011) and could be developed as value-added products in addition to the distilled essential oil. Using ultrasonic-aided extraction with 70% ethanol/water, a diverse chemical profile of phenolic acids, flavonols, glycosides, and flavones was extracted from *N. cataria* with the main constituents being rosmarinic acid and hyperoside (2413.1 and 1606.3 μg/g dry weight, respectively) (Mihaylova et al. 2013). Using free radical scavenging and FRAP assays, the extracts of the plant showed moderate antioxidant capabilities (Mihaylova et al. 2013).

Nepeta species also exhibit anxiolytic properties from ethanol (80%) extracts. Using the EPM model of anxiety testing, which tests against rodents' natural fear of elevated and open spaces, the ethanolic extracts of the aerial parts of *Nepeta persica* Boiss. significantly reduced anxiety at a dose of 50 mg/kg intraperitoneal (IP). Unlike diazepam, there were no marked effects on sedation or sleep. In fact, there was a stimulating effect for the mice when dosed at 100 mg/kg. Yet, the same doses did not statistically change ketamine induced sleeping time or latency to sleep, showing a lack of sedative effects (Rabbani et al. 2008). They hypothesized that the nepetalactones in the extract were responsible for the biological activity, but no chemical analysis was presented. In contrast, Bernardi et al. (2010) showed no anxiolytic effects, but did produce anti-depressant effects in mice (Bernardi et al. 2010). Instead of IP administration, *Nepeta cataria* leaf apolar extracts were given orally once per day to the mice, for 7 days (48 mg/kg dose). Another treatment group was given enriched chow (10% *N. cataria* leaves) for the same time frame. Repeated administration of the apolar extract of *N. cataria* leaves showed positive results in the BDT model of depression. In this model, the time and latency of immobility were reduced, similar to that of the antidepressant fluoxetine, a selective serotonin uptake inhibitor (SSRI) (Bernardi et al. 2010). An important distinction to make between these two studies is the parts of the plant used. In the former, the aerial parts of the plant were extracted, the latter used the leaves. While no chemical analysis was given in either study, this could show that there is a difference in the phytochemistry of the plant parts. Differences in administration could also account for the contrasting pharmacological results.

Nepeta cataria was tested for possible use in treating hyperactive gut and respiratory disorders (Gilani et al. 2009). The researchers found that the essential oil of field-grown catnip from Pakistan was comprised mainly of 1,8-cineol (21%), α-humulene (14.44%), α-pinene (10.43%), and geranyl acetate (8.21%) with no detectable nepetalactone present. Whether the genetic material was devoid of nepetalactone or not is unknown. However, when tested on rabbit jejunum tissue, the essential oil inhibited spontaneous contractions. The authors suggested the mechanism of action was through blockage of calcium ion channels, as the essential oil was compared to papaverine and verapamil, two pharmaceutical drugs. Dosages of 0.2 ± 0.10 mg/ml, 7.63 ± 0.04 µM, and 0.61 ± 0.07 µM were found to be effective for the oil, papaverine, and verapamil, respectively. The results showed comparable activity of the essential oil to papaverine. Similarly, the researchers tested the oil on guinea pig tracheal tissue to study its myorelaxant effect. Once again, the results showed efficacy in the research model with doses of 0.05 ± 0.14 mg/ml, 4.01 ± 0.12 µM, and 0.28 ± 0.08 µM. These tests also confirmed a possible dual inhibition of phosphodiesterases (PDE) and calcium ion channels of the essential oil (Gilani et al. 2009). With this knowledge it is critical to test cardiac stimulation, as PDE inhibitors are known to produce this side effect (Raeburn et al. 1993). No effects on the heart at doses therapeutic for potential use were noted (Gilani et al. 2009). Follow-up studies identifying the bioactive compounds in that particular essential oil is needed as many of the non-nepetalactone aromatic volatiles are present in many other aromatic plants of this family.

Other studies have produced results indicating that *Nepeta glomerata* has cytotoxic activity against two human cancer cell lines. Hydrodistilled essential oil showed 48% inhibition of renal adenocarcinoma (ACHN) at 100 µg/ml and 28% inhibition of amelanotic melanoma (C32) at 100 µg/ml. Moreover, the anti-inflammatory abilities of the oil were evaluated in the same study. Treatment with the essential oil yielded 50% inhibition at 78.1 µg/ml, which is comparable to the reference drug, indomethacin: 50% inhibition at 52.8 µg/ml (Rigano et al. 2011). Much like the oil from *N. cataria* grown in Pakistan, the *N. glomerata* cultivar did not show the presence of nepetalactones in GC/MS analysis. Instead, α-pinene (13.4%), spathulenol (8.6%), carvacrol (5.4%), and hexadacenoic acid (4.2%) were the most abundant compounds.

Rigano et al. (2011) suggested there are two main chemotypes for the essential oils: the nepetalactone chemotype and the 1,8-cineole chemotype. Each contributing diverse phytochemistry, thus changing the aroma and medicinal impact of the plants. Despite this, the authors report only 0.1% 1,8-cineole in their sample, which is quite unique when compared to the reported literature (Reichert et al. 2016; Bourrel et al. 1993; Malizia et al. 1996).

There is a third chemotype of catnip, *Nepeta cataria* var. *citriodora*, commonly referred to as lemon catnip. Lemon catnip presents a different biochemistry than the traditional nepetalactone-dominant plants (Wesołowska et al. 2011). Lemon catnip plants produce essential oil composed mostly of citral (geranial and neral), citronellol, geraniol, and nerol, which confer to the plants a lemon-like scent, often containing the same monoterpenoids found in *Melissa officinalis* L., known as lemon balm (Chalchat and Lamy 1997; Klimek and Modnicki 2005; Kolalite 1998; Modnicki et al. 2007; Said-Al Ahl et al. 2018). Cats are not attracted to this chemotype as nepetalactone is usually not found (Chalchat and Lamy 1997; Modnicki et al. 2007; Saeidnia et al. 2008; Suschke et al. 2007; Wesołowska et al. 2011), or found in smaller amounts when compared to the other major components of the essential oil (Suschke et al. 2007). Lemon catnip has been reported as a commercial source of citral (Kolalite 1998) and exhibited trypanocidal, cytotoxic, and antibacterial activities (Saeidnia et al. 2008; Suschke et al. 2007).

11.4 Chemistry and Biochemistry of *Nepeta cataria*

Catnip produces a range of secondary metabolites including polyphenols (Reichert et al. 2018) and essential oils or aromatic volatiles. The essential oils are accumulated and stored within microscopic glandular hairs (trichomes) in the leaves and flowers (Herron 2003). Essential oil yields for catnip can vary widely depending on genotype, time of harvest and various environmental factors. The yields tend to average 1.0% (Park et al. 2007; Saharkhiz et al. 2016). Nepetalactone can be found and potentially isomerize into eight different stereoisomers and four different diastereoisomers along with their corresponding enantiomers (Aničić et al. 2018). Nepetalactones are iridoid monoterpenes that originate from the non-mevalonate pathway or the methylerythritol phosphate (MEP) pathway (Mint Evolutionary Genomics Consortium 2018). The complete biosynthetic pathway of nepetalactone has not yet been elucidated. Researchers have however, isolated several of the enzymes responsible for nepetalactone formation (Lichman et al. 2019).

The two essential metabolic precursors for all terpenes are the five-carbon isomers: isopentenyl diphosphate (IPP) and dimethylallyl diphosphate (DMAPP) (Kuzuyama and Seto 2012). These two monomers are then synthesized into geranyl pyrophosphate (GPP) by the enzyme geranyl pyrophosphate synthase (GPPS) (Sherden et al. 2018). Subsequent conversion of GPP into geraniol is catalyzed by the geraniol synthase (GES) enzyme (Iijima et al. 2004; Sherden et al. 2018). Geraniol is then hydroxylated by geraniol 8-hydroxylase (G8H) to form 8-hydroxygeraniol (Hallahan et al. 1992; Collu et al. 2001; Sherden et al. 2018) and oxidized by 8-hydroxygeraniol oxidoreductase (8HGO) to form 8-oxogeranial (Miettinen et al. 2014; Sherden et al. 2018). The iridoid synthase (ISY) enzyme then converts 8-oxogeranial to its closed ring form nepetalactol and to its open-ring form iridodial (Geu-Flores et al. 2012; Sherden et al. 2018). It has been theorized that an oxidoreductase enzyme then finishes the conversion of nepetalactol to nepetalactone (Fig. 11.3); however, the exact mechanism is still unclear (Hallahan et al. 1998; Sherden et al. 2018).

Researchers have recently identified three new enzymes involved in the cyclization and oxidation of nepetalactols (Lichman et al. 2019). This group discovered a new class of nepetalactol-related short-chain dehydrogenase enzymes (NEPS) in *Nepeta mussinii* (Lichman et al. 2019). The NEPS enzymes are responsible for forming the distinct nepetalactol stereoisomers while also providing the final nepetalactone compound (Lichman et al. 2019). Future work is still needed to further clarify the exact regulatory pathways and genes involved in nepetalactone biosynthesis. The origins of stereocontrol in the nepetalactone isomers are important since the different isomers may exhibit differential bioactivities.

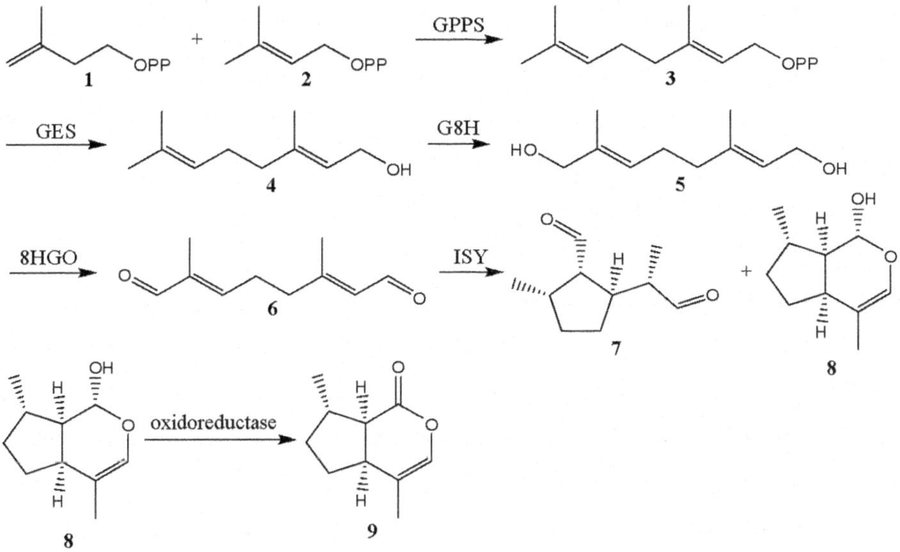

Fig. 11.3 Biosynthesis of *cis-trans*-nepetalactone 1: IPP; 2: DMAPP; 3: GPP; 4: geraniol; 5: 8-hydroxygeraniol; 6: 8-oxogeranial; 7: nepetalactol; 8: iridodial; 9: *cis-trans*-nepetalactone

Within the non-iridoid-producing Nepetoideae subfamily, *Nepeta* is a consistent producer of iridoids (El-Gazzar and Watson 1970; Kooiman 1972; Hegnauer 1989; Taskova et al. 1997; Wink 2003; Mint Evolutionary Genomics Consortium 2018). Metabolome studies have resulted in data showing that iridoids seemed to have disappeared in Nepetoideae, and then re-appeared in *Nepeta* specifically (Mint Evolutionary Genomics Consortium 2018). Other species within Nepetoideae also show great monoterpene diversity yet lack the complexity found in iridoids (Mint Evolutionary Genomics Consortium 2018).

Species within *Nepeta* produce several different stereoisomers of nepetalactone contained in the essential oils (De Pooter et al. 1988; Javidnia et al. 2011; Sharma and Cannoo 2013; Reichert et al. 2019). The identified and characterized nepetalactones within the genus include: 4aα,7α,7aα-nepetalactone, 4aα,7α,7aβ-nepetalactone, 4aβ,7α,7aβ-nepetalactone, 4aβ,7α,7aα-nepetalactone, 4aα,7β,7aβ-nepetalactone, and 4aα,7β,7aα-nepetalactone (McElvain and Eisenbraun 1957; Bellesia et al. 1979; Eisenbraun et al. 1980; Heuskin et al. 2009; Formisano et al. 2011; Süntar et al. 2018). The ratio of the available isomers may impact the efficacy of the essential oil as an insect repellant (Reichert et al. 2019). Yet essential oil samples containing various amounts of the purified and crude nepetalactone isomers have still been able to achieve optimal insect repellency (Reichert et al. 2019). Nepetalactones are also available in their hydrogenated forms such as α-dihydronepetalactone, β-dihydronepetalactone, 5,9-dehydronepetalactone, nepetalic acid, iridomyrmecin, and isoiridomyrmecin (Moghaddam and Hosseini 1996; Kalpoutzakis et al. 2001; Javidnia et al. 2011; Eom et al. 2005; Suschke et al. 2007; Heuskin et al. 2009; Süntar et al. 2018).

The composition of volatile catnip oil varies depending on the species and variety, growing site, climatic conditions, and analysis method in addition to seasonal and environmental factors (Baranauskiene et al. 2003; Setzer 2016). The relative content of nepetalactones within the essential oil can range from non-detectable to very high and the range varies across *N.* spp. In *N. cataria*, two new cultivars, CR9 and CR3 have been released that are richer in total essential oil and with elevated levels of E,Z-nepetalactone (4aα,7α,7aβ-nepetalactone) upwards of 45–75% of the total aromatic volatiles significantly higher than available commercial varieties (Reichert et al. 2018; Simon et al. 2019). Yet, Aydin et al. (1998) reported an essential oil from *Nepeta causarea* to contain >90% nepetalactones. While it is likely that such a concentration of nepetalactones was from a concentrated essential oil rather than the steam distilled essential oil, that level is what would be a target in plant breeding efforts. Other aromatic volatile components include β-pinene, β-caryophyllene, humulene, caryophyllene oxide, α-pinene, and carvone at differential concentrations as well as citral in the lemon catnips (Reichert et al. 2019).

11.5 Production of Catnip

Catnip is produced as a garden herb across much of the United States where it is grown as an ornamental and an herbal tea, for its dried leaves and flowers. Small farmers have reported limited acreage in the east coast, midwest and Pacific Northwest regions. The larger-scale production of catnip for essential oil has historically largely come from the Canadian provinces by those involved in the production of other essential oil crops, including mint. The USDA National plant Germplasm System maintains the catnip and *Nepeta* spp. collection at the USDA/ARS site in Ames, Iowa (www.ars-grin.gov).

11.5.1 Plant Propagation

Catnip can be either sexually propagated or cloned through vegetative cuttings and/or micropropagation techniques. The most common way to propagate the plant for commercial purposes is by seeds, which can be both directly sown or sown in trays to produce seedlings that will later be transplanted to the field. A study performed in two locations in the state of New Mexico concluded that despite the extra costs related to transplanting, the increase in biomass yield is enough to make it more profitable than direct sowing (Falk et al. 2000). The authors emphasize, however, that both direct sowed and transplanted catnip can generate a positive economic return for growers in that state (Falk et al. 2000), yet as with any minor specialty crop, the profitability determination must be based upon final marketable yields, costs of production and prices for that product in a given year.

To produce transplants, catnip seeds are in general sown into trays in green-houses or into warm plant beds 60–65 days prior to transplanting, which in the northern temperate zone in North America occurs between March and April such as in the State of North Carolina (Ferguson et al. 1990). The same authors reported that air temperature should not exceed than 24 °C for two consecutive days during seedling growth and coverings should be done when temperatures are below 7 °C. Once seedlings start to grow they need to be thinned to ensure proper growth. Seedlings are suitable for transplanting when they reach 15–20 cm of height (Ferguson et al. 1990). Transplants can also be produced under greenhouse conditions using commercial soil media and polypropylene plug trays. According to our experience at the New Jersey Agricultural Experiment Station, 45 days are enough to produce 20 cm high transplants in such conditions.

Our experiences indicate that good seed emergence and the establishment of a strong stand of catnip can be challenging. In part this appears to be due to low seed viability of catnip, inadequate soil preparation or lack of proper irrigation to use post seeding. Seeding rate and soil emergence are among the most initial critical issues in catnip production when field produced by seed. Commercial growers need to use high quality catnip seed, sown at high density with specialty planters to maintain a shallow sowing depth as the seed size is very small (0.7–1.4 mm in diameter). Emergence can vary over a long time period (from several weeks to over a month) as well. Proper irrigation to maintain moist soil surface and using nozzles that provide more of a mist appears to improve stand establishment. According to Ferguson et al. (1990), *N. cataria* seeds germinate rapidly and produce healthy seedlings at temperatures between 20 and 30 °C. The type of soil media can also affect the germination percentage of catnip seeds. Seeds sown in a mixture of sand and clay loamy soil (2 volumes of sand soil +1 volume of clay loam) performed better than seeds planted in pure sand soil or in a mixture of sand soil and clay loam in equal amounts (Ibrahim et al. 2017). In the same experiment, the authors also concluded that the most appropriate months for sowing the crop are February and March, when germination percentages reached up to 80% (Ibrahim et al. 2017). Asgari et al. (2015) reported that 1 to 2-month chilling treatment (4 °C) on seeds can increase germination percentage, seedling vigor and germination speed in *N. cataria* and other *Nepeta* spp. In another study, treatment with 1 mM solution of GA$_3$ (gibberelic acid) containing 500 mg L^{-1} nystatin was reported to induce seed germination for *in vitro* cultures of catnip (Dmitrović et al. 2016). Studies on strategies to improve catnip seed germination rates, speed and uniformity (e.g.: gibberellic acid application, thermic treatments, priming, etc.) are lacking.

Vegetative propagation of catnip, although not commonly used for commercial purposes due to the expense of large-scale cloning and transplanting, can be useful for cloning select genotypes for germplasm testing and breeding programs. Reichert et al. (2019) reported successfully propagating catnip by terminal node cuttings, briefly dipping the base of the propagules in 0.3% talc solution of indole-3-butyric acid (IBA) and subsequently placing the cuttings in mist houses until roots develop. In a different experiment, terminal and subterminal, single-node *N. cataria* cuttings were also rooted under mist chamber and reported to accumulate more root biomass if treated with IBA in talc solution at concentrations up to 0.8% (St. Hilaire et al. 2002). Rim and Jang (2017) recommended the use of saprolite as soil media and treatment with Rootone® (1-naphthaleneacetamide and tetra methyl thiuramdisulfide) for maximum rooting percentages of catnip cuttings. Regarding the type of propagule, terminal cuttings (9 cm long with at least 3 buds), single-node apical cuttings (3 cm long with one pair of leaves) and tip cuttings (3 cm long with terminal bud and a pair of leaves) can be successfully used for catnip cloning with survival and rooting rates over 90% (St. Hilaire 2003).

Members of the catnip family (Lamiaceae) are also commonly multiplied *in vitro* (i.e.: micropropagated) for mass distribution of disease-free plants or germplasm cloning (Tisserat and Vaughn 2004). For *N. cataria*, studies on micropropagation protocols are limited. Stimart (1986) reported that axillary shoot proliferation can be achieved from shoot tips but provided no description of culture media and other specifications of the culture system. Tisserat and Vaughn (2004) described a successful establishment of catnip shoot culture in Murashige and Skoog media with pH adjusted to 5.7, and emphasized the use of alternative tissue culture techniques such as high CO_2 and the automated plant culture system (APCS) as promising to improve plant growth and morphogenesis *in vitro*. Seed surface sterilization was reported to be done with 20% solution of commercial bleach (0.8% active chlorine) for 10 min and rinsing 5 times in sterile distilled water (Dmitrović et al. 2016) or with 70% (v/v) ethanol for 30 s and 2% (v/v) sodium hypochlorite solution for 10 min, then rinsed three times in sterilized water (Lee et al. 2010). Protocols for nodal segments or shoot tip surface sterilization, still need to be studied to improve cloning of selected genotypes. Research on acclimatization, *ex vitro* rooting of micropropagated shoot segments and field performance of *in vitro* propagated plants have yet to be examined in detail for catnip.

11.5.2   *Environmental Requirements and Plant Nutrition*

Catnip has been reported to grow naturally in temperate and subtropical areas, with soil pH ranging between 5.8 and 7.5 and average annual temperatures between 7 and 19 °C (Duke 1976). Temperatures during various periods of growth are among the main factors that can affect catnip productivity (Manukyan and Schnitzler 2006). In an experiment designed to assess the effects of different air temperatures on *N. cataria* var. *citriodora* (lemon catnip), 15 °C and 25 °C were observed to induce the highest contents of essential oil, while 20 °C was considered the ideal temperature for biomass and essential oil yield per plant (Manukyan and Schnitzler 2006). The increase of temperature from 15 °C to 25 °C reduced the contents of major compounds such as geraniol, geranial and trans-3-hexen-1-ol, while augmenting the contents of citronellol and nerol (Manukyan and Schnitzler 2006).

Studies on catnip nutrition and ideal soil conditions are also limited and as such present a future line of research to pursue to increase productivity. Until more specific research is available, nutrition, fertility and water management guidelines for mint can be applied. Catnip responds well to high nitrogen, well-drained soil and organic soil amendments. Park et al. (2007) reported successfully growing catnip in well drained sandy soil supplemented with 34 kg of nitrogen per hectare, applied via irrigation system. Exceptional nepetalactone yields, up to 1.54 g/plant, were achieved with the new cultivar CR9 planted in disc plowed cultivated land, with drip irrigation, plastic mulch and 900 lbs./acre of 15–15–15 fertilizer prior to planting (Reichert et al. 2016). Recently, the foliar application of potassium humate (400 ppm) has also been reported to increase essential oil and flavonoids contents in *N. cataria* and *N. cataria* var. *citriodora* (Mohamed et al. 2018).

A more specific and controlled attempt to understand nutritional aspects of lemon catnip is reported by Manukyan (2005). While the maximum content of essential oil (0.246% of fresh biomass) was obtained with N-P-K at a ratio of 15:70:15, the highest essential oil yield (1.56 ml/plant) was obtained with $N_{70}P_{15}K_{15}$, under hydroponic conditions. Mathematical modeling predicted that the optimum ratio of N-P-K for hydroponically grown lemon catnip is 53:40:7 atom%, with an estimated yield of 2.49 mL/plant (Manukyan 2005). In another nutritional study conducted under controlled environmental conditions to investigate the effects of nitrogen concentrations (50, 150 and 300 mg/l in nutrient solution) on lemon catnip growth and secondary metabolite production, the greatest biomass accumulation and essential oil yield as well as the greatest contents of citronellol, nerol, caffeic acid, rosmarinic acid and p-coumaric acid were observed when applying the highest nitrogen concentration (Manukyan 2011b). Future studies on catnip nutrition, specifically those addressing elemental composition and rates of nutrient absorption, will provide more precise recommendations, which can be adapted to the correction of different soils based both on soil conditions and plant necessity and ability to absorb and assimilate mineral nutrients.

Another fundamental feature for catnip growth and productivity relates to the availability of photosynthetically active radiation (PAR) (Manukyan 2013). In a comparison among field growing conditions, greenhouse with no artificial lighting and a treatment with additional PAR (15 mol m^{-2} day^{-1}), lemon catnip plants presented maximum essential oil content (%) and yield (ml/plant) when grown under additional PAR (Manukyan 2013). As for the oil composition, relative contents of neral and geranial increased while contents of citronellol and nerol were slightly reduced when plants were treated with no additional PAR (Manukyan 2013).

Given the sensitivity of the species to levels of PAR, plant spacing can also be an important aspect related to catnip productivity, since it affects plant competition for light, in addition to access to water and nutrients (Lambers et al. 2008). Ferguson et al. (1990) recommended that catnip transplants must be placed 23–30 cm apart at row widths of 76–97 cm. Park et al. (2007) tested different plant densities for catnip in the state of New Jersey and concluded that the highest yields of biomass and essential oil were obtained from the highest plant population (80,742 plants/ha). In that study, essential oil yields more than doubled from the lowest (26,909 plants/ha) to the highest plant population, increasing from 3.4 to 8.9 kg/ha (Park et al. 2007). Additional studies on ideal plant densities are still needed to improve land use and catnip productivity and may also represent a strategic approach for cultural weed control.

11.5.3 Plant Physiological Responses to Environmental Stress

Relative to other members of the Lamiaceae family, *N. cataria* is susceptible to several environmental stresses (Reichert et al. 2016). While these stresses can hinder biomass accumulation and, therefore, jeopardize productivity, deliberately applying stresses under ideal conditions, can in some cases increase secondary metabolite production and improve essential oil yield and quality (Al-Gabbiesh et al. 2015). In one study with 30 ornamental plants, catnip was ranked as one of the most dehydration tolerant herbs, with a lethal leaf water potential of -5.17 MPa (Augé et al. 2003). Despite being resistant to dehydration as a species, as a crop, catnip plants are reported to produce less biomass, lower essential oil yields (ml/plant) and lower contents of caffeic, rosmarinic and p-coumaric acid under drought stress (Manukyan 2011a). An experiment conducted in a controlled environment concluded that growth characteristics (plant height, number of branches and biomass) and essential oil yield (ml/plant) were significantly decreased with the rise in water stress levels, while essential oil content (% of leaves biomass) was stimulated in response to water stress for both *N. cataria* and *N. cataria* var. *citriodora* (Said-Al et al. 2016). In the same experiment, foliar application of salicylic acid (200 mg/l) was observed to be a practical approach for improving the performance of catnip plants under drought stress, increasing biomass and essential oil accumulation (Said-Al et al. 2016).

Another frequent and important environmental stress is salinity, a major threat to modern agriculture, causing inhibition and impairment of crop growth and development (Isayenkov and Maathuis 2019). Catnip appears sensitive to salinity stress since saline conditions have been reported to reduce vegetative growth, starch content, membrane stability, chlorophyll contents and photosynthetic rates, while increasing soluble sugars, proline concentrations and essential oil content (Mohammadi et al. 2017). Foliar applications of polyamines, specifically spermidine, may present a possible strategy to prevent sodium and chlorine toxicity, promoting better growth and preventing leaf abscission in *N. cataria* (Mohammadi et al. 2017).

11.5.4 Catnip as a Potential Soil Remediator

Catnip was reported to be relatively tolerant to heavy metal stress and has even been reported as a potential agent for phytoremediation. A mix of six herbs, including *N. cataria*, showed promising results on cadmium sequestration from soil (Yang et al. 2016). Catnip was also reported to grow spontaneously on sites affected by chromite mining with extreme levels of Cr and Ni, being recommended as a potential species for remediation of such areas (Nawab et al. 2015). Additionally, catnip has been reported to efficiently uptake Pb, Zn and Cd from contaminated soil without transferring such metals to the essential oil during plant growth or processing (Angelova 2012). Soil contamination with petroleum hydrocarbons, can cause phytotoxicity and inhibit the growth and development of higher plants (Potashev et al.

2014). While catnip seed germination has been reported to be significantly damaged by kerosene and *n*-tridecane contaminated soils (Sharonova and Breus 2012; Potashev et al. 2014), rhizosphere soil from catnip was reported to mineralize the herbicide atrazine, suggesting the use of the crop as a potential agent to facilitate microbial detoxification of unwanted organic compounds in soils (Anderson and Coats 1995). The possibility of using this crop as a phytoremediation agent is highly speculative given the wider array of other plants known to be phytoremediators from which to select. Production of catnip in soils with heavy metals and/or other toxic compounds is not recommended when the use of the dry, processed or essential oil is to be used for health, nutrition and/or as an insect repellent agent.

11.6   Harvesting

Catnip usually reaches a peak of essential oil yield when in full bloom (Figs. 11.1 and 11.2), after this stage essential oil yields decrease (Chalchat and Lamy 1997; Ferguson et al. 1990; Ibrahim et al. 2017; Mohammadi and Saharkhiz 2011). The first harvest usually commences in early to mid-August from direct seeded crops sown that same season in April/early May and for fields established from transplants, in late May/early July in the northern hemisphere. Catnip in the northern temperate zones is grown commercially either as an annual crop or short-lived perennial (2–3 growing seasons only, due to winter injury, loss of stand and increased weed pressure). Thus, second- and third-year catnip is normally harvested earlier than first-year catnip, which tends to occur in June. A few newer varieties including catnips, CR3 and CR9, were developed to allow a second harvest each year (Reichert et al. 2016, Simon et al. 2019).

Several reports describe the differences not only on yield but also on the chemical composition of essential oils from catnip plants harvested at different phenological stages (Bourrel et al. 1993; Hornok et al. 1992; Ibrahim et al. 2017; Mohammadi and Saharkhiz 2011; Schultz et al. 2004). For *N. cataria* var. *citriodora,* for example, maximum essential oil content was observed at the beginning of flowering for 1-year-old plants and at full flowering for 2-year-old plants, with gradual increases in neral and citronellol up to the stage of full flowering (Hornok et al. 1992). Catnip plants grown in Egypt yielded more essential oil (0.45% of leaves dry mass) at the stage of full flowering when compared to vegetative and seed formation stages (Ibrahim et al. 2017). Similarly, for plants grown in Iran, essential oil yields (w/w %) were 0.3, 0.5, 0.9, and 0.4% at vegetative, floral budding, full flowering, and fruit set stages, respectively, with the highest proportions of nepetalactones as major oils constituents at fruit set and full flowering (Mohammadi and Saharkhiz 2011). Increased nepetalactone content while flowering was also observed in catnip plants originating from the Province Center of France, with significant reductions of β-caryophyllene relative contents as flowering progressed (Bourrel et al. 1993).

Seasonality effect is also reported to affect the ratio of nepetalactone isomers in the plants as well as the repellent activity of catnip essential oil. Plants collected at the end of May showed greater $Z,E:E,Z$ nepetalactone ratios when compared to plants harvested in June and July in the state of Iowa (Schultz et al. 2004). The same study reports that plants collected in July presented the highest repellency to cockroaches (*Blattella germanica*) at a concentration of 0.1% in comparison to plants collected at different dates. The authors emphasize that no clear relationship was observed between the fluctuations of the $Z,E:E,Z$ nepetalactone ratio and the repellency (Schultz et al. 2004).

The time of day when catnip is harvested was reported to significantly influence secondary metabolite content. Plants collected at 11 am and 4 pm that were subsequently shade-dried and extracted with 70% ethanol,presented different contents of polyphenols and flavones, although no clear pattern was established on what is the best time to harvest due to the large variation between collection dates (Duda et al. 2015b). Studies on basil (Filho et al. 2006) and roses (Younis et al. 2009), for example, demonstrated that different times of harvesting within the same day can influence yields and quality of essential oils. In this regard, studies are still needed to determine the best time of day to harvest catnip to achieve maximum essential oil and nepetalactone contents.

Harvesting of catnip is performed by clipping the stems about 10 cm above the crown, thus allowing for regrowth and, if done in proper timing, for a second harvest in the same growing season (Ferguson et al. 1990). Ibrahim et al. (2017) reported two harvests in the same season for catnip plants grown in Egypt, where longer growing seasons are noted as compared to northern temperate zones found in North America. According to the authors, the first cut yielded greater essential oil content (0.25%) than the second cut (0.19%), although the second cut produced a larger amount of leaves biomass. Both harvests combined yielded 37.6 liters of oil per hectare, with nepetalactone contents varying from 22.6% in the first cut to 24.67% in the second cut (Ibrahim et al. 2017). Ferguson et al. (1990) stressed, however, that at higher elevations, cutting a second time within the same season could result in significant injuries over the winter and, therefore, may jeopardize the survival and use of catnip plants as a perennial/semi-perennial crop. We have observed that late autumn harvests even at sea level elevations can lead to increased winter injury the following growing season. Indeed, catnip is known to be sensitive to winter injury and has been observed to present poor re-growth after the first season in northern temperate zones (Park et al. 2007; Reichert et al. 2016). As a result, the crop is sometimes grown as an annual rather than a perennial in some cases (Park et al. 2007) particularly where winter survival and regrowth is problematic. Rutgers' cultivars CR9 and CR3, however, survived winter conditions in the state of New Jersey and exhibited the least winter injury and dieback compared with the commercial catnips evaluated, performing consistently as a perennial horticultural crop (Reichert et al. 2016; Simon et al. 2019).

According to Ferguson et al. (1990), a stand of catnip should last for 3 years, after which the quality of the crop decreases. This statement is in agreement with our observations with catnip in North America. Hornok et al. (1992) also reported that 2-year-old plantations tend to produce more biomass than 1-year-old ones, with an average 79% increase. Given the increase in yield and the costs associated with transplants, continuing development of new winter-resistant cultivars as well as the study of agronomic strategies that improve winter survival and catnip regrowth for a second and third year are important approaches that can lead to better economic return of the crop.

11.7 Postharvest Handling

Postharvest handling systems are based upon the intended end-product. For dry catnip, such as those plant materials in catnip toys and herbal products, the plant is normally cut by hand or mechanically and the harvested leaves and/or leaves and flowers are either allowed to sun dry or immediately moved into a dryer to reduce the moisture content to 10%. When using dryers, special attention must be directed to the temperature applied, since excessively high temperatures can volatilize and decrease the oil content of the dried catnip (Ferguson et al. 1990). Reichert et al. (2016) reported yields up to 1.54 g/plant in the new catnip cultivar CR9 after drying at 37 °C in a walk-in forced air commercial tobacco dryer converted to the drying of herbs and botanicals.

In a comparison between freshly harvested plants, sun-drying, shade-drying, oven-drying at 35, 45, and 55 °C and microwave-drying at 100 and 200 W, the highest essential oil yield after hydrodistillation in Clevenger-type apparatus was observed in 55 °C oven-dried *N. cataria*. Total nepetalactone content was equal in the fresh catnip and oven-dried plants at 55 °C, although some differences in the proportion of isomers was observed (Mohammadizad et al. 2017).

For the production of essential oils, the catnip can be collected and steam distilled when fresh, partially dried or completely dry (Chalchat and Lamy 1997; Gonzalez et al. 2012; Mohammadizad et al. 2017; Park et al. 2007). The freshly cut biomass (leaves and leaves plus flowers) can be distilled immediately or allowed to partially sun-dry as practiced with mint for essential oil production. Both dried and fresh plant material can be extracted by hydrodistillation or steam distillation, the most common methods of obtaining catnip essential oil for commercial purposes.

For *N. cataria* var. citriodora, the essential oil yield obtained using steam distillation was reported to be better than the yield from hydrodistillation, (1.5% and 0.5% of dry biomass, respectively), but the concentration of neral and geranial in the oil was higher when extracted by hydrodistillation (Wesołowska et al. 2011).

Park et al. (2007) reported successfully using a pilot-scale 500-liter portable steam distillation unit coupled with a special separator to capture the two fractions observed when catnip essential oil is distilled (Fig. 11.4a). Catnip essential oil can elute in two fractions, one lighter (Fig. 11.4b) and the other heavier (Fig. 11.4c) than water, a fact observed both in steam distillation and hydrodistillation when using the Clevenger-type apparatus (Fig. 11.4) (Clevenger 1928). While the specific gravity of essential oil of catnip essential oil is close to that of water, altering the essential oil of catnip to be higher in nepetalactone content

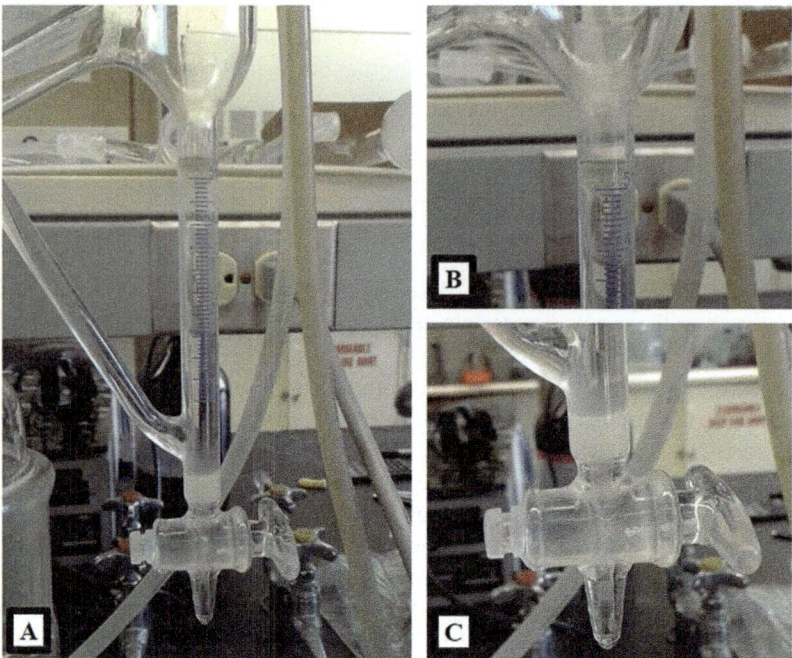

Fig. 11.4 Essential oil of Catnip obtained by hydrodistillation in Clevenger-type apparatus. (**a**) Both fractions of catnip essential oil; (**b**) Lighter fraction; (**c**) Heavier fraction

will results in a the larger oil fraction separating and needing to be collected under the water layer.

In 2012 a patent was awarded to improve the methods of *N. cataria* steam distillation (Gonzalez et al. 2012). According to the authors, catnip essential oil does not readily form a separate oil phase from the condensed water in the traditional steam distillation process and nepetalactone hydrates at high temperatures and is converted in unwanted byproducts. The new method overcomes those disadvantages by the addition of salts to the aqueous phase, which reduces the solubility of the oil in the water and allows the oil to be less dense than the aqueous phase, avoiding the above mentioned problems of the separation in two different phases. The author's approach also reduces the amount of water used in the process as well as decreases the steam temperature under vacuum (Gonzalez et al. 2012).

In addition to distillation methods, catnip oil can also be separated by supercritical CO_2 extraction and solvent extractions. Supercritical CO_2 extraction yielded from 4.6 to 5.7% of oil, while the traditional solvent extraction method generated material that represented 11.2% of catnip biomass, both superior to the essential oil yields obtained by conventional distillation methods (Louey et al. 2001). Louey et al. (2001) emphasize however, that solvent and supercritical fluid extracts can contain large amounts of impurities, therefore demanding further methods of purification adding to the cost of the final product. Other solvent extractions have also been reported using Soxhlet apparatus for 8 h, with hexane, trichloro-1,1,2-trifluoro-1,2,2-ethane (T.T.E.), dichloromethane, chloroform and ethanol (Bourrel et al. 1993). Maximum nepetalactone yield (57.24 g/kg of plant material) was reported when using ethanol, while the lowest content (3.92 g/kg) was obtained with T.T.E (Bourrel et al. 1993). Additional methods including solvent extraction with ethylic ether (Saeidnia et al. 2008), simultaneous distillation–solvent extraction, static headspace, and solid phase microextraction have been reported to isolate aroma volatiles in *N. cataria* and *N. cataria* var. *citriodora* (Baranauskiene et al. 2003).

Commercial catnip oil in North America and globally is generally produced using steam or hydro-distillation with the same commercial equipment and operations as employed in the distillation and separation of other aromatic essential oils including basil, coriander seed, dill, lavender, oregano, peppermint and spearmint. Growers of essential oils with their distillation equipment are in a unique position to commercially produce catnip essential oil given their expertise and the possible advantage of utilizing their distillation facilities either prior to or after their core essential crops are harvested and distilled. Growers producing catnip for the dried market only are generally located in the western states in North America or the Canadian prairies where the drying conditions for all herbs and agronomic crops are predictable and characterized by longer sunny days, low humidity, and little to no rainfall during the time when catnip is harvested.

The final quality of catnip essential oil is determined by its chemical composition, usually assessed by gas chromatography coupled to mass spectrometry (GC-MS) (Baranauskiene et al. 2003; Reichert et al. 2016). Nepetalactones can also be analyzed by reversed phase HPLC coupled with UV and MS detection, in a simple and reliable method as described in detail by Wang et al. (2007). Quality factors considered in addition to the essential oil yield and composition are similar to those factors in all essential oils including the polarity, viscosity and clarity of the oil as well as freedom from pesticides and other contaminants.

The major types of nepetalactones include: Z, E-nepetalactone and E, Z-nepetalactone (Fig. 11.5). By identifying the retention times of isolated standards and comparing them to the methanol extracts of *N. cataria*, the content of nepetalactones in samples can be identified and quantitated. Further use of ^{13}C-NMR allows

Fig. 11.5 Left: Z, E-nepetalactone (4aα,7α,7aα-nepetalactone; *cis*, *trans*-nepetalactone); Right: E, Z-nepetalactone (4aα,7α,7aβ-nepetalactone; *trans*, *cis*-nepetalactone)

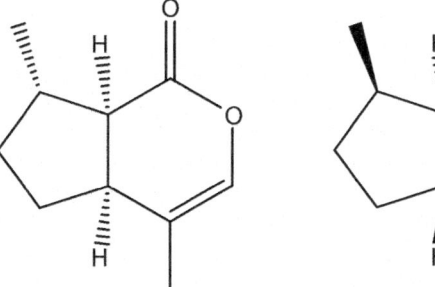

for the elucidation of stereochemistry. HPLC/MS with UV detection is a powerful, high throughput tool for large and small sample sizes and can be used for rapid and sensitive detection of the nepetalactones. Gas chromatography is often used for essential oil composition determination.

11.8   Weed Control

Weed competition remains a major issue for the production of medicinal plants, causing significant yield and quality decrease (Carrubba 2017). Weeds are the primary pest for catnip production, especially because of the very limited number of herbicides labeled for use on this crop (Ferguson et al. 1990). The lack of effective herbicides may limit the longevity of catnip stands to 3 years and significantly increase the costs of production, and with high weed pressure the catnip yield and quality of the crop will decrease (Ferguson et al. 1990). Lack of weed control reduced catnip yield by 38–53% in comparison to a hand-weeded control but did not significantly affect the concentration of nepetalactone (Duppong et al. 2004). Despite the fact that weeds are a significant obstacle to economically sustainable catnip commercial production and may affect the purity and quality of catnip essential oil, very little research is available regarding this topic.

Ninety commercial herbicide formulations are labeled for weed control on catnip in the United States and four in Canada, totalizing only six active ingredients (caprylic acid, carfentrazone, clethodim, fluazifop-*p*-butyl, glyphosate, pendimethalin) (Agrian Label Database 2019). By contrast, 18 active ingredients have commercial labels for use on spearmint (*Mentha spicata* L.), including ten active ingredients that can be applied preemergence versus a single compound (pendimethalin) on catnip (Agrian Label Database 2019). The lack of effective preemergence weed control options on catnip increase reliance on postemergence herbicides that may cause severe injury to catnip plants or result in unacceptable amount of herbicide residues in harvested parts of the plant. Currently, only one new active ingredient for postemergence weed control (bentazon) is being considered for review by the IR-4 Project (IR-4 Food Crop Database 2019). Thus, in the United States, there is a need for evaluating catnip response to various preemergence herbicides that have already received a label for use on other crops of the Lamiaceae family.

Duppong et al. (2004) have addressed the effect of flax straw, oat straw, and wool natural mulches on weed suppression and biochemical performances of catnip. Regardless of constituents, natural mulches in these studies provided on average 98% reduction of weed density on a variety of grasses that included giant foxtail (*Setaria faberi* Herrm), yellow foxtail (*S. glauca* L.), green foxtail (*S. viridis* L.), large crabgrass [*Digitaria sanguinalis* (L.) Scop.], witchgrass (*Panicum capillare* L.), and barnyardgrass [*Echinochloa crus-galli* (L.) P. Beauv.]. Density of broadleaf weeds such as Virginia pepperweed (*Lepidium virginicum* L.), common purslane (*Portulaca oleracea* L.), smooth groundcherry [*Physalis longifolia* Nutt. var. *subglabrata* (Mack. & Bush) Cronquist], and horsenettle (*Solanum carolinense* L.) was not affected by natural mulches. Dry biomass yield and nepetalactone content of plants mulched with flax straw or wool mulch were similar to results obtained by hand-weeding (Duppong et al. 2004). Additional studies of other mulching systems (such as plastic, wood chips) or cover crops are still needed to evaluate their impact on weed control as well as catnip productivity. Cultivation with a rolling-tine weeder early season followed by hand-hoeing as needed was sufficient for controlling weeds and successfully establishing transplanted or direct-seeded catnip in New Mexico (Kleitz et al. 2008). However, it has yet to be seen if mechanical weed control could be well adapted for catnip grown in less arid regions where multiple weed emergence flushes occur during the season and where abundant rainfalls may prevent the timely use of mechanical equipment. Hand-removal is the most-effective solution for controlling weeds, but is labor- and time-consuming with 200 to 600 manpower hours per hectare estimated for many aromatic and medicinal plants (Pank 1992).

Weed management of catnip is most important during the establishment phase of the crop. While catmint (*Nepeta x faasenii*) was shown to reduce light reaching the soil surface by more than 80% 2 years after transplantation, resulting in extremely low production of weed biomass in plots where catmint was planted (Eom et al. 2005), catnip is a poor competitor to weeds and both pre-emergence and post-emergence weed control is needed over the course of the entire growing season. Future research on weed management strategies for catnip production should primarily focus on weed control within catnip rows during the establishment phase of the crop and subsequently address the need for weed control between rows.

11.9   Main Diseases and Pests

11.9.1   Diseases

The need for an effective integrative pest management system is still needed and the lack of one has hindered further commercial production. Catnip is susceptible to numerous diseases. For example, with fungal diseases alone, the USDA GRIN Fungal Database lists 22 fungal-host interactions for *N. cataria* (Farr and Rossman 2019). Those interactions include Leaf Spot caused by *Ascochyta nepetae*, *Cercospora nepetae* and *Phyllosticta decidua* among many others (Alfieri et al. 1984; Cash 1953; Chupp 1954; Greene 1944, 1945; United States department of Agriculture (USDA) 1960). Other literature also reference *Phoma nepeticola* as causing leaf spot and stem lesions to catnip and other *Nepeta* species (Boerema et al. 2004; De Gruyter et al. 2002; Zimowska 2008).

Xanthomonas campestris was also identified to cause bacterial leaf spot on greenhouse grown catnip in California with host specificity, as no other Lamiaceae plants were able to become inoculated in their study (Koike et al. 2001). Symptoms began as small brown flecks on either side of the leaves, that later developed into larger, darker, more angular spots (Koike et al. 2001). This decreases the marketability for ornamental catnips based upon aesthetics since size and quality are affected (Koike et al. 2001).

Powdery mildew, widely affects members of the Lamiaceae family, including catnip (Braun 1987; Jamal et al. 2011). In the Ukraine, *Erysiphe biocellata*, was identified as the fungus causing the powdery mildew symptoms (Braun 1995). Catnip plants from other European countries have been known to be infected with other fungi. In France, Germany, Italy, Poland, Romania, the USSR and Yugoslavia, catnip was infected by *Erysiphe galeopsidis* (Amano 1986; Braun 1995). In Poland, catnip was also found to be infested by *Cladosporium* spp. and *Oidium* spp., but only *Oidium* spp. negatively affected the catnip leading to causing disease symptoms (Jadczak and Pinzon 2017).

Most recently in the literature, root rot caused by *Macrophomina phaseolina* was discovered in a catnip field in India (Nishad et al. 2018). In that study, identification using sequence analysis of internal transcribed spacer (ITS) region of rDNA was utilized and inoculation of healthy plants was achieved in a greenhouse study. Symptoms in the field included wilting along with dark brown necrotic lesions on the roots, black microsclerotia and even total vascular crumble (Nishad et al. 2018). Root rot can also be caused by *Rhizoctonia solani* in catnip, as noted by a report from the United States Department of Agriculture (1960).

The common mint rust, *Puccinia menthae* has been reported to infect catnip both in the field and greenhouse (Jamal et al. 2011). Symptoms included leaf fall, with dark brown uredial pustules on the leaves prior (Jamal et al. 2011). Other well-known fungal diseases that affect catnip include *Fusarium* sp. wilt and *Sclerotium rolfsii* Southern blight (United States Department of Agriculture 1960).

Given the range of plant diseases that can infect catnip, continual monitoring of the crop is needed. Table 11.1 presents a list of reported diseases on catnip.

Table 11.1 List of plant diseases reported on *Nepeta* spp.

Disease	Reported locations	Sources
Leaf spot (*Ascochyta nepetae*)	Wisconsin	Greene (1945), USDA (1960)
Leaf spot (*Cercospora nepetae*)	Florida, Illinois, Texas	Alfieri Jr. et al. (1984), Chupp (1954); USDA (1960)
Leaf spot (*Phyllosticta decidua*)	Ohio, Wisconsin	Cash (1953), Greene (1945), USDA (1960)
Leaf spot (*Xanthomonas campestris*)	California	Koike et al. (2001)
Leaf spot (*Phoma nepeticola*)	Europe, Netherlands, North America	Boerema et al. (2004), De Gruyter et al. (2002), Zimowska (2008)
Powdery Mildew (*Erysiphe biocellata*)	Ukraine	Braun (1995)
Powdery Mildew (*Erysiphe galeopsidis*)	France, Germany, Italy, Poland, Romania, USSR, Yugoslavia	Amano (1986), Braun (1995)
Powdery Mildew (*Oidium* spp.)	Poland	Jadczak and Pinzon (2017)

(continued)

Table 11.1 (continued)

Disease	Reported locations	Sources
Powdery Mildew (*Golovinomyces biocellatus*)	Russia, Ukraine	Dudka et al. (2004), Rusanov and Bulgakov (2008)
Root Rot (*Macrophomina phaseolina*)	India	Nishad et al. (2018)
Root Rot (*Rhizoctonia solani*)	Texas	USDA (1960)
Mint Rust (*Puccinia menthae*)	Pakistan	Jamal et al. (2011)
Wilt (*Fusarium* sp.)	Georgia	USDA (1960)
Southern Blight (*Sclerotium rolfsii*)	Texas	USDA (1960)
Gray Mold (*Botrytis cinerea*)	Alaska	USDA (1960)
Didymella catariae	New Jersey, Ohio	USDA (1960)
Phialea cyathoidea	Canada	Ginns (1986)
Phoma exigua	Ukraine	Dudka et al. (2004)
Phoma exigua var. *exigua*	Poland	Mulenko et al. (2008)
Puccinia menthae	Poland	Majewski (1979), Mulenko et al. (2008)
Septoria catariae	Romania	Radulescu et al. (1973)
Septoria nepetae	Poland, Texas, Wisconsin	Cash (1954), Mulenko et al. (2008), USDA (1960)
Sphaeria catariae	New Jersey	Cash (1954)
Teichospora nepetae	Canada	Cash (1954)
Xenodidymella catariae	Netherlands	Chen et al. (2017)
Cucumber Mosaic Virus (CMV)	Italy, Bulgaria	Davino et al. (2012), Dikova (2009)
Alfalfa Mosaic Virus (AMV)	Bulgaria	Dikova (2009)
Broad Bean Wilt Virus 1 (BBWV1)	Bulgaria	Dikova (2009)
Tomato Spotted Wilt Virus (TSWV)	Bulgaria	Dikova (2011), Khanzada et al. (2012)
Impatiens Necrotic Spot Virus (INSV)	Minnesota	McDonough et al. (1999)

11.9.2   Pests

Beyond plant pathogens, nematodes and other pests have also been reported to attack catnip. *Nepeta* spp. are affected by a wide range of nematodes. For example, *Nepeta savi is* affected by *Aphelenchoides fragariae* (Kohl 2011). Phytonematodes primarily are responsible for infections of the root, as they are present in the rhizosphere (Khanzada et al. 2012). In a study conducted in Pakistan on the nematode fauna associated with the rhizosphere of several members of the mint family, catnip was found to be associated with the phytoparasitic nematodes *Helicotylenchus* sp. and *Hoplolaimus* sp. (Khanzada et al. 2012). In terms of bacterial feeder nematodes, *Cephalobid* sp. and *Rhabditis* sp. were present with catnip (Khanzada et al. 2012). Nematodes are capable of initiating diseases as leaf and bud lesions and root-knot, determined by their mode of action (Khanzada et al. 2012).

Insect pests can also cause damage by themselves to catnip or serve as carriers of other diseases. Thrips have been known to be carriers of Tomato Spotted Wilt Virus (TSWV) and Impatiens Necrotic Spot Virus (INSV) (Dikova 2011; Khanzada et al. 2012; McDonough et al. 1999) Symptoms of TSWV range from typical mosaic, ring spots, localized necrotic spots to systemic chlorotic spots (Dikova 2011). Another thrips-trasmitted disease, Cucumber Mosaic Virus (CMV), is transmitted by almost 60 aphid species, and has a broad host range including catnip (Davino et al. 2012; Dikova 2009). In addition to CMV, Alfalfa Mosaic Virus (AMV) and Broad Bean Wilt Virus 1 (BBWV1) were also reported in Bulgaria (Dikova 2009). This study noted direct effect on essential oils composition and quality. The author mentioned that CMV affected the composition of the major constituents in the root extract of purple coneflower, as well as the yield of extracted oils from BBWV1 infected sage plants since they produced 2/3 the yield of healthy plants (Dikova 2009). The quality and yield of catnip oil could then be affected.

A few leafhopper species, the Ligurian leafhopper (*Eupteryx decemnotata*) and the mint leafhopper (*Eupteryx melissae* Curtis) have been reported to cause severe stippling damage on the leaves of catnip and other mint species throughout Europe and more recently in the United States (Halbert et al. 2009; Lowery et al. 2007; Nickel and Holzinger 2006; Rung et al. 2009). Their damage can be easily mistaken with that of thrips, lacebugs and mites but their high density causes more significant damage; in a case of catnip at a discount store in Florida, hundreds of cast skins were left behind indicating a large population (Halbert et al. 2009). The growing popularity of catnip could be in part leading to the expansion of these pest populations. For example, *Eupteryx decemnotata* likely spread from the trade of ornamental catnip from within Europe, especially in Germany (Nickel and Holzinger 2006). With the expansion of catnip production for essential oil as an insect repellent, an expansion of population of damaging pests should be expected and plans for the control developed.

Other insect pests of interest for catnip growers include the Cotton Bollworm *(Helicoverpa armigera)* and numerous species of thrips. During a study of aromatic plant hosts in India, *Helicoverpa armigera* was found to damage the leaves of field-grown catnip (Nadda 2013). Although they have only been reported to feed on catnip in India thus far, they can easily spread similar to that observed for the abovementioned species of leafhopper.

Thrips spp. can be found in the field, or in greenhouses with catnip. They can be found in high densities, particularly *T. fuscipennis* and *T. albopilosus* for catnip and other *Nepeta* species (Pobożniak and Anna 2011). Additionally, there are many other thrips species in which adults can be found associated with catnip, they

Table 11.2 List of insect pests reported on *Nepeta* spp.

Pest	Damage	Reported locations	Sources
Ligurian leafhopper (*Eupteryx decemnotata*)	Severe stippling on lower leaves	Europe: Austria, Britain, France, Germany, Greece, Italy, Portugal, Slovenia, Switzerland United States: California, Florida	Halbert et al. (2009), Nickel and Holzinger (2006), Rung et al. (2009)
Mint leafhopper (*Eupteryx melissae* Curtis)	Severe stippling of leaves	British Columbia, Canada	Lowery et al. (2007)
Thrips (*T. fuscipennis, T. albopilosus*)	Damages flowers and young leaves, can affect flowering, transmission of TSWV and INSV	Europe: Poland Canada United States: Indiana, Texas, New York, New Jersey, Maryland, Virginia, Colorado, California, District of Columbia, Ohio, New Mexico, Mississippi, Connecticut, Idaho, Illinois, Michigan, Louisiana, Tennessee, Pennsylvania, Florida, North Carolina, Arkansas, Iowa, Washington, Oregon	Chittenden (1919), McDonough et al. (1999), Poboźniak and Anna (2011)
Cotton Bollworm (*Helicoverpa armigera*)	Damages Leaves	India	Nadda (2013)

include: *Aeolothrips intermedius, Chirothrips manicatus, Frankliniella intonsa, T. atratus, T. flavus, T. major, T. physapus, T. tabaci, Haplothrips acanthoscelis, H. aculeatus,* and *H. statices* (Chittenden 1919; McDonough et al. 1999; Poboźniak and Anna 2011). With such a broad range, thrips can easily become problematic and even carry diseases, notably viruses, as discussed earlier. For example, catnip may be susceptible to a virus carried by *Frankliniella intonsa* or the Western Flower Thrip, as well as tospoviruses, Tomato Spotted Wilt Virus (TSWV) and Impatiens necrotic spot virus (INSV) (McDonough et al. 1999).

From a biodiversity study of thrips on 34 herbs in Poland, catnip was among one of the flowering herbs most greatly infested by thrips species (Poboźniak and Anna 2011). Thrips can normally locate their host by their pollen, as pollen is a major food source for them, but they also feed on flowers, inflorescences and young leaves (Poboźniak and Anna 2011). Thrips can be extremely damaging and affect further plant development. Table 11.2 presents a list of insect pests reported on catnip.

11.10   Conclusion

The primary applications for *Nepeta* spp. include the use as ornamental plants, and uses of their bioactive polyphenols and essential oils for health and nutrition. The fresh and dry leaves are used in traditional medicines and as herbal teas, while the dry leaves and essential oils have been used to attract domestic cats in cat toys. The applications of catnip oil, as a promising natural product for the repellency of arthropod, insects and other pests is of increasing interest as a tool to use in support of global health issues relative to mosquitoes, ticks and other pests that transmit human and animal diseases.

Catnip essential oils also show promise in the possible development of new medicinal and health applications given their antimicrobial, antioxidant, anti-inflammatory, sedative, relaxant, cholesterol lowering, anti-asthmatic, carminative, diuretic, diaphoretic, febrifuge, vermifuge, herbicidal, anxiolytic, and opioid-like effects (Aydin et al. 1998; Firoozabadi et al. 2015; Salehi et al. 2018; Zomorodian et al. 2012). *Nepeta cataria* has also been shown to produce significant amounts of iridodial that has been reported to attract goldeneyed lacewing, (*Chrysopa oculata*), which commonly eats pests like mites and aphids (Aldrich et al. 2016). These are just some illustrative applications of the potential uses of catnip.

The behavioral effects of catnip on felines are thought to be due to the interaction of nepetalactone with their opioid receptors (Espín-Iturbe et al. 2017). Additionally, there is evidence to support that the stereoisomer $4a\alpha,7\alpha,7a\alpha$-nepetalactone from the essential oil of *Nepeta caesarea* has specific opioid receptor subtype agonistic activity, conferring to such natural product significant analgesic and sedative activities (Aydin et al. 1998). Yet another feature of nepetalactone qualifying the molecule as a new prototype opioid analgesic is the fact that it has a lipophilic structure, which enables it to pass through the blood-brain barrier and be active in the central nervous system (Aydin et al. 1998). Given the current international opioid epidemic (Humphreys 2017), natural products from catnip show phenomenal potential as alternatives for pain relief and mitigation of opioid abuse, although more research is needed to determine the specific receptor or receptors subtypes acted on to develop new analgesic medicines, making it a challenging and exciting topic for future studies (Aydin et al. 1998).

The methanol extract of lemon catnip has shown strong positive permeation enhancing properties, as a result of the synergistic action of oleanolic and ursolic acids with other substances present in the dry extract, such as β-sitosterol 3-O- β-D-glucopyranoside (Winnicki et al. 2013). The abovementioned ursolic acid, reported to be present in the lemon-scented chemotype in contents up to 1.3% of dry biomass (Klimek and Modnicki 2005), is a triterpenoid of outstanding pharmacological importance, with potential anti-microbial, anti-oxidant, anti-inflammatory, anti-cancer and anti-angiogenesis effects (Kashyap et al. 2016).

Studies on murine models have also shown several potential applications and future opportunities for research and development of pharmaceutical products and dietary supplements based on *N. cataria* (Bernardi et al. 2010, 2011; Massoco et al. 1995). Some examples are the antidepressant effects of n-hexane extract of *N. cataria* (Bernardi et al. 2010;) and the modified dopaminergic-related behaviors with an amphetamine-like effect when catnip dried leaves were administered in the diet of mice (Massoco et al.,1995). The administration of dried leaves of *N. cataria* var. *citriodora* in the diet is also reported to increase penile erection and slightly improve male rat sexual behavior by an action on dopaminergic systems (Bernardi et al. 2011).

Chemotyping has revealed that many of these compounds (Salehi et al. 2018) vary significantly by species, populations/landraces and varieties. Given the essential oil has such promising potential as an insect repellent and as a biological pest control method (Reichert et al. 2019), current constraints that limit expanded production reside in the high cost of product due to the historically low biomass and essential oil yields of the available catnip and the low content of nepetalactone in the essential oils. New developments in crop improvement (Reichert et al. 2016; Simon et al. 2019) that can increase significantly total essential oil yields and increase the relative content of nepetalactone in the essential oil can stimulate larger-scale production, thus overcoming the limited supply and high cost of the 'raw essential oil'. New product development can move ahead when there is a secure oil supply. Coupled to the development of improved production systems that would lead to economic and effective weed control and pest management, growers could look toward catnip as an additional essential oil crop to complement their established essential oil crops, while simultaneously using the same distillation equipment and their agricultural and processing expertise prior to or following the processing of their core aromatic crops.

Acknowledgements We thank the New Jersey Farm Bureau, the New Jersey Agricultural Experiment Station Project NJ12158, the New Jersey Health Foundation and the Deployed Warfighter Protection (DWFP) Program in support of the research project, War FighterProtection from Arthropods Utilizing Naturally Sourced Repellents (Grant W911QY1910007) for their support of our research on *Nepeta* spp. We also recognize the Brazilian Federal Agency for Support and Evaluation of Graduate Education (CAPES) (DOC_PLENO/ proc. n° 88881.129327/2016-01) for providing a Graduate Student Fellowship to the senior author (E. Gomes) as this work is part of his dissertation studies.

References

Adiguzel A, Ozer H, Sokmen M, Gulluce M, Sokmen A, Kilic H, Sahin F, Baris O (2009) Antimicrobial and Antioxidant Activity of the Essential Oil and Methanol Extract of *Nepeta cataria*. Pol J Microbiol 58:69–76

Agrian Database (2019) Herbicide Catnip. Retrieved October 7, 2019, from https://www.agrian.com/labelcenter/results.cfm

Aldrich JR, Chauhan K, Zhang QH (2016) Pharmacophagy in green lacewings (Neuroptera: Chrysopidae: Chrysopa spp.)? PeerJ 4:e1564

Alfieri SA Jr, Langdon KR, Wehlburg C, Kimbrough JW (1984) Index of plant diseases in Florida. Florida Department of Agriculture & Consumer Services, Division of Plant Industry, Tallahassee

Al-Gabbiesh A, Kleinwächter M, Selmar D (2015) Influencing the contents of secondary metabolites in spice and medicinal plants by deliberately applying drought stress during their cultivation. Jordan J Biol Sci 147:1–10

Amano K (1986) Host range and geographical distribution of the powdery mildew fungi. Japan Science Society Press, Tokyo

Amer A, Mehlhorn H (2006) Repellency effect of forty-one essential oils against *Aedes, Anopheles*, and *Culex* mosquitoes. Parasitol Res 99:478–490

Anderson TA, Coats JR (1995) Screening rhizosphere soil samples for the ability to mineralize elevated concentrations of atrazine and metolachlor. J Environ Sci Health Part B 30(4):473–484

Angelova V (2012) Potential of some medicinal and aromatic plants for phytoremediation of soils contaminated with heavy metals. Agrarni Nauki 4(11):61–66

Aničić N, Matekalo D, Skorić M, Pećinar I, Brkušanin M, Živković JN, Mišić D (2018) Trichome-specific and developmentally regulated biosynthesis of nepetalactones in leaves of cultivated *Nepeta rtanjensis* plants. Ind Crop Prod 117:347–358

Asgari M, Nasiri M, Ashrafe Jafari A, Flah Hoseini L (2015) Investigation of chilling effects on characteristics of seed germination, vigor and seedling growth of *Nepeta* spp. J Rangeland Sci 5:313–324

Augé RM, Stodola AJ, Moore JL, Klingeman WE, Duan X (2003) Comparative dehydration tolerance of foliage of several ornamental crops. Sci Hortic 98:511–516

Aydin S, Beis R, Öztürk Y, Hüsnü K, Baser C (1998) Nepetalactone: a new opioid analgesic from *Nepeta caesarea* Boiss. J Pharm Pharmacol 50(7):813–817

Bandh SA, Kamili AN, Ganai BA, Lone BA, Saleem S (2011) Evaluation of antimicrobial activity of aqueous extracts of *Nepeta cataria*. J Pharm Res 4:3141–3142

Baranauskiene R, Venskutonis RP, Demyttenaere JC (2003) Sensory and instrumental evaluation of catnip (*Nepeta cataria* L.) aroma. J Agric Food Chem 51(13):3840–3848

Bellesia F, Pagnoni UM, Trave R, Andreetti GD, Bocelli G, Sgarabotto P (1979) Synthesis and molecular structures of (1S) -cis, cis-iridolactones. J Chem Soc Perkin Trans 2:1341–1346

Bernardi MM, Kirsten TB, Salzgeber SA, Ricci EL, Romoff P, Guilardi Lago JH, Lourenço LM (2010) Antidepressant-like effects of an apolar extract and chow enriched with *Nepeta cataria* (catnip) in mice. Psychol Neurosci 3(2):251–258

Bernardi MM, Kirsten TB, Lago JHG, Giovani TM, de Oliveira Massoco C (2011) *Nepeta cataria* L. var. *citriodora* (Becker) increases penile erection in rats. J Ethnopharmacol 137(3):1318–1322

Bernier UR, Furman KD, Kline DL, Allan SA, Barnard DR (2005) Comparison of contact and spatial repellency of catnip oil and N, N-diethyl-3-methylbenzamide (DEET) against mosquitoes. J Med Entomol 42(3):306–311

Birkett MA, Hassanali A, Hoglund S, Pettersson J, Pickett JA (2011) Repellent activity of catmint, *Nepeta cataria*, and iridoid nepetalactone isomers against Afro-tropical mosquitoes, ixodid ticks and red poultry mites. Phytochemistry 72:109–114

Boerema GH, De Gruyter J, Noordeloos ME, Hamers MEC (2004) Phoma identification manual: differentiation of specific and infra-specific taxa in culture. CABI Publishing, Wallingford

Bol S, Caspers J, Buckingham L, Anderson-Shelton GD, Ridgway C, Buffington CT, Bunnik EM (2017) Responsiveness of cats (Felidae) to silver vine (*Actinidia polygama*), Tatarian honeysuckle (*Lonicera tatarica*), valerian (*Valeriana officinalis*) and catnip (*Nepeta cataria*). BMC Vet Res 13:70

Bourrel C, Perineau F, Michel G, Bessiere JM (1993) Catnip (*Nepeta cataria* L.) essential oil: analysis of chemical constituents, bacteriostatic and fungistatic properties. J Essent Oil Res 5(2):159–167

Braun U (1987) A monograph of the *Erysiphales* (powdery mildews). Beihefte zur Nova Hedwigia 89:1–700

Braun U (1995) The Powdery mildews (Erysiphales) of Europe. Gustav Fischer Verlag, Jena

Carrubba A (2017) Weeds and weeding effects on medicinal herbs. In: Ghorbanpour M, Varma A (eds) Medicinal plants and environmental challenges. Springer, Cham, pp 295–327

Cash EK (1953) A record of the fungi named by J.B. Ellis (Part 2). USDA National Agricultural Library, Beltsville

Cash EK (1954) A record of the fungi named by J.B. Ellis (Part 3). USDA National Agricultural Library, Beltsville

Chalchat JC, Lamy J (1997) Chemical composition of the essential oil isolated from wild catnip *Nepeta cataria* L. cv. *citriodora* from the Drôme region of France. J Essent Oil Res 9(5):527–532

Chen Q, Hou LW, Duan WJ, Crous PW, Cai L (2017) Didymellaceae revisited. Stud Mycol 87:105–159

Chittenden FH (1919) Farmers bulletin 1007: Control of the onion thrips. United States Department of Agriculture, Washington, DC

Chupp C (1954) Monograph of the fungus genus Cercospora. Published by the Author, Ithaca

Clapperton BK, Eason CT, Weston RJ, Woolhouse AD, Morgan DR (1994) Development and testing of attractants for feral cats, Felis catus L. Wildl Res 21:389–399

Claßen-Bockhoff R, Wester P, Tweraser E (2003) The staminal lever mechanism in Salvia L. (Lamiaceae)-a review. Plant Biol 5(01):33–41

Clevenger JF (1928) Apparatus for the determination of volatile oil. J Am Pharm Assoc 17(4):345–349

Collu G, Unver N, Peltenburg-Looman AM, van der Heijden R, Verpoorte R, Memelink J (2001) Geraniol 10 hydroxylase1, a cytochrome P450 enzyme involved in terpenoid indole alkaloid biosynthesis. FEBS Lett 508(2):215–220

Croteau R, Gershenzon J (1994) Genetic control of monoterpene biosynthesis in mints (Mentha: Lamiaceae). In: Genetic engineering of plant secondary metabolism. Springer, Boston, pp 193–229

Daskalova T (2004) Histological structure of the microsporangia, microsporogenesis and development of the male gametophyte in *Nepeta cataria* (Lamiaceae). Phytologia Balcanica 10(2–3):241–246

Davino S, Panno S, Rangel EA, Davino M, Bellardi MG, Rubio L (2012) Population genetics of cucumber mosaic virus infecting medicinal, aromatic and ornamental plants from northern Italy. Arch Virol 157(4):739–745

De Gruyter J, Boerema GH, Van Der HA (2002) Contributions towards a monograph of Phoma (Coelomycetes) – VI. 2. Section Phyllostictoides: Outline of its taxa. Persoonia 18:1–53

De Pooter HL, Nicolai B, De Laet J, De Buyck LF, Schamp NM, Goetghebeur P (1988) The essential oils of five *Nepeta* species. A preliminary evaluation of their use in chemotaxonomy by cluster analysis. Flavour Fragr J 3(4):155–159

Dikova B (2009) Establishment of some viruses-Polyphagues on economically important essential oil-bearing and medicinal plants in Bulgaria. Biotechnol Biotechnol Equip 23(Sup1):80–84

Dikova B (2011) Tomato spotted wilt virus on some medicinal and essential oil-bearing plants in Bulgaria. Bulgarian J Agric Sci 17(3):306–313

Dmitrovic S, Skoric M, Boljevic J, Anicic N, Bozic D, Misic D, Opsenica D (2016) Elicitation effects of a synthetic 1, 2, 4, 5-tetraoxane and a 2, 5-diphenylthiophene in shoot cultures of two *Nepeta* species. J Serb Chem Soc 81(9):999–1012

Duppong LM, Delate K, Liebman M, Horton R, Romero F, Kraus G, Petrich J, Chowdbury PK (2004) The effect of natural mulches on crop performance, weed suppression and biochemical constituents of catnip and St. John's wort Crop Sci 44(3):861–869

Duda SC, Mărghitaş LA, Dezmirean DS, Bobiş O, Duda MM (2015a) *Nepeta cataria*: medicinal plant of interest in phytotherapy and bee keeping. Hop Med Plants 23:34–38

Duda SC, Mărghitaş LA, Dezmirean DS, Duda M, Mărgăoan R, Bobiş O (2015b) Changes in major bioactive compounds with antioxidant activity of *Agastache foeniculum, Lavandula angustifolia, Melissa officinalis* and *Nepeta cataria*: effect of harvest time and plant species. Ind Crop Prod 77:499–507

Dudka IO, Heluta VP, Tykhonenko YY, Andrianova TV, Hayova VP, Prydiuk MP, Dzhagan VV, Isikov VP (2004) Fungi of the Crimean Peninsula (Translated from Russian). M.G. Kholodny Institute of Botany, National Academy of Sciences of Ukraine, Kiev

Duke JA (1976) Perennial weeds as indicators of annual climatic parameters. Agric Meteorol 16(2):291–294

Eisenbraun EJ, Browne CE, Irvin-Willis RL, McGurk DJ, Eliel EL, Harris DL (1980) Structure and stereochemistry of 4aα, 7α, 7aβ-nepetalactone from *Nepeta mussini* and its relationship to the 4aα, 7α, 7aα-nepetalactone and 4aβ, 7α, 7aβ-nepetalactone from *Nepeta cataria*. J Org Chem 45(19):3811–3814

El Gazzar A, Watson L (1970) Some economic implications of the taxonomy of Labiatae essential oils and rusts. New Phytol 69(2):487–492

Eom SH, Senesac AF, Tsontakis-Bradley I, Weston LA (2005) Evaluation of herbaceous perennials as weed suppressive groundcovers for use along roadsides or in landscapes. J Environ Hortic 23(4):198–203

Espín-Iturbe LT, Yañez BAL, García AC, Canseco-Sedano R, Vázquez-Hernández M, Coria-Avila GA (2017) Active and passive responses to catnip (Nepeta cataria) are affected by age, sex and early gonadectomy in male and female cats. Behav Process 142:110–115

Falk CL, Voorthuizen HV, Wall MM, Guldan SJ, Martin CA, Kleitz KM (2000) An economic analysis of transplanting versus direct seeding of selected medicinal herbs in New Mexico. J Herbs Spices Med Plants 7(4):15–29

Farr DF, Rossman AY (2019) Fungal databases, U.S. National Fungus Collections, ARS, USDA. Retrieved September 25, 2019, from https://nt.ars-grin.gov/fungaldatabases/

Feaster JE, Scialdone MA, Todd RG, Gonzalez YI, Foster JP, Hallahan DL (2009) Dihydronepetalactones deter feeding activity by mosquitoes, stable flies, and deer ticks. J Med Entomol 46(4):832–840

Ferguson JM, Weeks WW, Fike WT (1990) Production of catnip in North Carolina. In: Janick J, Simon JE (eds) Advances in new crops. Timber Press, Oregon, pp 527–528

Filho JLS, Blank AF, Alves PB, Ehlert PA, Melo AS, Cavalcanti SC, Silva-Mann R (2006) Influence of the harvesting time, temperature and drying period on basil (Ocimum basilicum L.) essential oil. Rev Bras 16(1):24–30

Firoozabadi A, Zarshenas MM, Salehi A, Jahanbin S, Mohagheghzadeh A (2015) Effectiveness of Cuscuta planiflora Ten. and Nepeta menthoides Boiss. & Buhse in major depression: a triple-blind randomized controlled trial study. J Evid Based Complement Alternat Med 20(2):94–97

Formisano C, Rigano D, Senatore F (2011) Chemical constituents and biological activities of Nepeta species. Chem Biodivers 8:1783–1818

Geu-Flores F, Sherden NH, Courdavault V, Burlat V, Glenn WS, Wu C, O'Connor SE (2012) An alternative route to cyclic terpenes by reductive cyclization in iridoid biosynthesis. Nature 492:138–142

Gilani AH, Shah AJ, Zubair A, Khalid S, Kiani J, Ahmed A, Ahmad VU (2009) Chemical composition and mechanisms underlying the spasmolytic and bronchodilatory properties of the essential oil of Nepeta cataria L. J Ethnopharmacol 121(3):405–411

Ginns JH (1986) Compendium of plant disease and decay fungi in Canada 1960–1980. Canadian Government Publishing Centre, Ottawa

Gonzalez Y, Jackson SC, Manzer LE (2012) U.S. Patent No. 8,329,229. U.S. Patent and Trademark Office, Washington, DC

Greene HC (1944) Notes on Wisconsin parasitic fungi V. Wis Acad Sci 36:225–268

Greene HC (1945) Notes on Wisconsin parasitic fungi VII. Am Midl Nat 34(1):258–270

Halbert SE, Rung A, Ziesk DC, Gill RJ (2009) A leafhopper pest of plants in the mint family, Eupteryx Decemnotata Rey, Ligurian Leafhopper, New To North America and intercepted in Florida on plants from California (PDF file). Retrieved September 26, 2019, from https://www.fdacs.gov/content/download/68538/file/Pest_Alert_-_Eupteryx_Decemnotata,_Ligurian_Leafhopper.pdf

Hallahan DL, Dawson GW, West JM, Wallsgrove RM (1992) Cytochrome P-450 catalysed monoterpene hydroxylation in Nepeta mussinii. Plant Physiol Biochem 30(4):435–443

Hallahan DL, West JM, Smiley DW, Pickett JA (1998) Nepetalactol oxidoreductase in trichomes of the catmint Nepeta racemosa. Phytochemistry 48(3):421–427

Hegnauer (ed) (1989) Chemotaxonomie der Pflanzen. Lehrbücher und Monographien aus dem Gebiete der Exakten Wissenschaften. Birkhäuser, Basel

Herron S (2003) Catnip, Nepeta cataria, a morphological comparison of mutant and wild type specimens to gain an ethnobotanical perspective. Econ Bot 57:135–142

Heuskin S, Godin B, Leroy P, Capella Q, Wathelet JP, Verheggen F et al (2009) Fast gas chromatography characterisation of purified semiochemicals from essential oils of

Matricaria chamomilla L. (Asteraceae) and *Nepeta cataria* L. (Lamiaceae). J Chromatogr A 1216:2768–2775

Hornok L, Domokos J, Hethelyi E (1992) Effect of harvesting time on the production of *Nepeta cataria* var. *citriodora* Balb. Acta Hortic 306:290–295

Humphreys K (2017) Avoiding globalization of the prescription opioid epidemic. Lancet 390:437–439

IBISWorld (2019) Insect repellent manufacturing. Retrieved October 7, 2019 from https://clients1. ibisworld.com/reports/us/industry/ataglance.aspx?entid=4948

Ibrahim ME, El-Sawi SA, Ibrahim FM (2017) *Nepeta cataria* L, one of the promising aromatic plants in Egypt: seed germination, growth and essential oil production. J Mat Environ Sci 8:1990–1995

Iijima Y, Gang DR, Fridman E, Lewinsohn E, Pichersky E (2004) Characterization of geraniol synthase from the peltate glands of sweet basil. Plant Physiol 134:370–379

IR-4 Food Crop Database (2019) Herbicide Catnip. Retrieved October 7, 2019, from http://ir4app. rutgers.edu/ir4FoodPub/simpleSch.aspx

Isayenkov SV, Maathuis FJ (2019) Plant salinity stress: many unanswered questions remain. Front Plant Sci 10:1–11

ITIS Report (2019) *Nepeta cataria* L. (n.d.). Retrieved October 5, 2019, from https://www.itis.gov/ servlet/SingleRpt/SingleRpt?search_topic=TSN&search_value=32623#null

Jadczak P, Pizoń K (2017) Identification of taxa of microscopic fungi occurring on selected herbal plants and possible methods of their elimination. World Sci News 69:1–17

Jamal A, Naeemullah M, Masood MS, Zakria M, Tahira R, Iqbal U, Iqbal SM (2011) Screening of mint germplasm under field and glasshouse conditions. Mycopathologia 9:29–32

Jamzad Z, Ingrouille M, Simmonds MS (2003) Three new species of *Nepeta* (Lamiaceae) from Iran. Taxon 52(1):93–98

Javidnia K, Mehdipour AR, Hemmateenejad B, Rezazadeh SR, Soltani M, Khosravi AR, Miri R (2011) Nepetalactones as chemotaxonomic markers in the essential oils of *Nepeta* species. Chem Nat Compd 47(5):843–847

Kaiser C, Ernst M (2019) Catnip. Retrieved August 17, 2019, from http://www.uky.edu/ccd/sites/ www.uky.edu.ccd/files/catnip.pdf

Kalpoutzakis E, Aligiannis N, Mentis A, Mitaku S, Charvala C (2001) Composition of the essential oil of two Nepeta species and in vitro evaluation of their activity against *Helicobacter pylori*. Planta Med 67:880–883

Kashyap D, Tuli HS, Sharma AK (2016) Ursolic acid (UA): a metabolite with promising therapeutic potential. Life Sci 146:201–213

Khanzada SA, Naeemullah M, Munir A, Iftikhar S, Masood S (2012) Plant parasitic nematodes associated with different *Mentha* species. Pak J Nematol 30(1):21–26

Kleitz KM, Wall MM, Falk CL, Martin CA, Remmenga MD, Guldan SJ (2008) Stand establishment and yield potential of organically grown seeded and transplanted medicinal herbs. Hortic Technol 18(1):116–121

Klimek B, Modnicki D (2005) Terpenoids and sterols from *Nepeta cataria* L. var. *citriodora* (Lamiaceae). Acta Pol Pharm 62(3):231–235

Kohl LM (2011) Astronauts of the nematode world: an aerial view of foliar nematode biology, epidemiology, and host range. Retrieved September 26, from https://www.apsnet.org/edcenter/ apsnetfeatures/Pages/foliarnematodes.aspx

Koike ST, Azad HR, Cooksey DA (2001) *Xanthomonas* leaf spot of catnip: a new disease caused by a pathovar of *Xanthomonas campestris*. Plant Dis 85(11):1157–1159

Kolalite MR (1998) Comparative analysis of ultrastructure of glandular trichomes in two *Nepeta cataria* chemotypes (*N. cataria* and *N. cataria* var. *citriodora*). Nord J Bot 18(5):589–598

Kooiman P (1972) The occurrence of iridoid glycosides in the Labiatae. Acta Botanica Neerlandica 21(4):417–427

Kuzuyama T, Seto H (2012) Two distinct pathways for essential metabolic precursors for isoprenoid biosynthesis. Proce Jpn Acad Ser B Phys Biol Sci 88(3):41–52

Lambers H, Chapin FS III, Pons TL (2008) Plant physiological ecology. Springer, New York
Lee SY, Lee CY, Eom SH, Kim YK, Park NI, Park SU (2010) Rosmarinic acid production from transformed root cultures of *Nepeta cataria* L. Sci Res Essays 5(10):1122–1126
Lewis WH, Elvin-Lewis MP (1982) Medical botany: plants affecting man's health. Wiley, New York
Lichman BR, Kamileen MO, Titchiner GR, Saalbach G, Stevenson CE, Lawson DM, O'Connor SE (2019) Uncoupled activation and cyclization in catmint reductive terpenoid biosynthesis. Nat Chem Biol 15(1):71–79
Linnaeus C (1800) Species plantarum, vol 4. Impensis GC Nauk, Berlin
Louey J, Petersen N, Salotti D, Shaeffer H, James KD (2001) Oil of catnip by supercritical fluid extraction. Pap Am Chem Soc 221:215–221
Lowery DT, Triapitsyn SV, Judd GJ (2007) Leafhopper host plant associations for *Anagrus parasitoids* (Hymenoptera: Mymaridae) in the Okanagan Valley, British Columbia. J Entomol Soc BC 104:9–16
Majewski T (1979) Fungi of Poland (Mycota), basidiomycetes, uredinles II, vol 11 (Translated from Polish). State Science Public House, Kraków
Malizia RA, Molli JS, Cardell DA, Retamar JA (1996) Volatile consituents of the essential oil of Nepeta cataria L. grown in Cordoba Province (Argentina). J Essent Oil Res:565–567
Manukyan AE (2005) Optimum nutrition for biosynthesis of pharmaceutical compounds in celandine and catmint under outside hydroponic conditions. J Plant Nutr 28(5):751–761
Manukyan A (2011a) Effect of growing factors on productivity and quality of lemon catmint, lemon balm and sage under soilless greenhouse production: I. drought stress. Med Aromat Plant Sci Biotechnol 5(2):119–125
Manukyan A (2011b) Effect of growing factors on productivity and quality of lemon catmint, lemon balm and sage under soilless greenhouse production: II. Nitrogen stress. Med Aromat Plant Sci Biotechnol 5(2):126–132
Manukyan AE (2013) Effects of PAR and UV-B radiation on herbal yield, bioactive compounds and their antioxidant capacity of some medicinal plants under controlled environmental conditions. Photochem Photobiol 89(2):406–414
Manukyan AE, Schnitzler WH (2006) Influence of air temperature on productivity and quality of some medicinal plants under controlled environment conditions. Eur J Hortic Sci 71(1):26–35
Marmy RMS, Norulakmal NH, Faridah G (2018) Optimization of *Nepeta cataria* essential oil extraction yield by ultrasonic-soxhlet extraction method using response surface methodology. IOP Conf Ser Mater Sci Eng 440:1–10
Martin E, Altinordu F, Özcan T, Dirmenci T (2013) Karyomorphological study in *Nepeta viscida* Boiss.(Lamiaceae) from Turkey. J Appl Biol Sci 7(3): 26–30. E-ISSN: 2146-0108
Massoco CO, Silva MR, Gorniak SL, Spinosa MS, Bernardi MM (1995) Behavioral effects of acute and long-term administration of catnip (*Nepeta cataria*) in mice. Vet Hum Toxicol 37(6):530–533
McDonough MJ, Gerace D, Ascerno ME (1999) Western flower thrips feeding scars and tospovirus lesions on petunia indicator plants. Retrieved September 26, from https://conservancy.umn.edu/bitstream/handle/11299/51810/1/7375.pdf
McElvain SM, Eisenbraun EJ (1957) The interconversion of nepetalic acid and isoiridomyrmecin (iridolactone). J Org Chem 22:976–977
McElvain SM, Bright RD, Johnson PR (1941) The constituents of the volatile oil of catnip. I. Nepetalic acid, nepetalactone and related compounds. J Am Chem Soc 63:1558–1563
Miettinen K, Dong L, Navrot N, Schneider T, Burlat V, Pollier J, Woittiez L, van der Krol S, Lugan R, Ilc T, Verpoorte R, Oksman-Caldentey KM, Martinoia E, Bouwmeester H, Goossens A, Memelink J, Werck-Reichhart D (2014) The seco-iridoid pathway from *Catharanthus roseus*. Nat Commun 5:3606
Mihaylova D, Georgieva L, Pavlov A (2013) In Vitro antioxidant activity and phenolic composition of Nepeta cataria L. extracts. Int J Agric Sci Technol 1(4):74–79

Mint Evolutionary Genomics Consortium (2018) Phylogenomic mining of the mints reveals multiple mechanisms contributing to the evolution of chemical diversity in Lamiaceae. Mol Plant 11(8):1084–1096

Missouri Botanical Garden (2019) Plant finder – *Nepeta cataria*. Retrieved April 24, 2019, from http://www.missouribotanicalgarden.org/PlantFinder/PlantFinderDetails. aspx?kempercode=e433

Modnicki D, Tokar M, Klimek B (2007) Flavonoids and phenolic acids of *Nepeta cataria* L. var. *citriodora* (Becker) Balb.(Lamiaceae). Acta Pol Pharm 64(3):247–252

Moghaddam FM, Hosseini M (1996) Composition of the essential oil from Nepeta crassifolia Boiss. & Buhse. Flavour Fragr J 11:113–115

Mohamed HF, Mahmoud AA, Alatawi A, Hegazy MH, Astatkie T, Said-Al Ahl HA (2018) Growth and essential oil responses of *Nepeta* species to potassium humate and harvest time. Acta Physiol Plant 40:1–8

Mohammadi S, Saharkhiz MJ (2011) Changes in essential oil content and composition of catnip (*Nepeta cataria* L.) during different developmental stages. J Essent Oil Bearing Plants 14(4):396–400

Mohammadi S, Saharkhiz MJ, Javanmardi J (2017) Evaluation of interaction effects of spermidine and salinity on physiological and morphology trait of catnip (*Nepeta cataria* L.). Zeitschrift Fur Arznei-& Gewurzpflanzen 22(3):104–109

Mohammadizad HA, Mehrafarin A, Naghdi Badi H (2017) Qualitative and quantitative evaluation of essential oil of Catnip (*Nepeta cataria* L.) under different drying conditions. J Med Plants 1(61):8–20

Mountain Rose Herbs (2019) Mountain rose herbs: *Catnip*. Retrieved April 19, 2019, from https://www.mountainroseherbs.com/products/catnip/profile

Mułenko W, Majewski T, Ruszkiewicz-Michalska M (2008) A preliminary checklist of micromycetes in Poland. (Kraków W). Szafer Institute of Botany, Polish Academy of Sciences, Poland

Nadda G (2013) Medicinal and aromatic crops as hosts of *Helicoverpa armigera* Hübner (Lepidoptera: Noctuidae). J Trop Asian Entomol 2:44–46

Naghibi F, Mosaddegh M, Mohammadi MM, Ghorbani A (2010) Labiatae family in folk medicine in Iran: from ethnobotany to pharmacology. Iran J Pharm Res 4:63–79

Nawab J, Khan S, Shah MT, Khan K, Huang Q, Ali R (2015) Quantification of heavy metals in mining affected soil and their bioaccumulation in native plant species. Int J Phytoremediation 17(9):801–813

Nickel H, Holzinger WE (2006) Rapid range expansion of Ligurian leafhopper, *Eupteryx decemnotata* Rey, 1891 (Hemiptera: Cicadellidae), a potential pest of garden and greenhouse herbs, in Europe. Russ Entomol J 15:295–301

Nishad I, Srivastava AK, Saroj A, Babu BK, Samad A (2018) First report of root rot of *Nepeta cataria* caused by *Macrophomina phaseolina* in India. Plant Dis 102(11):2380

Pank F (1992) The influence of chemical weed control on quality characters of medicinal and aromatic plants. Acta Hortic 306:145–154

Park CH, Tannous P, Juliani HR, Wu QL, Sciarappa WJ, VanVranken R, Simon JE (2007) Catnip as a source of essential oils. Creating markets for economic development of new crops and new uses (ed: Whipkey A), pp 311–315

Patience GS, Karirekinyana G, Galli F, Patience NA, Kubwabo C, Collin G, Boffito DC (2018) Sustainable manufacture of insect repellents derived from *Nepeta cataria*. Sci Rep 8:22–35

Payandeh M, Bordbar F, Mirtadzadini M, Khaniki GRB (2015) New chromosome counts for *Nepeta* (Lamiaceae) from flora of Iran. Biol Divers Conserv 8:70–73

Peterson CJ, Coats J (2001) Insect repellents-past, present and future. Pestic Outlook 12:154–158

Peterson CJ, Coats JR (2011) Catnip essential oil and its nepetalactone isomers as repellents for mosquitoes. In: Paluch GE, Coats JR (eds) Recent developments in invertebrate repellents. American Chemical Society, Washington, DC, pp 59–65

Peterson CJ, Nemetz LT, Jones LM, Coats JR (2002) Behavioral activity of catnip (Lamiaceae) essential oil components to the German cockroach (Blattodea: Blattellidae). J Econ Entomol 95:377–380

Pobożniak M, Anna S (2011) Biodiversity of thrips species (Thysanoptera) on flowering herbs in Cracow, Poland. J Plant Prot Res 51(4):393–398

Polsomboon S, Grieco JP, Achee NL, Chauhan KR, Tanasinchayakul S, Pothikasikorn J, Chareonviriyaphap T (2008) Behavioral responses of catnip (*Nepeta cataria*) by two species of mosquitoes, *Aedes aegypti* and *Anopheles harrisoni*, in Thailand. J Am Mosq Control Assoc 24:513–520

Potashev K, Sharonova N, Breus I (2014) The use of cluster analysis for plant grouping by their tolerance to soil contamination with hydrocarbons at the germination stage. Sci Total Environ 485:71–82

Proestos C, Boziaris IS, Nychas GJE, Komaitis M (2006) Analysis of flavonoids and phenolic acids in Greek aromatic plants: Investigation of their antioxidant capacity and antimicrobial activity. Food Chem 95:664–671

Rabbani M, Sajjadi SE, Mohammadi A (2008) Evaluation of the anxiolytic effect of Nepeta persica Boiss. in mice. eCAM 5:181–186

Radulescu E, Negru A, Docea E (1973) Septoriozele din Romania. Editura Academiei Republicii Socialiste România, Bucarest

Raeburn D, Souness JE, Tomkinson A, Karlsson J-A (1993) Isozyme-selective cyclic nucleotide phosphodiesterase inhibitors: Biochemistry, pharmacology and therapeutic potential in asthma. Prog Drug Res 40:9–32

Reichert WJ (2019) The phytochemical investigation, breeding and arthropod repellent efficacy of Nepeta cataria. Doctoral dissertation, Rutgers University-School of Graduate Studies

Reichert W, Park HC, Juliani HR, Simon JE (2016) 'CR9': a new highly aromatic catnip *Nepeta cataria* L. cultivar rich in Z, E-Nepetalactone. HortScience 51:588–591

Reichert W, Villani T, Pan MH, Ho CT, Simon JE, Wu QL (2018) Phytochemical analysis and anti-inflammatory activity of *Nepeta cataria* accessions. J Med Active Plants 7(1):19–27

Reichert W, Ejercito J, Guda T, Dong X, Wu Q, Ray A, Simon JE (2019) Repellency assessment of *Nepeta cataria* essential oils and isolated nepetalactones on *Aedes aegypti*. Sci Rep 9(1):1524

Rigano D, Arnold NA, Conforti F, Menichini F, Formisano C, Piozzi F, Senatore F (2011) Characterisation of the essential oil of *Nepeta glomerata* Montbret et Aucer ex Bentham from Lebanon and its biological activities. Nat Prod Res:614–626

Rim JA, Jang EJ (2017) Effects of substrate and Rootone on the rooting of *Mentha spicata*, *Mentha x piperita*, and *Nepeta cataria*. J People Plants Environ 20(5):511–520

Rung A, Halbert SE, Ziesk DC, Gill RJ (2009) A leafhopper pest of plants in the mint family, *Eupteryx dec emnotata* Rey (Hemiptera: Auchenorrhyncha: Cicadellidae), Ligurian leafhopper, new to North America. Insecta Mundi 88:1–4

Rusanov VA, Bulgakov TS (2008) Powdery mildew fungi of Rostov region. Mikol Fitopatol 42(4):314–322

Saeidnia S, Gohari AR, Hadjiakhoondi A (2008) Trypanocidal activity of oil of the young leaves of *Nepeta cataria* L. obtained by solvent extraction. J Med Plants 1:54–57

Saharkhiz MJ, Zadnour P, Kakouei F (2016) Essential oil analysis and phytotoxic activity of catnip (*Nepeta cataria* L.). Am J Essent Oils Nat Prod 4(1):40–45

Said-Al Ahl HA, Sabra AS, Hegazy MH (2016) Salicylic acid improves growth and essential oil accumulation in two *Nepeta cataria* chemotypes under water stress conditions. Ital J Agrometeorol 21(1):25–36

Said-Al Ahl H, Naguib NY, Hussein MS (2018) Evaluation growth and essential oil content of catmint and lemon catnip plants as new cultivated medicinal plants in Egypt. Ann Agric Sci 63:201–205

Salehi B, Valussi M, Jugran AK, Martorell M, Ramírez-Alarcón K, Stojanović-Radić ZZ, Setzer WN (2018) *Nepeta species*: From farm to food applications and phytotherapy. Trends Food Sci Technol 80:104–122

Schultz G, Simbro E, Belden J, Zhu J, Coats J (2004) Catnip, *Nepeta cataria* (Lamiales: Lamiaceae) – a closer look: seasonal occurrence of nepetalactone isomers and comparative repellency of three terpenoids to insects. Environ Entomol 33(6):1562–1569

Setzer WN (2016) Catnip essential oil: there is more to it than making your cat go crazy. Am J Essent Oils Nat Prod 4(4):12–15

Sharma A, Cannoo DS (2013) Phytochemical composition of essential oils isolated from different species of genus *Nepeta* of Labiatae family: a review. Pharmacophore 4(6):181–211

Sharonova N, Breus I (2012) Tolerance of cultivated and wild plants of different taxonomy to soil contamination by kerosene. Sci Total Environ 424:121–129

Shen D, Pan MH, Wu QL, Park CH, Juliani HR, Welch CR, Simon JE (2010) Identification of the anti-inflammatory bioactive compounds in oregano (*Origanum* spp.) and their simultaneous quantitation by LC/MS. J Agric Food Chem 58(12):7119–7125

Shen D, Pan MH, Wu QL, Park CH, Juliani HR, Ho CT, Simon JE (2011) A rapid LC/MS/MS method for the analysis of non-volatile anti-inflammatory agents from *Mentha* spp. J Food Science 76(6):900–908

Sherden NH, Lichman B, Caputi L, Zhao D, Kamileen MO, Buell CR, O'Connor SE (2018) Identification of iridoid synthases from *Nepeta* species: iridoid cyclization does not determine nepetalactone stereochemistry. Phytochemistry 145:48–56

Sih A, Baltus MS (1987) Patch size, pollinator behavior, and pollinator limitation in catnip. Ecology 68(6):1679–1690

Simon JE, Reichert W, Wu QL (2019) Catnip cultivar 'CR3'. US Patent 10,512,231 B2, December 24, 2019

Simpson MG (2010) Plant systematics. Academic, Amsterdam

Smitherman LC, Janisse J, Mathur A (2005) The use of folk remedies among children in an urban black community: remedies for fever, colic, and teething. Pediatrics 115:e297–e304

St. Hilaire R (2003) Propagation of catnip by terminal and single-node cuttings. J Environ Hortic 21(1):20–23

St. Hilaire R, Hockman AW, Chavez SM (2002) Propagation and irrigation regime affect the development of catnip. Acta Hortic 629:321–327

Stimart DP (1986) Commercial micropropagation of florist flower crops. In: Tissue culture as a plant production system for horticultural crops. Springer, Dordrecht, pp 301–315

Süntar I, Nabavi SM, Barreca D, Fischer N, Efferth T (2018) Pharmacological and chemical features of Nepeta L. genus: Its importance as a therapeutic agent. Phytother Res 32(2):185–198

Suschke U, Sporer F, Schneele J, Geiss HK, Reichling J (2007) Antibacterial and cytotoxic activity of *Nepeta cataria* L., *N. cataria* var. citriodora (Beck.) Balb. and Melissa officinalis L. essential oils. Nat Prod Commun 2:1277–1286

Taskova R, Mitova M, Evstatieva L, Ancev M, Peev D, Handjieva N, Popov S (1997) Iridoids, flavonoids and terpenoids as taxonomic markers in Lamiaceae, Scrophulariaceae and Rubiaceae. Bocconea 5(2):631–636

Technavio. Global mosquito repellent market 2018–2022. Retrieved August 17, 2019 from https://www.technavio.com/report/global-mosquito-repellent-market-analysis-share-2018

Tisserat B, Vaughn SF (2004) Techniques to improve growth, morphogenesis and secondary metabolism responses from Lamiaceae species in vitro. Acta Hortic 629:333–340

Transparency Market Research (TMR). Catnip essential oil market – global industry analysis, size, share, growth, trends, and forecast 2017–2025. Retrieved August 17, 2019 from https://www.transparencymarketresearch.com/catnip-essential-oil-market.html

United States Department of Agriculture (USDA) (1960) Index of plant diseases in the United States. USDA, Washington, DC

United States Department of Agriculture (USDA) (2019) *Nepeta cataria* L. Catnip. Retrieved August 17, 2019, from https://plants.usda.gov/core/profile?symbol=NECA2#

Waller GR, Price GH, Mitchell ED (1969) Feline attractant, cis, trans-nepetalactone: Metabolism in the domestic cat. Science 164(3885):1281–1282

Wang M, Cheng KW, Wu Q, Simon JE (2007) Quantification of nepetalactones in catnip (*Nepeta cataria* L.) by HPLC coupled with ultraviolet and mass spectrometric detection. Phytochem Anal 18(2):157–160

Wesołowska A, Jadczak D, Grzeszczuk M (2011) GC-MS analysis of lemon catnip (*Nepeta cataria* L. var. *citriodora* Balbis) essential oil. Acta Chromatogr 23(1):169–180

Wink M (2003) Evolution of secondary metabolites from an ecological and molecular phylogenetic perspective. Phytochemistry 64(1):3–19

Winnicki AM, Śmieszek JM, Partyka D, Modnicki D (2013) Permeation-enhancing properties of *Nepeta cataria* var. *citriodora* dry extract. Herba Polonica 59(3):5–16

Yang Y, Wang XY, Wang J, Zhao TT, Cheng SY, Shao DD, Xu J (2016) Effects of species diversity on plant growth and remediation of Cd contamination in soil. Acta Sci Circumst 36:2103–2113

Younis A, Riaz A, Khan MA, Khan AA (2009) Effect of time of growing season and time of day for flower harvest on flower yield and essential oil quality and quantity of four *Rosa* cultivars. Floricult Ornamental Biotechnol 3:98–103

Zhu J, Zeng X (2006) Adult repellency and larvicidal activity of five plant essential oils against mosquitoes. J Am Mosq Control Assoc 22(3):515–523

Zhu JJ, Zeng XP, Berkebile D, Du HJ, Tong Y, Qian K (2009) Efficacy and safety of catnip (*Nepeta cataria*) as a novel filth fly repellent. Med Vet Entomol 23:209–216

Zhu JJ, Berkebile DR, Dunlap CA, Zhang A, Boxler D, Tangtrakulwanich K, Brewer G (2012) Nepetalactones from essential oil of *Nepeta cataria* represent a stable fly feeding and oviposition repellent. Med Vet Entomol 26:131–138

Zimowska B (2008) Biodiversity of fungi colonizing and damaging selected parts of motherwort (*Leonurus cardiaca* L.). Herba Pollonica 54(2):30–40

Zomorodian K, Saharkhiz MJ, Shariati S, Pakshir K, Rahimi MJ, Khashei R (2012) Chemical composition and antimicrobial activities of essential oils from *Nepeta cataria* L. against common causes of food-borne infections. ISRN Pharm 2012:1–7

Zomorodian K, Saharkhiz MJ, Rahimi MJ, Shariatifard S, Pakshir K, Khashei R (2013) Chemical composition and antimicrobial activities of essential oil of *Nepeta cataria* L. against common causes of oral infections. J Dent 10:329–337

Chapter 12
25 Years of DSHEA: Impact on Supply, Conservation and Sustainability, GACPs and Regulatory Compliance of Botanical Ingredients

Josef A. Brinckmann, Edward J. Fletcher, Rupa Das, and Trish Flaster

Abstract Prior to October 1994 in the United States, botanical substances were regulated as ingredients of drug products, both over-the-counter (OTC) and prescription (Rx), or as components of food products. Most of the thousands of medicinal plant species in global commerce were not expressly permitted for use in food products. This fact, coupled with the reality that the U.S. Food and Drug Administration (FDA) had established panels to review all old drugs and establish new monographs, led to the eventual re-classification of most botanicals as non-monograph. Thus, by the early 1990s many botanicals had no safe harbor, as they were determined to be unlawful for use as components of both foods and drugs.

The tension caused by these facts is a backdrop, that leads to the U.S. Congress establishing a new regulatory framework, between food and drug, the Dietary Supplement Health and Education Act of 1994 (DSHEA). With the passage of DSHEA, most herbs of commerce, if intended for oral ingestion, became "old dietary ingredients" and could now be marketed with "nutrient content" or "structure function" claim statements. Botanicals intended for topical application were not protected by DSHEA however, and, for the most part, transitioned to use in

Presented at the 19th Annual Oxford International Conference on the Science of Botanicals 8 April 2019.

J. A. Brinckmann (✉)
Traditional Medicinals, Sebastopol, CA, USA
e-mail: jbrinckmann@tradmed.com

E. J. Fletcher
Herbal Ingenuity, Wilkesboro, NC, USA

R. Das
BI Nutraceuticals, Rancho Dominguez, CA, USA

T. Flaster
Botanical Liaisons, LLC, Boulder, CO, USA

non-drug cosmetic products, with the exception of a few that remained in the FDA monographs.

Post-DSHEA, the U.S. market for herbal supplements boomed and continues to grow. This paper examines the impacts of DSHEA, 25-years on, on the ever-increasing market demand for botanical ingredients, including legal and regulatory issues (state, federal, and international) affecting production, import, trade and use of botanical supplement ingredients, and the necessary developments of (a) quality management systems such as good agricultural and collection practices (GACPs) linked to good manufacturing practices (GMPs); (b) ingredient quality standards such as pharmacopoeial monographs that can serve as the basis of specifications for testing composition, identity, purity, and strength; (c) conservation status assessments of popular species, as many are obtained through wild-collection; and (d) standards (national and voluntary) for sustainable agriculture and wild collection practices in order to meet increasing market demand without detriment to biodiversity.

Keywords Biodiversity · Botanical supplements · Conservation · Quality management · Sustainable production

Abbreviations

ABC	American Botanical Council
AHP	American Herbal Pharmacopoeia
AHPA	American Herbal Products Association
CBD	Convention on Biological Diversity
CFR	Code of Federal Regulations
CITES	Convention on International Trade in Endangered Species
CR	Critically Endangered
DSHEA	Dietary Supplement Health and Education Act of 1994
E	Endangered
EPA	U.S. Environmental Protection Agency
FCC	Food Chemicals Codex
FDA	U.S. Food and Drug Administration
FD&C Act	Federal Food, Drug, and Cosmetic Act
FEMA	Flavor and Extract Manufacturers Association
FLO	Fair Trade International
FR	Federal Register
FSMA	Food Safety Modernization Act
FWF	FairWild Foundation
JECFA	The Joint Expert Committee on Food Additives
GACP	Good Agricultural and Collection Practices

GMP   Good Manufacturing Practices
GRAS   Generally Recognized as Safe
GRASE   Generally Recognized as Safe & Effective
HARPC   Hazard Analysis and Risk-Based Preventive Controls
IUCN   International Union for Conservation of Nature
LC   Least Concern
MAP   Medicinal and Aromatic Plants
MPWG   Medicinal Plant Working Group (of USFWS)
MPSG   Medicinal Plant Specialist Group (of IUCN)
NOP   National Organic Program
NT   Near Threatened
OTC   Over-the-counter drug product
R   Rare
UEBT   Union for Ethical BioTrade
UpS   United Plant Savers
USC   U.S. Congress
USDA   U.S. Department of Agriculture
USFWS   U.S. Fish and Wildlife Service
USP-NF   U.S. Pharmacopeia and National Formulary
VSS   Voluntary Sustainability Standards
VU   Vulnerable
WHO   World Health Organization.

12.1 Introduction

12.1.1 Pre-DSHEA Regulatory Landscape

In the United States (U.S.), prior to the passage of the Dietary Supplement Health and Education Act of 1994 (DSHEA) (USC 1994), botanicals were regulated as either botanical drug ingredients (over-the-counter (OTC) or prescription (Rx)), as non-drug cosmetic ingredients, or as conventional food ingredients. There was no middle ground between drug and food.

A famous case that begged for a middle ground began in 1988, when a shipment of black currant (*Ribes nigrum* L.) seed oil from England was seized by U.S. Marshals. The U.S. Food and Drug Administration (FDA) argued that by merely filling the seed oil into gelatin capsules, the botanical substance would become an unsafe food additive and therefore an adulterated food, i.e. the gelatin capsule being the food. In 1991, the U.S. District Court for the Central District of Illinois ruled in favor of the importing company and dismissed the case. In 1992, the FDA appealed the decision to the United States Court of Appeals for the Seventh Circuit. In January 1993, the appeals court affirmed the lower court decision, again, in favor of the importing company (Blumenthal 1994a).

For a botanical to have been permitted for use as a food ingredient, it required FDA classification as either a (1) Generally Recognized as Safe (GRAS) substance (listed in FDA regulations, Title 21 Code of Federal Regulations [21 CFR], Parts 182 and 184); or, as a (2) food additive or color additive (listed in 21 CFR Parts 172, 173 and Parts 73, 74, respectively), and/or flavoring substance evaluated by the Flavor and Extract Manufacturers Association (FEMA) and the Joint Expert Committee on Food Additives (JECFA); or, as a (3) prior-sanctioned substance, approved for specific uses in foods prior to September 6, 1958 (listed in 21 CFR Part 181) (FDA 2018a).

While many botanical substances had been classified as old drugs, i.e. as active ingredients of medicinal products sold from the early nineteenth century through the late twentieth century, in the mid-1970s, the FDA established expert review panels for the purposes of assessing the available evidence of safety and efficacy for all existing OTC drug active ingredients and establishing new labeling standards monographs. This process included the review of hundreds of "old" botanical drugs that were still on the pharmacy shelves in the 1970s. This is an important fact to consider in the historical context that led to the creation of DSHEA in the 1990s. From the mid-1970s through the mid-1990s, most botanical substances listed in FDA's proposed and tentative final monographs were removed, due to the agency's inability (at that time) to determine sufficient levels of evidence to support the claimed therapeutic uses. It is worth noting, however, that during the same time frame, other national health authorities (notably the German Federal Health Office (BGA)), were publishing positive therapeutic monographs for many of the same botanical drugs. Levels of evidence were defined differently in different countries. FDA's reclassification of many important botanicals as "non-monograph" created a tension between industry and government. Suddenly non-monograph botanicals had no safe-harbor for market access, as they were determined to be unlawful for use as components of both foods and drugs.

In 1992, as a strategy to address the issue, the European-American Phytomedicines Coalition (EAPC), a group of European and American phytomedicine manufacturers petitioned the FDA to expand its review of OTC drugs "to include products that have a long and safe history of use as nonprescription drugs in Western Europe." To no avail, the EAPC subsequently filed two citizens' petitions with the FDA, to allow the sale of (1) valerian as an OTC nighttime sleep aid; and (2) ginger as an OTC antiemetic drug product (Blumenthal 1994a, b, 1995).

Also in 1992, the American Herbal Products Association (AHPA) listed about 550 botanical species in their *Herbs of Commerce*, the majority of which had evidence of acceptable uses in food products (e.g. species listed in 21CFR, Parts172.5100,182.10 and 182.20), although the AHPA list also included species that were used only as botanical drugs (Foster 1992). In 1997, the FDA incorporated the AHPA list into regulation (21 CFR §101.4), requiring that the common or usual name of ingredients of dietary supplements (algal, botanical, and fungal) shall be consistent with the names standardized in the 1992 version of *Herbs of Commerce* (FDA 1997). The second edition of AHPA's *Herbs of Commerce* increased the estimated number of species in U.S. commerce to 2048 (McGuffin et al. 2000), and, in 2007, the

International Union for Conservation of Nature (IUCN) estimated that about 3000 medicinal and aromatic plant (MAP) species were in import-export trade (IUCN 2007). The point being that, of the thousands of herbs of commerce, most were not expressly listed as permitted for use in conventional food products. And, given the fact that only 2 new botanical drugs (out of over 600 applications) have been approved by the FDA in the 25 years since DSHEA, most botanicals must continue to navigate the middle ground between drug and food, as botanical dietary supplements. Both of the recently approved botanical drugs are prescription-only medicines with narrow indications for use, (1) an ointment containing a partially purified extract of green tea (*Camellia sinensis* (L.) Kuntze) leaf for treating genital/perianal warts, approved in 2006, and (2) a preparation of dragon's blood (*Croton lechleri* Müll.Arg.) latex for treating HIV-related diarrhea, approved in 2012 (Wu 2017).

After the passage of DSHEA, most oral ingestion herbal medicinal products transitioned to labeling as per the newly established dietary supplement subset of food regulations. However, the new regulatory framework provided no safe harbor for topical application botanical substances. While several of these did remain as Generally Recognized as Safe and Effective (GRASE) active ingredients in FDA tentative-final monographs in the *Federal Register* (FR) or final monographs in the *Code of Federal Regulations* (CFR), e.g. Capsicum Oleoresin USP, Cocoa Butter NF, Colloidal Oatmeal USP, Topical Starch USP, and Witch Hazel USP, most external application botanical products in the market transitioned to non-drug cosmetic labeling.

12.1.2 Post-DSHEA Market Boom

The botanical ingredient sector experienced an initial market boom due to the fact that DSHEA removed considerable market access barriers and afforded the opportunity to make claim statements on labeling, including many formerly OTC drug monograph structure function claim statements. At first, the boom benefited a range of clinically-tested herbal extract products that already had marketing authorizations as OTC medicines in European countries and in neighboring Canada. These same herbal medicinal products quickly poured into the U.S. market, although labeled as dietary supplement products rather than as medicines. At the same time, scores of lower-cost look-alike products entered the market making it difficult for consumers to differentiate the good from the bad. It wasn't unusual in the 1990s to read market reports of ginseng (*Panax* spp.) product retail sales increasing by nearly 250% over previous year (Anon 1994) or St. John's wort (*Hypericum perforatum* L.) product sales increasing by over 2800% from 1997 to 1998 (Brevoort 1998).

Although the top botanical products of the 1990s had relatively stable markets established in Europe, the new rapidly increasing market demand in the U.S. stimulated by the passage of DSHEA triggered considerable supply chain challenges, in terms of both quality and quantity. Suppliers of botanical raw materials spent the 1990s scaling up production to keep up with spikes in demand, significantly

Table 12.1 Top-selling botanical supplements – U.S. Mass Market – 1998 vs. 2017 ranking

Primary ingredient(s) of supplement	1998 ranking	2017 ranking
Ginkgo (*Ginkgo biloba* L.)	1	21
St. John's wort (*Hypericum perforatum* L.)	2	37
Ginseng (Panax ginseng C.A.Mey. or *P. quinquefolius* L.)	3	25
Garlic (*Allium sativum* L.)	4	18
Echinacea (*Echinacea angustifolia* DC., *E. pallida* (Nutt.) Nutt., or *E. purpurea* (L.) Moench)	5	2
Saw palmetto (*Serenoa repens* (W.Bartram) small)	6	14
Grape seed (*Vitis vinifera* L.)	7	Not in top-40
Kava-kava (*Piper methysticum* G.Forst.)	8	Not in top-40
Evening primrose (*Oenothera biennis* L., *O. lamarckiana* Ser.)	9	40
Echinacea/Goldenseal (*Echinacea* spp. and *Hydrastis canadensis* L.)	10	Not in top-40
Cranberry (*Vaccinium macrocarpon* Aiton)	11	3
Valerian (*Valeriana officinalis* L.)	12	16

Source: Data extrapolated from Brevoort (1998) and Smith et al. (2018)

increasing farm acreage and wild collection areas. But, then, booms do lead to busts, which hit the sector hard by the end of the decade. While the top-selling botanical supplements of the 1990s are still important today, their rankings have dropped and stabilized as new entries have displaced them. For example, turmeric (*Curcuma longa* L.) rhizome supplements are top-sellers today but were unheard of in the 1990s. Table 12.1 shows the top-selling botanical supplement products in the U.S. mass market in 1998 compared with their rankings in 2017.

12.1.3 Botanical Supply Chain Challenges

While the botanical ingredients market was booming, feeding the new dietary supplement finished product sector, the fundamentals to support it were lacking and yet to be developed. At that time, quality assurance and management systems such as good agricultural and collection practice (GACP) guidelines and implementable standards for sustainable production, suitable for cultivated and/or wild-collected botanical crops, were non-existent.

The U.S. Department of Agriculture (USDA) organic regulations, inclusive of a wild-crop harvesting practice standard, of particular relevance to botanicals, came into force in 2001 (USDA 2000). Guidance on how to comply with the new wild-crop standard, however, was not published for another decade, in July of 2011

(USDA 2011). In the meantime, in 2006, a nonprofit organization, the American Herbal Products Association (AHPA) published GACPs for herbal raw materials (relevant to both farmed and wild-collected botanicals), in collaboration with another nonprofit, the American Herbal Pharmacopoeia (AHP). As members of the AHPA Botanical Raw Materials Committee, authors of this paper were deeply involved with the development of the GACP document (AHPA 2006).

Authoritative or official quality standards from which to establish ingredient specifications for testing and verifying botanical identity, composition, strength and purity, were also lacking. From the first publication of the *Pharmacopoeia of the United States of America* in 1820 (USPC 1820) until 1920, around 875 botanical monographs were published in the USP (USPC 2016). including botanical drug substances and botanical preparations. However, with late nineteenth century advances in chemistry, the USP began to withdraw many botanical drug monographs. In the early twentieth century, only 170 botanical monographs remained, and by 1995, there were fewer than 40 USP monographs for botanicals and their preparations (Schiff et al. 2006). In 1995, two important initiatives commenced, (1) the United States Pharmacopeial Convention (USPC) adopted a resolution that led to the election of a 5-year term (1995–2000) subcommittee to explore candidate botanicals for dietary supplement monograph development work; and (2) the nonprofit organization AHP was established for the express purpose of developing comprehensive monographs for quality control testing of botanical raw materials and extracts used in dietary supplement products.

Another considerable challenge for the rapidly growing botanical supplement sector was the fact that many of the most popular botanical ingredients were procured, for the most part, from wild harvesting networks in rural and remote areas often through informal, non-documented cash trade lacking transparency and traceability. It wasn't until the early 2000s that the new USDA organic regulations enabled chain-of-custody and traceability of wild crops, due to the documentation and mapping requirements for inspection and certification. Popular high-demand wild harvested botanicals at that time included (European) bilberry (*Vaccinium myrtillus* L.) fruit and leaf, black cohosh (*Actaea racemosa* L.) rhizome, cascara sagrada (*Frangula purshiana* Cooper) bark, (Peruvian) cat's claw (*Uncaria tomentosa* (Willd. ex Schult.) DC.) bark, (European) dog rose (*Rosa canina* L.) hip, echinacea (*Echinacea angustifolia* DC.) root, (Siberian) eleuthero (*Eleutherococcus senticosus* (Rupr. & Maxim.) Maxim.) root, goldenseal (*Hydrastis canadensis* L.) rhizome, licorice (*Glycyrrhiza glabra* L., *G. uralensis* Fisch.) root, saw palmetto (*Serenoa repens* (W.Bartram) Small) berry, slippery elm (*Ulmus rubra* Muhl.) bark, St. John's wort (*Hypericum perforatum* L.) flowering tops, and wild yam (*Dioscorea villosa* L.) rhizome, among many others.

The conservation status of most medicinal plants had yet to be assessed, and the development of international standards for sustainable wild collection and ecosystem management were still a decade away. While the World Health Organization (WHO) 1993 "*Guidelines on the Conservation of Medicinal Plants*" existed (WHO, IUCN and WWF 1993), which provided a framework with recommended actions, they had little impact on industry. It was probably not until 1996 that the industry really sat up and took notice, when the World Wildlife Fund-US (WWF) suggested that the U.S. Fish & Wildlife Service (USFWS) propose listing the popular botanical goldenseal onto Appendix II of the Convention on International Trade in Endangered Species (CITES) (USFWS 1996).

And, it wasn't until the early 2000s, that rigorous international standards and corresponding guidance documents for sustainable wild collection of botanicals were being developed and test-implemented globally for feasibility, in collaboration with herbal companies (some American) and nature conservation organizations (e.g. WWF), with the financial and technical support of European governmental agencies. This included the (1) German-government supported *"International Standard for the Sustainable Wild-Collection of Medicinal and Aromatic Plants"* (ISSC-MAP) Draft 1 (Leaman 2004), Draft 2 (Leaman and Salvador 2005), Working Draft (MPSG 2006), and Version 1.0 (2007) (MPSG 2007); and (2) the Swiss-government supported *"FairWild Standard,"* Version 1.0 (Meinshausen et al. 2006) and Version 2.0 (FWF 2010), the latter which incorporated the ISSC-MAP as the result of the two initiatives merging in October 2008 (Brinckmann 2009).

12.2 Managing Quality and Sustainability

When DSHEA came into being, the U.S. botanical ingredients and products sectors, for the most part, did not yet have the requisite standards and tools developed to cope with, not only exponential market growth and strains on capacity, but also a series of regulations that would require development and implementation of standards for quality management and sustainability of botanical ingredients from the field to post-harvest processing, and on to finished product manufacturing. This chapter will tie the thread of quality management from implementation GACPs and sustainable production standards in the field to GMPs in the processing facilities with application of pharmacopoeial quality standards in the testing laboratories as the basis of specifications for ensuring the proper composition, identity, purity, and strength of the botanical ingredient. Table 12.2 provides a chronology of significant post-DSHEA developments related to managing the quality and sustainability of botanical supplement ingredients in the U.S.

12.2.1   1995–2000: AHP and USP Begin to Develop Botanical Monographs

Due to the passage of DSHEA, the advisability of developing botanical supplement monographs was deliberated at the 1995 Quinquennial Meeting of the United States Pharmacopeial Convention (USPC). The Convention adopted a resolution to encourage USP to explore the feasibility and advisability of establishing standards and developing information concerning dietary supplements. A subcommittee on

Table 12.2 Historical timeline of significant post-DSHEA events impacting quality and sustainability of botanical supply

Date	Event
1994, Oct. 25	U.S. Congress "Dietary Supplement Health and Education Act of 1994" (DSHEA) becomes law. The act afforded legal recognition to USP-NF standards for dietary supplement quality.
1995, Mar. 9–12	U.S. Pharmacopeial Convention (USPC) adopted Resolution No. 12, a resolution that "encouraged the USP to explore the feasibility and advisability of establishing standards and developing information concerning dietary supplements."
1995	Nonprofit organization, the American Herbal Pharmacopoeia (AHP) is established with a purpose to produce comprehensive monographs that outline the quality control criteria needed for ensuring the identity, purity, and quality of botanical raw materials.
1995	Nonprofit organization, the United Plant Savers (UpS) is founded with a mission "to protect native medicinal plants of the United States and Canada and their native habitat while ensuring an abundant renewable supply of medicinal plants for generations to come."
1997, Sept. 18	Goldenseal (*Hydrastis canadensis*) enters Appendix II of the Convention on International Trade in Endangered Species (CITES) as proposed by the U.S. Fish and Wildlife Service (USFWS).
1997, Dec. 16	USDA proposes establishment of a national organic program to include wild-crop harvesting.
1999	The Medicinal Plant Working Group (MPWG) was established under the umbrella of the Plant Conservation Alliance (PCA), facilitated by the USFWS.
2000	Nonprofit UpS publishes at-risk and to-watch list of native American botanicals.
2000, Dec. 21	USDA National Organic Program (NOP) final rule, which, of particular relevance to sustainable botanical supply, includes a "Wild-crop Harvesting Practice Standard" (7 CFR § 205.207).
2004, Nov.	Nonprofit organization, the American Herbal Products Association (AHPA) publishes "Background on California Proposition 65: Issues related to heavy metals and herbal products."
2006, Dec.	AHPA publishes "AHPA-AHP Good Agricultural and Collection Practice for Herbal Raw Materials," prepared by members of the AHPA Botanical Raw Materials Committee.
2007, Jun. 25	U.S. FDA final rule "Current Good Manufacturing Practice in Manufacturing, Packaging, Labeling, or Holding Operations for Dietary Supplements (21 CFR Part 111).
2008, Dec. 29	USDA Forest Service final rule on "Special Forest Products and Forest Botanical Products."
2010, Oct. 29	United Nations treaty affecting access & trade of medicinal plants adopted: "Nagoya Protocol on Access to Genetic Resources and the Fair and Equitable Sharing of Benefits Arising from their Utilization to the Convention on Biological Diversity," and came into force Oct. 12, 2014.
2011, Jan. 4	U.S. Congress "FDA Food Safety Modernization Act" becomes law.
2011, Jul. 22	USDA NOP issues guidance on organic "Wild Crop Harvesting."

(continued)

Table 12.2 (continued)

Date	Event
2011, Nov. 2	Three nonprofit organizations, the American Botanical Council (ABC), the AHP, and the University of Mississippi's National Center for Natural Products Research (NCNPR) initiated the "ABC-AHP-NCNPR Botanical Adulterants Program" to educate members of the herbal and dietary supplement industry about ingredient and product adulteration.
2012	USPC published first edition of the Dietary Supplements Compendium (DSC), which brought together monographs and general chapters of FCC, NF and USP.
2015, Sept. 17	U.S. FDA final rule "Current Good Manufacturing Practice, Hazard Analysis, and Risk-Based Preventive Controls for Human Food" (21CFR Part 117).
2016, Jan. 15	USDA NOP issues guidance on "Natural Resources and Biodiversity Conservation" for organic operations (farms and wild-collection operations).
2017, March	AHPA publishes updated "Good Agricultural and Collection Practices and Good Manufacturing Practices for Botanical Materials."

natural products was elected for a five-year term (1995–2000). The subcommittee, with the assistance of an appointed advisory panel, began the work to select candidate botanicals for monograph and methods development (Schiff et al. 2006). Also in 1995, the nonprofit AHP was founded, and, in the meantime, has developed and published nearly 40 botanical monographs, possibly the most comprehensive compendium available providing standards for identity, analysis, and quality control.

While the use of official pharmacopoeial standards is mandatory for quality specifications of botanical drugs, their use is voluntary for botanical dietary supplements in the U.S.

The Dietary Supplement Health and Education Act of 1994 (DSHEA) amendments to the Federal Food, Drug, and Cosmetic Act (FD&C Act) named the United States Pharmacopeia and National Formulary (USP-NF) as official compendia for dietary supplement ingredient quality standards. At the same time, industry compliance with USP-NF standards was made voluntary. Nonetheless, the amendments stipulate that if an herbal dietary supplement that is covered by the specifications of an official compendium is represented as conforming to the specifications of an official compendium, and fails to conform to such specifications, it shall be deemed misbranded. Thus, compliance with a USP monograph becomes mandatory only in cases where the product label specifies that it contains components with a USP quality designation (Brinckmann 2011).

One of the co-authors of this paper (JAB) submitted comments on FDA's draft cGMPs suggesting that use of authoritative quality standards, such as those published in pharmacopoeial monographs, should be a required minimum basis for the establishment of botanical supplement ingredient specifications, in that it could be disingenuous or fraudulent for companies to market products with claim statements, where the evidence to support the claim is based on the use of a specified pharmacopoeial quality, if, in fact, the company was using inferior quality grades of botanical raw materials. In the preamble of the final rule, while FDA stated that companies may use validated methods that can be found in official references, such as AOAC International, USP, and others, the agency rejected the argument that pharmacopoeial quality should be a minimum requirement for ingredients that are the subject of structure function claim statements (FDA 2007).

In a critique of the FDA cGMPs, published in *Food and Drug Law Journal*, representatives of the USPC stated:

> Unfortunately, these cGMPs establish no minimum requirement for quality. The manufacturer is entitled to use its best judgment in establishing a standard. Moreover, although the cGMPs will help ensure that one manufacturer's product is similar from batch to batch, the specifications for similar articles can vary widely from manufacturer to manufacturer. Effectively, the cGMPs call for standards without standardization. A manufacturer may create essentially private standards for a particular dietary ingredient or dietary supplement, but the manufacturer need not make these standards public. One manufacturer's standards may bear little or no resemblance to the standards created by other manufacturers for the same ingredient or product. Thus multiple manufacturers may establish different standards of identity, strength, quality, and purity for articles that are introduced into commerce under the same name (Miller et al. 2008).

Regardless of the fact that it was not mandatory for manufacturers of supplement products to base their ingredient quality specifications on pharmacopoeial standards, considerable progress has been made over the past 25 years. Monographs relevant to specifying the quality of botanical supplement ingredients have been made available through the American Herbal Pharmacopoeia (AHP), Food Chemicals Codex (FCC), National Formulary (NF) and United States Pharmacopeia (USP). Furthermore, the USPC, since 2012, publishes a separate Dietary Supplements Compendium (DSC), which provides monographs and general chapters of FCC, NF and USP along with supplemental information such as color plates and illustrations (USPC 2019). Table 12.3[1] provides a list of monographs available for botanical raw materials and extracts and oils obtained from them.

12.2.2   2000: USDA Final Rule for NOP (7 CFR Part 205)

In 2000, the USDA NOP final rule was published, which, of particular relevance to sustainable botanical supply, included the "Wild-crop Harvesting Practice Standard" (7 CFR § 205.207).

Experience has shown that the implementation of sustainability standards may also contribute to compliance with certain aspects of FDA's cGMPs, especially where they pertain to management and control of contamination. For example, the USDA NOP soil fertility and crop nutrient management practice standard (7 CFR §205.203) requires producers to manage plant and animal materials to maintain or improve soil organic matter content in a manner that does not contribute to contamination of crops, soil, or water by plant nutrients, pathogenic organisms, heavy metals, or residues of prohibited substances. Furthermore, if a botanical supplement component is certified organic, 7 CFR §205.670(c) requires pre-harvest or

---

[1] Due to its large size, Table 12.3 is presented under Annex I

post-harvest tissue test sample collection to be performed by an inspector or certifying agent.

> Sample integrity must be maintained throughout the chain of custody, and residue testing must be performed in an accredited laboratory. Chemical analysis must be made in accordance with AOAC methods or other current applicable validated methodology determining the presence of contaminants in agricultural products (USDA 2000).

12.2.3   2006: AHPA-AHP GACPs

The genesis of the development of GACPs, specifically designed for MAP crops (farmed and wild), can be traced to an initial 1983 round-table discussion at the 4th International Society for Horticultural Science (ISHS) Symposium in Angers, France (Máthé and Franz 1996). It would however be another 20 years before an elaboration of general international GACP guidelines for medicinal plants would be published by the World Health Organization (WHO 2003a). The "*WHO Guidelines on Good Agricultural and Collection Practices (GACP) for Medicinal Plants*" were developed in response to Resolution WHA56.31 from the Fifty-sixth World Health Assembly on Traditional Medicine which requested the Director-General to provide technical support for development of methodology to monitor or ensure product quality, efficacy and safety, preparation of guidelines, and promotion of exchange of information (WHO 2003b). The WHO GACP guidelines were primarily intended to provide "general" yet globally applicable technical guidance with the awareness that the GACPs would need to be adjusted according to each country's actual situation. One of the main objectives of these general guidelines was to provide the basis for the formulation of national and/or regional GACP guidelines, species-specific GACP monographs, and related standard operating procedures (SOPs).

While many national and international governmental organizations have developed and implemented national GACPs for MAPs, for example, the China Food and Drug Administration (CFDA 2002), Japanese Ministry of Health, Labour and Welfare (WHO 2003a), European Medicines Agency (EMA) (HMPC 2006), and India's National Medicinal Plants Board (NMPB 2009a, b), the U.S. has not yet. While GACP compliance is mandatory for botanical drugs it is voluntary for botanical supplements. In their 2004 guidance for industry on botanical drug products (FDA 2004), the FDA deferred to the EMA GACPs:

> Starting materials of botanical origin that are used to produce a botanical drug substance should be evaluated for quality. The use of appropriate starting materials and the drug substance manufacturer's ability to control the source depend on appropriate specifications (tests, analytical procedures, and acceptance criteria). In addition to establishing specifications, manufacturers can achieve adequate quality control of starting materials by applying the principles outlined in FDA's botanical guidance and by following good agricultural and good collection practice for starting materials of herbal origin (e.g., European Medicines Evaluation Agency HMPWP/31/99) (FDA 2004).

However, in the FDA's revised botanical drug guidance (FDA 2016), the agency deferred to both the EMA and WHO GACPs:

> To assess quality and therapeutic consistency, it is important to select representative raw material batches (i.e., raw material from three or more representative cultivation sites or farms) for the manufacturing of the clinical drug substance for multiple batch Phase 3 studies. The sponsor should establish large growing regions with three or more cultivation sites or farms whose locations are purposefully selected to be representative of the regions for each of the botanical raw materials following the principles of Good Agricultural and Collection Practices (GACP)." [Note: FDA footnotes this statement with "See the World Health Organization's *"WHO guidelines on good agricultural and collection practices (GACP) for medicinal plants"* and the European Medicines Agency's *"Guideline on Good Agricultural and Collection Practice (GACP) for Starting Materials of Herbal Origin"*] (FDA 2016).

And, while the USPC has, in the past, proposed to develop a new General Chapter section on GACPs, USP General Chapter <2030> *"Supplemental Information for Articles of Botanical Origin"* provides general guidance deferring to the WHO GACP:

> It is recommended that, at a minimum, growers and others involved in the handling and distribution of botanical products should become familiar with and follow the *WHO Guidelines on Good Agricultural and Collection Practices (GACP) for Medicinal Plants* (USPC 2018a).

Thus, for the U.S. context of quality management of botanical crops, in lieu of official GACPs from either the FDA or USP, two nonprofit organizations, AHPA and AHP collaborated in the production of the *"AHPA-AHP Good Agricultural and Collection Practice for Herbal Raw Materials,"* initiated under the auspices of the AHPA Botanical Raw Materials Committee (AHPA 2006). While it has no legal standing in the U.S., AHPA states that its GACP has relevance to herbal raw materials used in all herbal products, including cosmetics, dietary supplements, drugs, and foods. However, it should also be noted that the Canadian government officially recognized the AHPA-AHP GACP in their *"Quality of Natural Health Products Guide"* as acceptable for botanical medicinal ingredients of licensed natural health products in Canada (NNHPD 2015).

12.2.4   2007: U.S. FDA Final Rule cGMPs (21 CFR Part 111)

FDA's 2007 final rule on cGMP in manufacturing, packaging, labeling, or holding operations for dietary supplements (21 CFR Part 111) required the establishment of quality specifications with scientifically valid methods for ensuring the composition, identity, purity and strength of dietary supplement components, as well as the establishment of limits on types of contamination that may adulterate or may lead to adulteration (FDA 2007). Experience of some companies has shown that compliance with the cGMP can be accomplished by implementing other standards earlier in the botanical supply chain. For example, GACP compliance generally requires written agreements between producers and buyers of botanicals with regard to quality such as content of active principle, macroscopical and olfactory properties, limit values for microbial contamination, chemical residues and heavy metals, and that specifications should be based on recognized national pharmacopoeial monographs (HMPC 2006). And, basing one's quality specification on quality standards monographs such as those of the USP ensure that scientifically valid methods and limits will be established for determining composition, identity, purity, and strength.

Subsequently, in 2011, also impacting the production and trade of botanical dietary supplement ingredients, Congress enacted the Food Safety Modernization Act (FSMA)

> in response to dramatic changes in the global food system and in our understanding of foodborne illness and its consequences, including the realization that preventable foodborne illness is both a significant public health problem and a threat to the economic wellbeing of the food system. FSMA is transforming the nation's food safety system by shifting the focus from responding to foodborne illness to preventing it (FDA 2018b).

As part of the FSMA, Congress has enacted several rules:

1. Preventive Controls for Human Food
2. Foreign Supplier Verification Program for Importers of Food for Human and Animals
3. Sanitary Transportation of Human and Animal Food
4. Mitigation Strategies to Protect Food against Intentional Adulteration
5. Accredited Third Party Certification
6. Preventive Controls for Food for Animals
7. Standards for Produce Safety

While these rules are now final, all corresponding guidance documents are not yet finalized.

12.3   Control of Adulterants and Contaminants

This section discusses the problems and solutions concerning adulteration of botanical ingredients, whether intentional (economic) or unintentional (look-alike species growing in proximity to target species). Additionally unavoidable contamination of botanical ingredients has become one of the most disruptive quality problems negatively impacting trade and quality of products. Our planet is not getting cleaner. Anthropogenic pollution and related contamination to the natural environment is now global and pervasive. Types of pollution affecting botanical crops, whether farmed or wild-collected, include heavy metals contamination from industrial pollution, nonpoint source pesticide pollution from conventional agriculture operations negatively impacting certified organic crops and wild crops, pharmaceutical drugs in irrigation water, plastic pollution, polycyclic aromatic hydrocarbons (PAHs), and radioactive residues.

12.3.1 Economic Adulteration

In 2011, three nonprofit organizations, the ABC, the AHP, and the University of Mississippi's National Center for Natural Products Research (NCNPR) initiated the "*ABC-AHP-NCNPR Botanical Adulterants Program*" to educate members of the herbal and dietary supplement industry about ingredient and product adulteration. To date, 17 comprehensive Botanical Adulterant Bulletins have been published and seven laboratory guidance documents (Table 12.4), among other materials useful for the industry.

Table 12.4 Botanical Adulterants Program bulletins and laboratory guidance documents list

Date	Title
2015, January	Skullcap Adulteration Laboratory Guidance Document
2015, August	Bilberry Fruit Extract Laboratory Guidance Document
2015, November	Black Cohosh Adulteration Laboratory Guidance Document
2016, April	Adulteration of Bilberry (*Vaccinium myrtillus*) Extracts
2016, April	Adulteration of Grape Seed Extract (*Vitis vinifera*)
2016, April	Adulteration of Skullcap (*Scutellaria lateriflora*)
2016, June	Adulteration of Black Cohosh (*Actaea racemosa*)
2016, June	Adulteration of *Hydrastis canadensis* root and rhizome
2016, August	Adulteration of Arnica (*Arnica montana*)
2017, January	St. John's Wort (*Hypericum perforatum*) Adulteration
2017. March	Adulteration of Grapefruit Seed Extract (*Citrus paradisi*)
2017, May	Grapefruit Seed Extract Laboratory Guidance Document
2017. August	Tea Tree (Melaleuca alternifolia and M. linariifolia) Oil
2017, October	Adulteration of Rhodiola (*Rhodiola rosea*) Rhizome, Root, and Extracts
2017, December	Adulteration of Cranberry (*Vaccinium macrocarpon*)
2018, January	Adulteration of *Ginkgo biloba* Leaf Extract
2018, April	Pomegranate Products Laboratory Guidance Document
2018, May	Turmeric (*Curcuma longa*) Root and Rhizome, and Root and Rhizome Extracts
2018, June	Boswellia serrata Adulteration
2018, September	Adulteration of Maca (*Lepidium meyenii*)
2018, September	Tea Tree Oil Laboratory Guidance Document
2018, October	Adulteration of Saw Palmetto (*Serenoa repens*)
2018, November	Cranberry Products Laboratory Guidance Document
2019, January	Adulteration of Ashwagandha (*Withania somnifera*) Roots, and Extracts
2019, February	Grape Seed Extract Laboratory Guidance Document

Source: American Botanical Council: http://cms.herbalgram.org/BAP/

12.3.2 Heavy Metal Pollutants in the Environment

Medicinal plant crops can be contaminated by absorbing heavy metals from soil, water and air, i.e. from industrial emissions, mining, smelting, landfills, conventional agriculture (fertilizers and pesticides; polluted water used for irrigation, sewage sludge), transportation (traffic-induced contaminated atmosphere), rainfall, and atmospheric dust:

- Air: Aerial plant parts accumulate heavy metals by uptake of airborne pollution relative to proximity to industrial zones and traffic (Serbula et al. 2013).
- Soil: "Absorption of heavy metal in medicinal plants is governed by soil characteristics such as pH, salinity, conductivity and organic matter content" (Rădulescu et al. 2013).
- Water: "Toxic elements from wastewater may contaminate agricultural soils, water supplies and environment… Elevated levels of heavy metals in plants are reported from the areas having long-term uses of treated or untreated wastewater" (Shaban et al. 2016).

While the United States Pharmacopoeia (USP) prescribes general limits for elemental impurities in general chapter ⟨561⟩ ARTICLES OF BOTANICAL ORIGIN (USPC 2018b) (see Table 12.5), the State of California enacted its own limits through implementation of the Safe Drinking Water and Toxic Enforcement Act of 1986, more commonly known as California Proposition 65.

The California Proposition 65 limits for heavy metals continue to impact the botanical dietary supplement trade, from increased testing for selection of California-compliant lots, to product brands lowering their daily serving size instructions on labeling in order to comply. That is, because the limits are not based on content but on daily ingestion levels. For example, the California maximum allowable daily level (MADL) for lead content in an herbal dietary supplement product is not-more-than 0.5 μg per day. Depending on the recommended daily serving size for a product, the USP limit of 5 μg per gram (in botanical material) may, or may not, prove to be compliant in California. In 2004, the American Herbal Products Association (AHPA) published its "*Background on California Proposition 65: Issues related to heavy metals and herbal products,*" (McGuffin and Norris 2004), which was revised in 2008, and in 2010, and again in 2017, retitled as "*Guidance on California Proposition 65 and Herbal Products*" (AHPA 2017). AHPA's document provides

Table 12.5 USP General Chapter ⟨561⟩ Articles of Botanical Origin – limits of elemental impurities

Element	Limit (μg/g)
Arsenic (inorganic)	2
Cadmium	0.5
Lead	5
Mercury (total)	1
Methylmercury (as Hg)	0.2

guidance for industry on regulatory and liability implications of Proposition 65 for constituents that may be present in herbal products sold in California.

12.3.3 Nonpoint Source Pesticide Contamination

In the decades since the passage of DSHEA, pesticide drift from conventional agriculture sites has become ubiquitous in the natural world. Residues can now be detected the world over, in the air, ice, snow, soil and water, and on crops grown without the intentional use of pesticides.

- Residues of "legacy" (e.g. DDT) and "current use pesticides" (CUPs) are now detected in Arctic ice caps (evidence of long range atmospheric transport).
- In rural and remote areas, there is widespread contamination of wildflowers and bee-collected pollen with agricultural pesticides.
- Nonpoint source pesticide detection is an increasing problem even with certified organically grown and/or wild-collected botanicals, where pesticides have not been applied.

As per FDA's 21 CFR Part 111 cGMPs for dietary supplements, specifications are required to ensure that a dietary supplement derived from a botanical source does not contain contaminants such as an unlawful pesticide. Pesticide tolerances, however, are established by the Environmental Protection Agency (EPA) on a crop-specific and/or crop-group basis. While FDA enforces the EPA limits that were established mainly for food crops, tolerances have been established for only a relatively small number of botanical crops such as certain aromatic or culinary herbs that are grown in the U.S. on a large scale, e.g., spearmint (*Mentha spicata*) and hops (*Humulus lupulus*). EPA tolerances have not been established for thousands of other botanicals in commerce. Thus, when *de minimis* levels of pesticide residues are found at above the limit of quantitation (LoQ), even on wild-collected botanicals or on certified organically grown botanicals, where no pesticides were intentionally applied, there is a risk of FDA detention or import refusal at the port of entry.

Although USP General Chapter ⟨561⟩ Articles of Botanical Origin provides reasonable limits for pesticide residues, in the U.S. the USP limits are applicable only to botanical drugs and not to botanical dietary supplements. That is because dietary supplements are regulated as a subset of foods. EPA establishes limits for food crops while USP has the authority to establish limits for botanical drug crops.

In 2016, a stimuli article titled "*Need for Clear Regulation of Pesticide Residue Limits for Articles of Botanical Origin*," was published in the *Pharmacopeial Forum* for public comment (USP Botanical Dietary Supplements and Herbal Medicines Expert Committee 2016). This was followed-up with a USP roundtable on pesticide residues in botanical dietary supplements with participants from governmental agencies, trade associations, and industry.

12.4   Conservation and Sustainability

This chapter looks at impacts of international conventions such as the Convention on Biological Diversity (CBD) and the Convention on International Trade in Endangered Species (CITES) on botanical ingredient production and import trade, as well as conservation status assessments of botanical species carried out by non-governmental organizations such as IUCN Red List assessments, and the emergence of voluntary standards for demonstrating economic, environmental and social sustainability in the botanical supply chain.

By the start of the twenty-first century, awareness grew in the global botanical trade that conserving biodiversity and improving rural economies appeared to be linked to access to traditional harvesting areas, sustainable production, trade and use of ever-increasing quantities of botanicals needed for the natural cosmetic, dietary supplement, botanical drug, and functional food sectors. This was evidenced by the fact that numerous sustainability standards (both national and private voluntary) as well as GACP standards designed for MAP crop production were being developed through multi-stakeholder processes and implemented with funding support from governmental agencies (e.g. German Federal Agency for Nature Conservation (BfN)), intergovernmental organizations (e.g. agencies of the United Nations), and non-governmental organizations (e.g. WWF).

> Implementation of voluntary sustainability standards (VSS) that include economic, environmental and social criteria and indicators is an emerging market-based approach to biodiversity conservation, sustainable harvesting and commercial use of MAPs, and to demonstrate compliance with international agreements. While such standards are voluntary, in that compliance is not required by governmental regulations, independent third-party inspection and certification organisations determine if companies are operating in compliance with the standard (Brinckmann 2017; Komives and Jackson 2014).

From 2010 to 2014, international botanical ingredient and finished product companies also needed to learn the (new) ropes for compliance with a United Nations multilateral treaty, adopted in 2010 for enforcement by 2014, the *"Nagoya Protocol on Access to Genetic Resources and the Fair and Equitable Sharing of Benefits Arising from their Utilization to the Convention on Biological Diversity"* (Secretariat of the Convention on Biological Diversity 2011). Some of the aforementioned voluntary sustainability standards provide ways for companies to demonstrate compliance with the access and benefit sharing (ABS) requirements of the protocol, for example, the Ethical BioTrade Standard, developed by the Union for Ethical BioTrade (UEBT), and the FairWild Standard (FWS), developed by the FairWild Foundation. Criterion 4.2 of the FWS requires that agreements made with local communities and/or indigenous peoples are executed in compliance with relevant international and national laws and ABS regulations including protection of traditional knowledge (FWF 2010). Written and mutually accepted fair and equitable agreements on use of medicinal plants and associated traditional knowledge must be in place to maintain FairWild certification (FWF 2013). Section 3 of the Ethical

BioTrade Standard *"Fair and Equitable Sharing of Benefits derived from the Use of Biodiversity"* includes comparable requirements (UEBT 2012).

Table 12.6 provides a listing of selected botanicals with information on their conservation status, for example if the species is listed in the CITES appendixes, or has been assessed according to the IUCN Red List criteria or the UpS ranking tool. Species prioritized in the AHPA tonnage survey are also included in the table if the species has been assessed according to UpS, IUCN or CITES.

CITES Appendices Appendix I: lists species that are the most endangered; threatened with extinction. CITES prohibits international trade in specimens of these species except when the purpose of the import is not commercial, for instance for scientific research. **Appendix II:** lists species that are not necessarily now threatened with extinction but that may become so unless trade is closely controlled. It also includes so-called "look-alike species", i.e. species whose specimens in trade look like those of species listed for conservation reasons. **Appendix III:** is a list of species included at the request of a Party that already regulates trade in the species and that needs the cooperation of other countries to prevent unsustainable or illegal exploitation.

IUCN Categories CR, critically endangered; DD, data deficient; EN, endangered; LC, least concern; NT, near threatened; VU, vulnerable.

UpS Ranking ATR, at-risk; TW, to watch.

12.4.1 AHPA Tonnage Survey

The American Herbal Products Association (AHPA) initiated and has conducted "Tonnage Surveys" since 1999 for the purpose of quantifying the annual harvests of specific North American botanicals in commerce. AHPA's Tonnage Survey is considered "a vital index of native U.S. botanical consumption," according to the Fish and Wildlife Service of the U.S. Department of the Interior. According to AHPA:

> Harvest information is a more powerful tool when harvest amounts are tabulated to include combined total usage. By working together, AHPA and the herbal products industry generate valuable information that helps ensure sustainable growth and stability. Participation in this survey also demonstrates the industry's commitment to sustainable harvests (AHPA 2019).

The first of these surveys addressed only the plant goldenseal (*Hydrastis canadensis*) and solicited information about both wild and cultivated harvests for 1998, as well as certain harvest and cultivation practices. The second survey extended this attention on goldenseal for the 1999 harvest year and also compiled information for the years 1997–1999 for a number of other plants. The 2000–2001

Table 12.6 Conservation status of selected botanicals used in botanical supplement products

Botanical name	CITES	IUCN red list	UpS
Agarwood (*Aquilaria* spp.)	II	CR, *Aquilaria crassna*; CR, *A. malaccensis*; VU, *A. sinensis*	
American ginseng (*Panax quinquefolius* L.)	II	Not assessed	ATR
Asian ginseng (*Panax ginseng* C.A.Mey.) from Russia	II	Not assessed	
Arnica (*Arnica* spp.)		LC, *Arnica montana*	TW
Bethroot (*Trillium erectum* L.)		Not assessed	ATR
Bletilla (*Bletilla striata* (Thunb.) Rchb.f.)	II	Not assessed	
Bloodroot (*Sanguinaria canadensis* L.)		Not assessed	ATR
Black cohosh (*Actaea racemosa* L.)		Not assessed	ATR
Blue cohosh (*Caulophyllum thalictroides* (L.) Michx.)		Not assessed	ATR
Brazilian rosewood (*Aniba rosaeodora* Ducke)	II	EN	
Candelilla (*Euphorbia antisyphilitica* Zucc.)	II	Not assessed	
Cape aloe (*Aloe ferox* Mill.)	II	Not assessed	
Cascara sagrada (*Frangula purshiana* Cooper)		LC	TW
Chun lan (*Cymbidium goeringii* (Rchb.f.) Rchb.f.)	II	Not assessed	
Common pleione (*Pleione yunnanensis* (Rolfe) Rolfe)	II	Not assessed	
Costus (*Saussurea costus* (Falc.) Lipsch.)	I	CR	
Dendrobium (*Dendrobium* spp.)	II	CR, *D. huoshanense*; CR, *D. officinale*;	
Desert broomrape (*Cistanche deserticola* Y.C.Ma)	II	Not assessed	
Echinacea (*Echinacea* spp.)		Not assessed	ATR
Elephant tree (*Bursera microphylla* A.Gray)		Not assessed	TW
Eyebright (*Euphrasia* spp.)		Not assessed	ATR
Fragrant rosewood (*Dalbergia odorifera* T.C.Chen)	II	VU	
False unicorn (*Chamaelirium luteum* (L.) A. Gray)		Not assessed	ATR
Gastrodia (*Gastrodia elata* Blume)	II	VU	
Gentian (*Gentiana* spp.)		CR, *G. kurroo*	TW
Goldenseal (*Hydrastis canadensis* L.)	II	VU	ATR
Goldthread (*Coptis* spp.)		EN, *C. teeta*	TW
Guaiacum (*Guaiacum* spp.)	II	EN, *G. officinale*; NT, *G. sanctum*	
Hoodia (*Hoodia gordonii* (Masson) Sweet ex Decne.)	II	Not assessed	
Indian rosewood (*Dalbergia sissoo* DC.)	II	Not assessed	
Jatamansi (*Nardostachys grandiflora* DC.)	II	CR	
Jivakah (*Malaxis acuminata* D.Don)	II	Not assessed	

(continued)

Table 12.6 (continued)

Botanical name	CITES	IUCN red list	UpS
Lady's slipper orchid (*Cypripedium* spp.)	II	LC, *C. acaule*; LC, *C. parviflorum*	ATR
Lobelia (*Lobelia* spp.)		LC, *L. cardinalis*; LC, *L. siphilitica*	TW
Lomatium (*Lomatium dissectum* (Nutt.) Mathias & Constance)		Not assessed	ATR
Oregon grape (*Mahonia aquifolium* (Pursh) Nutt., *M. nervosa* (Pursh) Nutt., *M. repens* (Lindl.) G. Don))		Not assessed	TW
Osha (*Ligusticum porteri* J.M.Coult. & rose, L. spp.)		Not assessed	ATR
Partridge berry (*Mitchella repens* L.)		Not assessed	TW
Perry's aloe (*Aloe perryi* Baker)	II	NT	
Picrorhiza (*Picrorhiza kurrooa* Royle)	II	Not assessed	
Pleurisy (*Asclepias tuberosa* L.)		Not assessed	TW
Pygeum (*Prunus africana* (Hook.f.) Kalkman)	II	VU	
Red saunders (*Pterocarpus santalinus* L.f.)	II	NT	
Riddhi (*Habenaria intermedia* D.Don)	II	Not assessed	
Sandalwood (*Santalum* spp.)		VU, *S. album*	ATR
Scythian lamb (*Cibotium barometz* (L.) J.Sm.)	II	Not assessed	
Skullcap (*Scutellaria lateriflora* L.)		LC	
Slippery elm (*Ulmus rubra* Muhl.)		LC	ATR
Sundew (*Drosera* spp.)		Not assessed	ATR
True unicorn (*Aletris farinosa* L.)		Not assessed	ATR
Venus' fly trap (*Dionaea muscipula* J.Ellis)		VU	ATR
Wild cherry (*Prunus serotine* Ehrh.)		LC	
Wild indigo (*Baptisia tinctoria* (L.) Vent.)		Not assessed	TW
Wild yam (*Dioscorea villosa* L., D. spp.)		Not assessed	ATR
Witch hazel (*Hamamelis virginiana* L.)		LC	
Yerba mansa (*Anemopsis californica* (Nutt.) Hook. & Arn.)		Not assessed	TW

survey identified all of the plants from these earlier efforts and a number of additional species. These surveys have continued to evolve by adding new species, more cultivation oriented, fresh and dried material numbers and other information based on the need of new data for today's market. Figure 12.1 provides a chart illustrating the annual harvested quantities of four high-volume American botanicals from 1999 through 2010.

According to AHPA, with regard to its 14 years of harvest data for goldenseal:

Annual variations in harvest quantities for certain of these commodities are consistent with market factors over the past several years. This review of data over the fourteen-year period reveals that patterns in market demand are reflected in annual variations of harvests of both wild and cultivated sources of these botanical commodities. Several species for which there are large market demands are cultivated to a sufficient degree so that some meaningful portion

CHART 1

Four High-Volume Commodities, 1999 – 2010 (Wild and Cultivated Combined)

Black cohosh root and slippery elm bark harvests rebound after depressed harvests in 2009, cascara sagrada reaches new peak, then returns to 2006 and 2007 harvest levels, while goldenseal root harvests remain relatively stable.

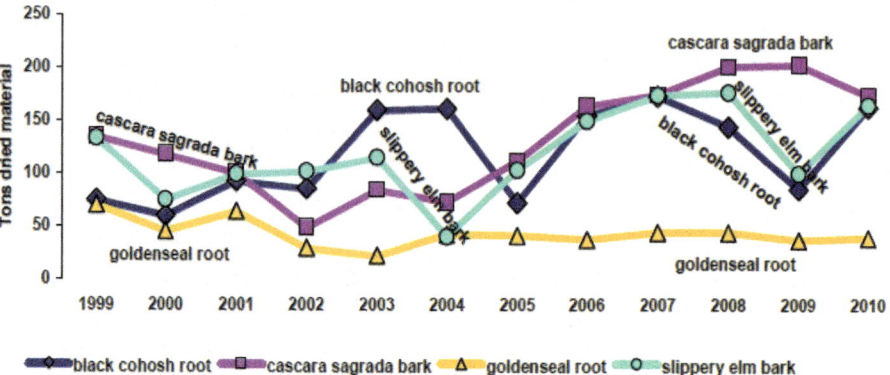

Fig. 12.1 Annual harvested quantities (dry weight basis) of four high-volume botanicals, 1999–2010, excerpted with permission from the "*Tonnage Surveys of Select North American Wild-Harvested Plants, 2006–2010*". (AHPA 2012)

of the total usage is provided by farmers rather than by harvesters of wild plants. The most recent surveys affirm that this continues to be the case for the goldenseal market… "However, market demand, not plant availability, seems to be the primary driver of total harvest quantities. As has been noted in other publications, information about the population dynamics of many of these species in their native habitats is not well understood" (AHPA 2012).

The latest survey (with submission deadline of 26 April 2019), continues to focus on selected North American botanicals that are produced in part from wild-harvested populations, but has been expanded to track 51 herbal commodities representing 44 species, including American ginseng, black cohosh, cascara sagrada, *Echinacea* spp., saw palmetto, slippery elm, and witch hazel (AHPA 2019).

Data from the AHPA tonnage surveys are the foundational information needed to both track and assess the sustainability of harvests of North American species in trade within the herbal dietary supplement and herbal medicine industries. It may still require a collaborative effort between industry, regulatory and nature conservation organizations to objectively and impartially study the data thoroughly enough to better understand the impact on native species relative to each species' life cycle, while considering other impacts such as human expansion and habitat depletion.

12.4.2 Medicinal Plant Working Group

Facilitated by the USFWS, the Medicinal Plant Working Group (MPWG) formed in 1999: "Recognizing that commercial demands may pose threats to native plants in the United States, representatives from industry, government, academia, tribes, and environmental organizations joined together to form the Medicinal Plant Working Group (PCA-MPWG) under the umbrella of the Plant Conservation Alliance (PCA)" (Lyke 2000). The primary focus of the MPWG has been to facilitate action in cases where the conservation status of a native medicinal plant is of concern.

12.4.3 UpS Ranking Tool

While UpS launched their medicinal plant species "At Risk" and "To Watch" lists in 2000, it was initially informed by input from UpS-member herbalists, ecologists, growers, and buyers. As the botanical industry continued to grow rapidly, UpS realized that a more transparent methodology was needed. As the result of a multi-year, multi-stakeholder process, in 2014, UpS launched what it called a *"Ranking Tool Created for Medicinal Plants."* The tool can be used to quantify and compare vulnerability to overharvest for wild collected medicinal plants, scoring species "according to their life history, the effects of harvest, their abundance and range, habitat, and demand" (Castle et al. 2014).

12.4.4 USDA NOP Biodiversity Conservation Guidance

The general natural resources and biodiversity conservation requirement of the USDA organic regulations (7 CFR §205.200) requires producers (both farms and wild collection operations) to "maintain or improve the natural resources of the operation, including soil and water quality." And the wild-crop harvesting practice standard (7 CFR §205.207) states that a "wild crop must be harvested in a manner that ensures that such harvesting or gathering will not be destructive to the environment and will sustain the growth and production of the wild crop" (USDA 2000). In 2011, USDA published guidance aiming to clarify the ways in which certified operations could demonstrate compliance with the standard (USDA 2011). Then, 16 years after the organic regulations came into force, guidance was finally published on the biodiversity conservation requirements of organic operations (USDA 2016a). In response to public comments, USDA stated:

> There are many ways for operations to meet the requirements of § 205.200... Wild harvest operations may voluntarily elect to follow the FairWild Foundation (FWF) standards in order to meet the NOP requirements for biodiversity conservation; however the change was not made to the final guidance (USDA 2016b).

12.4.5 Emergence of Voluntary Sustainability Standards

The end of the twentieth century saw the emergence of voluntary sustainability standards (VSS) relevant to the production and trade of medicinal and aromatic plants according to principles of economic, environmental, and social sustainability. For example, Fair Trade International (FLO) was founded in 1997 in Bonn, Germany, and soon afterwards developed fair trade standards for herbs, herbal teas, and spices. In 2000, the USDA National Organic Program (NOP) standards were published. In the first two decades of the twenty-first century, there has been a proliferation of VSS, with several hundreds of botanical species being traded globally with one or more certifications. Table 12.7 provides a list of selected VSS that are being applied to botanical crops. Implementation of VSS by botanical producers can, not only fill in some of the gaps in GACP guidelines and organic agriculture regulations, respectively, but also provide producers with the value-addition of certifications that demonstrate compliance with credible sustainability standards, that can strengthen their economic viability and market position (Brinckmann 2016).

12.4.6 Conclusion and Recommendations for the Future

This paper shows that, 25 years since DSHEA, in response to strains caused by a continually growing market demand for botanical ingredients and products in the United States, a considerable range of useful standards, guidance manuals, and tools have been developed by educational organizations (e.g. ABC-AHP-NCNPR), governmental organizations (e.g. FDA, USDA-NOP), medicinal plant conservation organizations (e.g. IUCN-MPSG, PCA-MPWG, UpS), quality standards setting organizations (e.g. AHP, USP), sustainability standards setting organizations (e.g. FLO, FWF, UEBT), and trade associations (e.g. AHPA). *While compliance with FDA regulations governing the manufacture and marketing of herbal dietary supplement products is mandatory, the use of pharmacopoeial quality standards, GACP standards, and sustainability standards (all of which support regulatory compliance), is voluntary.*

Table 12.7 Selected voluntary sustainability standards applicable to botanical crops

Name of standard	Owner of the standard	Economic sustainability	Environmental sustainability	Social sustainability
Agricultural Production Standard	Fair Trade USA (FTUSA)	✓	✓	✓
Biodynamic® Farm Standard	DEMETER Association Inc.		✓	
EcoSocial Fair	Instituto Biodinâmico de Desenvolvimento Rural (IBD)	✓	✓	✓
Ethical BioTrade Standard	Union for Ethical BioTrade (UEBT)	✓	✓	✓
Fair Choice Social and Fair Trade Standard	CONTROL UNION (CU)	✓	✓	✓
Fair For Life Standard	ECOCERT SA	✓	✓	✓
Fairtrade Standard for Herbs, Herbal Teas & Spices for Small Producer Organizations and traders	Fairtrade Labelling Organizations International, e.V. (FLO)	✓		✓
Fairtrade Standard for Herbs and Herbal Teas for Hired Labour and Traders	Fairtrade Labelling Organizations International, e.V. (FLO)	✓		✓
FairTSA Consolidated standards for the production of agricultural products, processed foods, wild collected plants, handicrafts and personal-care products	Fair Trade Sustainability Alliance (FairTSA)	✓	✓	✓
FairWild Standard	FairWild Foundation (FWF)	✓	✓	✓
National Standard for Organic Production	United States Department of Agriculture (USDA)		✓	
Non-GMO Project Standard	Non-GMO Project		✓	
Regenerative Organic Certification Standard	Regenerative Organic Alliance	✓	✓	✓
Sustainable Agriculture Standard	Rainforest Alliance and UTZ		✓	
UEBT-UTZ Sustainable Herbal Tea Standards	UEBT and UTZ	✓	✓	✓

Source: Table created by J.A. Brinckmann for the AHPA Sustainability Sub-committee

The way that botanical ingredients and products are regulated and marketed in the United States, for the most part, remains unique and possibly an anomaly in the world, certainly by comparison to the frameworks of our neighbors and trading partners, such as Canada, Mexico, the European Union, Australia, Japan, and China. In the aforementioned countries, the same types of ingredients are generally viewed today as active ingredients for use in licensed, listed, or registered medicinal products that require pre-marketing authorization. This means that implementation of suitable GACPs is expected for quality assurance of active ingredients and use of pharmacopoeial monographs, to serve as the basis for quality specifications, is mandated. In the United States, GACP implementation and compliance with pharmacopoeial quality standards is mandatory only for botanical drugs but voluntary for botanical dietary supplements. This difference leaves it up to the individual company whether, or not, they will make such commitments to quality that will increase the price of their finished products. It is difficult for the consumer in the United States to understand and sort out quality differences between similarly packaged dietary supplement products (and just about every company claims that they have the highest quality!).

Commitments to nature conservation, sustainable production and trade is a different matter and perhaps easier for American companies to communicate. For example, in the European Union, registered herbal medicinal products are not permitted to carry any certification logos on the retail packages. That is because of EU medicines policy stating that the conservation and sustainability status of medicinal ingredients and products have nothing to do with pharmaceutical quality or safety and efficacy of the medicine. That is an unfortunate policy (for European consumers concerned with ethical trade and sustainability). However, in the United States, it is not unusual to find herbal dietary supplement products with multiple credible certification marks on the labeling, demonstrating to the consumer that the product and its ingredients conform to a range of environmental, economic, and social sustainability standards.

Regardless of market trends and consumer preferences, companies intrinsically concerned with reproducible quality, safety and efficacy of their botanical products may need to engage more deeply in the ecological-economic viability of the rural and remote communities where so many medicinal plants are produced. With mass migration of youth to urban areas, happening at a scale never before seen in the world, labor is rapidly disappearing in rural farming and wild-collection regions. Non-sustainable rural livelihoods can lead to land-use changes, loss of ecosystem stewardship and biodiversity. The rapid loss of biodiversity in recent decades, not only impacts access to and quality of wild botanicals, but is reportedly an existential threat (IPBES 2019). There may be a correlation between sustainable resource management and equitable trade, with a companies' ability to procure increasing quantities of botanicals of a specified and consistent quality.

Annex I:

Table 12.3 Botanical, algal, fungal quality standards monographs available from AHP, FCC, NF, and USP

Monograph	AHP	FCC	NF	USP
Acacia (gum Arabic)		X	X	
Acacia syrup			X	
Agar		X	X	
Aloe (dried latex)				X
Aloe Vera leaf, leaf juice & inner leaf juice	X			
Ambrette seed oil		X		
American ginseng root	X			X
American ginseng root dry extract				X
Amla fruit	X			
Amyris oil, west Indian type		X		
Andrographis (leaf and stem)				X
Andrographis dry extract				X
Angelica root oil		X		
Angelica seed oil		X		
Anise oil		X	X	
Annatto extracts		X		
ARA from fungal (*Mortierella alpina*) oil		X		
Arabinogalactan		X		
Astaxanthin esters (from *Haematococcus pluvialis*)		X		X
Ashwagandha root	X			X
Ashwagandha root dry extract				X
Asian ginseng root				X
Asian ginseng root dry extract				X
Astragalus root	X			X
Astragalus root dry extract				X
Aztec Marigold Zeaxanthin extract				X
Bacopa (leaf and stem)				X
Bacopa dry extract				X
Balsam Peru oil, cold-pressed		X		
Banaba leaf				X
Banaba leaf dry extract				X
Basil oil, Comoros type		X		
Basil oil, European type		X		
Bay oil		X		
Belleric Myrobalan fruit	X			
Bergamot oil, cold-pressed		X		
Bilberry fruit	X			
Bilberry fruit dry extract				X
Birch tar oil, rectified		X		
Black cohosh (rhizome)	X			X
Black cohosh dry extract				X

(continued)

Table 12.3 (continued)

Monograph	AHP	FCC	NF	USP
Black cohosh fluidextract				X
Black haw bark	X			
Black pepper (fruit)				X
Black pepper dry extract				X
Black pepper oil		X		
Blue cohosh root/rhizome	X			
Bois de rose oil		X		
Borage seed oil				X
Boswellia serrata (Oleogum resin)				X
Boswellia serrata extract				
Cananga oil		X		
Candelilla wax		X	X	
Cannabis inflorescence	X			
Capsicum (ripe fruit)				X
Capsicum oleoresin				X
Capsicum tincture				X
Caraway fruit			X	
Caraway oil		X	X	
Cardamom seed			X	
Cardamom oil		X	X	
Cardamom tincture, compound			X	
Carnauba wax		X	X	
Carrageenan		X	X	
Carrot seed oil		X		
Casanthranol				X
Cascara fluidextract, aromatic				X
Cascara Sagrada (aged bark)				X
Cascara Sagrada dry extract				X
Cascara Sagrada fluidextract				X
Cascarilla oil		X		
Cassia oil		X		
Castor oil		X		X
Castor oil, aromatic				X
Castor oil emulsion				X
Cat's claw (stem inner bark)				X
Cat's claw dry extract				X
Cedar leaf oil		X		
Celery seed oil		X		
Centella asiatica (aerial parts)				X
Centella asiatica dry extract				X
Centella asiatica triterpenes				X
Chamomile (flower heads)				X

(continued)

Table 12.3 (continued)

Monograph	AHP	FCC	NF	USP
Chamomile oil, English type		X		
Chamomile oil, German type		X		
Chaste tree fruit	X			X
Chaste tree fruit dry extract				X
Chebulic Myrobalan fruit	X			
Cherry juice			X	
Cherry syrup			X	
Chia seed oil				X
Chinese Salvia (root and rhizome)				X
Chocolate			X	
Chocolate syrup			X	
Cinnamomum cassia twig				X
Cinnamon bark oil, Ceylon type		X		
Cinnamon leaf oil		X		
Clary oil		X		
Clove flower bud oil		X	X	
Clove leaf oil		X		
Clove stem oil		X		
Cocoa butter			X	
Coconut oil (refined)			X	
Coconut oil (Unhydrogenated)		X		
Coffee fruit dry extract				X
Cognac oil, green		X		
Coix seed				X
Copaiba oil		X		
Coriander oil		X	X	
Costus root oil		X		
Cramp bark	X			
Cranberry fruit	X			
Cranberry liquid preparation				X
Crypthecodinium cohnii oil				X
Cubeb oil		X		
Cumin oil		X		
Curcuminoids				X
Dammar gum		X		
Dang Gui root	X			
Dill seed oil, European type		X		
Dill seed oil, Indian type		X		
Dillweed oil, American type		X		
Echinacea angustifolia root	X			X
Echinacea angustifolia dry extract				X
Echinacea pallida root	X			X

(continued)

Table 12.3 (continued)

Monograph	AHP	FCC	NF	USP
Echinacea pallida dry extract				X
Echinacea purpurea aerial parts	X			X
Echinacea purpurea dry extract				X
Echinacea purpurea root	X			X
Eleuthero root and rhizome				X
Eleuthero root and rhizome dry extract				X
Eucalyptus oil		X	X	
European elder berry dry extract				X
Evening primrose oil				X
Fennel oil		X	X	
Fenugreek seed				X
Fenugreek seed dry extract				X
Feverfew aerial parts	X			
Feverfew leaf				X
Fir needle oil, Canadian type		X		
Fir needle oil, Siberian type		X		
Flax seed oil				X
Forskohlii (root)				X
Forskohlii dry extract				X
Galageenan			X	
Ganoderma lucidum fruiting body				X
Garcinia cambogia (pericarp)				X
Garcinia Hydroxycitrate dry extract				X
Garlic (bulb)				X
Garlic dry extract				X
Garlic fluidextract				X
Garlic oil		X		
Gellan gum		X	X	
Geranium oil, Algerian type		X		
Ginger (rhizome)				X
Ginger oil		X		
Ginger tincture				X
Ginkgo leaf	X			X
Ginkgo leaf dry extract	X			X
Goldenseal (root and rhizome)	X			X
Goldenseal dry extract				X
Grape seeds Oligomeric Proanthocyanidins				X
Grape skin extract		X		
Grapefruit oil, cold-pressed		X		
Green tea dry extract, decaffeinated				X
Guar gum		X	X	
Guggul (oleo-gum-resin)				X

(continued)

Table 12.3 (continued)

Monograph	AHP	FCC	NF	USP
Guggul native extract				X
Guggul purified extract				X
Gum Ghatti		X		
Gum guaiac		X		
Gymnema (leaf)				X
Gymnema native extract				X
Gymnema purified extract				X
Hawthorn berry	X			
Hawthorn leaf with flower	X			X
Holy basil leaf				X
Holy basil leaf dry extract				X
Hops oil		X		
Horse chestnut seed				X
Horse chestnut seed dry extract				X
Japanese honeysuckle flower				X
Japanese honeysuckle flower dry extract				X
Juniper berries oil		X		
Karaya gum		X		
Kelp		X		
Konjac flour		X		
Labdanum oil		X		
Laurel leaf oil		X		
Lavandin oil, Abrial type		X		
Lavender oil		X		
Lemon oil			X	
Lemon oil, cold-pressed		X		
Lemon oil, desert type, cold-pressed		X		
Lemon oil, distilled		X		
Lemon tincture			X	
Lemongrass oil		X		
Licorice (root, rhizome, stolon)				X
Licorice dry extract				X
Licorice fluidextract			X	
Lime oil, cold-pressed		X		
Lime oil, distilled		X		
Linaloe wood oil		X		
Locust (carob) bean gum		X		
Lovage root oil		X		
Lutein				X
Lycium (goji) berry	X			
Lycopene				X
Lycopene extract from tomato		X		

(continued)

Table 12.3 (continued)

Monograph	AHP	FCC	NF	USP
Lycopene from Blakeslea trispora		X		
Lycopene preparation				X
Mace oil		X		
Malabar-nut-tree leaf				X
Malabar-nut-tree leaf dry extract				X
Mandarin oil, cold-pressed		X		
Maritime pine (stem bark)				X
Maritime pine extract		X		X
Marjoram oil, Spanish type		X		
Marjoram oil, sweet		X		
Masticatory substances, natural		X		
Mentha arvensis oil, partially Dementholized		X		
meso-Zeaxanthin		X		X
meso-Zeaxanthin preparation				X
Milk thistle (ripe fruit)				X
Milk thistle dry extract				X
Monk fruit extract		X		
Motherwort aerial parts	X			
Mustard oil		X		
Myrrh (oleo-gum-resin)				X
Myrrh oil		X		
Nutmeg oil		X		
Olibanum oil		X		
Olive leaf				X
Olive leaf dry extract				X
Onion oil		X		
Orange oil			X	
Orange oil, bitter, cold-pressed		X		
Orange oil, cold-pressed		X		
Orange oil, distilled		X		
Orange Peel tincture, sweet			X	
Orange Spirit, compound			X	
Orange syrup			X	
Origanum oil, Spanish type		X		
Orris root oil		X		
Osha root	X			
Palm kernel oil			X	
Palm kernel oil (Unhydrogenated)		X		
Palm oil, hydrogenated			X	
Palm oil, refined			X	
Palm oil (Unhydrogenated)		X		
Palmarosa oil		X		

(continued)

Table 12.3   (continued)

Monograph	AHP	FCC	NF	USP
Papain				X
Parsley herb oil		X		
Parsley seed oil		X		
Pea starch		X	X	
Pectin				X
Pennyroyal oil		X		
Peppermint			X	
Peppermint oil		X	X	
Peppermint Spirit				X
Peppermint water			X	
Petitgrain oil, Paraguay type		X		
Phyllanthus amarus (aerial parts)				X
Pimenta fruit oil		X		
Pimenta leaf oil		X		
Pine needle oil, dwarf		X		
Pine needle oil, scotch type		X		
Plant Stanol esters		X		X
Plantago seed				X
Psyllium husk				X
Pygeum bark				X
Pygeum bark extract				X
Rebaudioside A		X		
Red clover aerial parts				X
Red clover aerial parts Isoflavone Aglycones dry extract				X
Red clover aerial parts dry extract				X
Red clover flowering tops, aerial parts, and dry extracts	X			
Reishi mushroom	X			
Rhodiola crenulata root and rhizome				X
Rhodiola crenulata root and rhizome dry extract				X
Rhodiola rosea root and rhizome				X
Rhodiola rosea root and rhizome dry extract				X
Rhodiola rosea root and rhizome tincture				X
Rose oil		X	X	
Rosemary extract		X		
Rosemary leaf				X
Rosemary leaf dry aqueous extract				X
Rosemary oil		X		
Rose water, stronger			X	
Rue oil		X		
Sage oil, Dalmatian type		X		
Sage oil, Spanish type		X		
Sandalwood oil, east Indian type		X		

(continued)

Table 12.3 (continued)

Monograph	AHP	FCC	NF	USP
Salix species bark				X
Salix species bark dry extract				X
Savory oil (summer type)		X		
Saw palmetto (ripe fruit)				X
Saw palmetto extract				X
Schisandra berry	X			
Schisandra fruit, northern				X
Schisandra fruit, northern, dry extract				X
Schizochytrium oil				X
Senna fluidextract				X
Senna leaf				X
Senna oral solution				X
Senna pods				X
Sennosides				X
Sheanut oil, refined		X		
Skullcap aerial parts	X			
Slippery elm inner bark	X			X
Solin oil		X		
Soy Isoflavones dry extract				X
Spearmint oil		X		
Spice oleoresins		X		
Spike lavender oil		X		
Spirulina		X		X
Star Anise fruit	X			
Steviol glycosides		X		
Stinging nettle herb	X			
Stinging nettle root	X			X
Stinging nettle root dry extract				X
St. John's Wort flowering top	X			X
St. John's Wort flowering top dry extract				X
Storax (balsam)				X
Tagetes extract		X		
Tangerine oil, cold-pressed		X		
Tangerine Peel				X
Tangerine Peel dry extract				X
Tannic acid (from nutgalls on *Quercus* tree twigs)		X		X
Tara gum		X		
Tarragon oil		X		
Tea polyphenols from green tea, decaffeinated		X		
Thyme oil		X		
Tolu balsam				X
Tolu balsam syrup			X	

(continued)

Table 12.3 (continued)

Monograph	AHP	FCC	NF	USP
Tolu balsam tincture			X	
Tapioca starch			X	
Tienchi ginseng root and rhizome				X
Tienchi ginseng root and rhizome dry extract				X
Tomato extract containing lycopene				X
Tragacanth		X	X	
Turmeric (rhizome)				X
Turmeric dry extract				X
Turmeric oleoresin		X		
Uva Ursi leaf	X			
Valerian root	X			X
Valerian root dry extract				X
Valerian root tincture				X
Vanilla (unripe fruit)			X	
Vanilla tincture			X	
Vegetable oil Phytosterol esters		X		
Wheat bran				X
Willow bark	X			
Wintergreen oil		X		

References

[AHPA] American Herbal Products Association (1999) Tonnage survey of select north American wild-harvested plants, 1999. American Herbal Products Association, Silver Spring

[AHPA] American Herbal Products Association (2006) AHPA-AHP good agricultural and collection practice for herbal raw materials. American Herbal Products Association, Silver Spring

[AHPA] American Herbal Products Association (2007) Tonnage survey of select north American wild-harvested plants, 2004–2005. American Herbal Products Association, Silver Spring

[AHPA] American Herbal Products Association (2012) Tonnage surveys of select north American wild-harvested plants, 2006–2010. American Herbal Products Association, Silver Spring

[AHPA] American Herbal Products Association (2017) Guidance on California proposition 65 and herbal products. American Herbal Products Association, Silver Spring

[AHPA] American Herbal Products Association (2019) Complete the AHPA Tonnage Survey by April 26 to support sustainable harvests. American Herbal Products Association, Silver Spring

Anon (1994) Herbs top growth category in supermarkets and drug stores! HerbalGram 31:9. Available at: http://cms.herbalgram.org/herbalgram/issue31/article899.html

Blumenthal M (1994a) Firm wins appeal in black currant oil case: court chides FDA's "Alice-in-Wonderland" approach. HerbalGram 29:38,39. Available at: http://cms.herbalgram.org/herbalgram/issue29/article879.html

Blumenthal M (1994b) European/American phytomedicines group moves to expand FDA OTC drug policy. HerbalGram 28:36. Available at: http://cms.herbalgram.org/herbalgram/issue28/article285.html

Blumenthal M (1995) EAPC files petitions for OTC drug use for valerian and ginger. HerbalGram:35–19. Available at: http://cms.herbalgram.org/herbalgram/issue35/article1153.html

Brevoort P (1998) The booming U.S. botanical market: a new overview. HerbalGram 44:33–46. Available at: http://cms.herbalgram.org/herbalgram/issue44/article1309.html

Brinckmann JA (2009) FairWild and ISSC-MAP wild collection standards under one roof. Org Stand 96:3–7

Brinckmann JA (2011) Reproducible efficacy and safety depend on reproducible quality: matching the various quality standards that have been established for botanical ingredients with their intended uses in cosmetics, dietary supplements, foods, and medicines. HerbalGram 91:40–55. Available at: http://cms.herbalgram.org/herbalgram/issue91/FEAT_reproducible.html

Brinckmann JA (2016) Sustainable sourcing: markets for certified Chinese medicinal and aromatic plants. International Trade Centre, Geneva. Available at: http://www.intracen.org/publication/Sustainable-Sourcing/

Brinckmann JA (2017) Owning up to owning traditional knowledge of medicinal plants. In: Callahan M, Rogers J (eds) A critical guide to intellectual property. Zed Books, London

Castle LM, Leopold S, Craft R, Kindscher K (2014) Ranking tool created for medicinal plants at risk of being overharvested in the wild. Ethnobiology Letters 5:77–88. Available at: https://unitedplantsavers.org/wp-content/uploads/2016/01/169-741-2-PB.pdf

[CFDA] China Food and Drug Administration (2002) Good agricultural practice for Chinese crude drugs (interim). China Food and Drug Administration, Beijing. Available at: http://eng.sfda.gov.cn/WS03/CL0768/61642.html#

[FDA] U.S. Food and Drug Administration (1997) 21 CFR Part 101. Food labeling; statement of identity, nutrition labeling and ingredient labeling of dietary supplements; compliance policy guide, revocation. Final Rule. Fed Regist 62(184):49826–49858. Available at: https://www.govinfo.gov/content/pkg/FR-1997-09-23/pdf/97-24739.pdf

[FDA] U.S. Food and Drug Administration (2004) Guidance for industry – botanical drug products. U.S. Food and Drug Administration, Rockville

[FDA] U.S. Food and Drug Administration (2007) 21 CFR Part 111. Current good manufacturing practice in manufacturing, packaging, labeling, or holding operations for dietary supplements; final rule. Fed Regist 72(121):34752–34958. Available at: https://www.govinfo.gov/content/pkg/FR-2007-06-25/pdf/07-3039.pdf

[FDA] U.S. Food and Drug Administration (2016) Botanical drug development guidance for industry. U.S. Food and Drug Administration, Silver Spring. Available at: https://www.fda.gov/downloads/Drugs/Guidances/UCM458484.pdf

[FDA] U.S. Food and Drug Administration (2018a) Substances added to food (formerly EAFUS). U.S. Food and Drug Administration, Silver Spring. Page last updated: 10/10/2018. Available at: https://www.accessdata.fda.gov/scripts/fdcc/?set=FoodSubstances

[FDA] U.S. Food and Drug Administration (2018b) FDA food safety modernization act (FSMA). U.S. Food and Drug Administration, Silver Spring. Page last updated: 11/15/2018. Available at: https://www.fda.gov/food/guidanceregulation/fsma/

Foster S (ed) (1992) Herbs of commerce. American Herbal Products Association, Bethesda

[FWF] FairWild Foundation (2010) FairWild Standard, Version 2.0. FairWild Foundation, Weinfelden. Available at: http://www.fairwild.org/documents/

[FWF] FairWild Foundation (2013) Guidance manual for implementation of social and fair trade aspects in FairWild operations (version 1.1 – December 2013). FairWild Foundation, Weinfelden. Available at: http://www.fairwild.org/documents/

[HMCP] Committee on Herbal Medicinal Products (2006) Guidelines on Good Agricultural and Collection Practice (GACP) for starting materials of herbal origin. European Medicines Agency, London. Available at: http://www.ema.europa.eu/docs/en_GB/document_library/Scientific_guideline/2009/09/WC500003362.pdf

[IPBES] Intergovernmental Science-Policy Platform on Biodiversity and Ecosystem Services (2019) Summary for policymakers of the global assessment report on biodiversity and ecosystem services of the Intergovernmental Science – Policy Platform on Biodiversity and Ecosystem Services – ADVANCE UNEDITED VERSION. Available at: https://www.ipbes.net/news/Media-Release-Global-Assessment

[IUCN] International Union for Conservation of Nature, Medicinal Plant Specialist Group, Species Survival Commission (2007) International Standard for Sustainable Wild Collection of Medicinal and Aromatic Plants (ISSC-MAP). Version 1.0. Bonn, Gland, Frankfurt, and Cambridge: Bundesamt für Naturschutz (BfN), MPSG/SSC/IUCN, WWF Germany, and TRAFFIC (BfN-Skripten 195). Available at: https://www.bfn.de/fileadmin/MDB/documents/service/skript195.pdf

Komives K, Jackson A (2014) Introduction to voluntary sustainability standard systems. In: Schmitz-Hoffmann C, Schmidt M, Hansmann B, Palekhov D (eds) Voluntary standard systems – a contribution to sustainable development. Springer, Berlin Heidelberg

Máthé Á, Franz Ch (1996) Good agricultural practice vs. the quality of phytomedicines. In: Workshop "Storage of Medicinal Plants," 43rd Annual Congress of the Society of Medicinal Plant Research, Halle/Saale, 4 September 1995

Leaman DJ (2004) Standards for Sustainable Wild Collection of Medicinal and Aromatic Plants. Discussion Draft 1. Bundesamt für Naturschutz (BfN), MPSG/SSC/IUCN, WWF Germany, and TRAFFIC, Bonn, Gland, Frankfurt, and Cambridge

Leaman DJ, Salvador S (2005) International Standard for Sustainable Wild Collection of Medicinal and Aromatic Plants (ISSC-MAP). Draft 2. Bundesamt für Naturschutz (BfN), MPSG/SSC/IUCN, WWF Germany, and TRAFFIC, Bonn, Gland, Frankfurt, and Cambridge

Lyke J (2000) The Medicinal Plant Working Group. Endanger Species Bull XXV(1–2):20–21

McGuffin M, Kartesz J, Leung A, Tucker A (2000) Herbs of commerce, 2nd edn. American Herbal Products Association, Silver Spring

McGuffin M, Norris T (2004) Background on California proposition 65: issues related to heavy metals and herbal products. American Herbal Products Association, Silver Spring

Meinshausen F, Winkler S, Bächi R, Staubli F, Dürbeck K (2006) *FairWild Standards*, version 1. Institute for Market Ecology, Weinfelden

Miller RK, Celestino C, Giancaspro GI, Williams RL (2008) FDA's dietary supplement cGMPs: standards without standardization. Food Drug Law J 63(4):929–942

[MPSG] Medicinal Plant Specialist Group (2006) International Standard for Sustainable Wild Collection of Medicinal and Aromatic Plants (ISSC-MAP). Working Draft. Bundesamt für Naturschutz (BfN), MPSG/SSC/IUCN, WWF Germany, and TRAFFIC, Bonn, Gland, Frankfurt, and Cambridge

[MPSG] Medicinal Plant Specialist Group (2007) International Standard for Sustainable Wild Collection of Medicinal and Aromatic Plants (ISSC-MAP). Version 1.0. Bundesamt für Naturschutz (BfN), MPSG/SSC/IUCN, WWF Germany, and TRAFFIC, Bonn, Gland, Frankfurt, and Cambridge (BfN-Skripten 195). Available at: https://www.bfn.de/fileadmin/MDB/documents/service/skript195.pdf

[NMPB] National Medicinal Plants Board (2009a) Guidelines on good field collection practices for Indian medicinal plants. NMPB, Department of AYUSH (Ayurveda, Yoga & Naturopathy, Unani, Siddha, and Homoeopathy), Ministry of Health and Family Welfare, Government of India, New Delhi. Available at: https://www.nmpb.nic.in/content/nmpb-publications

[NMPB] National Medicinal Plants Board (2009b) Good field collection practices standard for medicinal plants – requirements. NMPB, Department of AYUSH (Ayurveda, Yoga and Naturopathy, Unani, Siddha and Homoeopathy), Ministry of Health and Family Welfare, New Delhi. Available at: https://www.nmpb.nic.in/content/npb-publications

[NNHPD] Natural and Non-prescription Health Products Directorate (2015) Quality of Natural Health Products Guide, Version 3.1. Health Canada, Ottawa. Available at: https://www.canada.ca/content/dam/hc-sc/migration/hc-sc/dhp-mps/alt_formats/pdf/prodnatur/legislation/docs/eq-paq-eng.pdf

Rădulescu C, Stihi C, Popescu IV, Ionita I, Dulama I, Chilian A, Bancuta OR, Chelarescu ED, Let D (2013) Assessment of heavy metals level in some perennial medicinal plants by flame atomic absorption spectrometry. Romanian Reports in Physics 65(1):246–260

Schiff PL Jr, Srinivasan VS, Giancaspro GI, Roll DB, Salguero J, Sharaf MH (2006 Mar) The development of USP botanical dietary supplement monographs, 1995–2005. J Nat Prod 69(3):464–472

Secretariat of the Convention on Biological Diversity (2011) Nagoya Protocol on Access to Genetic Resources and the Fair and Equitable Sharing of Benefits Arising from their Utilization to the Convention on Biological Diversity. Convention on Biological Diversity, United Nations Environmental Programme, Montreal

Serbula SM, Kalinovic TS, Ilic AA, Kalinovic JV, Steharnik MM (2013) Assessment of airborne heavy metal pollution using *Pinus* spp. and *Tilia* spp. Aerosol Air Qual Res 13:563–573

Shaban NS, Abdou KA, Hassan NEHY (2016) Impact of toxic heavy metals and pesticide residues in herbal products. Beni-Suef Univ J Basic Appl Sci 5:102–106

Smith T, Kawa K, Eckl V, Morton C, Stredney R (2018) Herbal supplement sales in US increased 8.5% in 2017, topping $8 billion. HerbalGram 119:62–71. Available at: http://cms.herbalgram.org/herbalgram/issue119/hg119-herbmktrpt.html

[UEBT] Union for Ethical BioTrade (2012) UEBT (2012) STD01 – ethical BioTrade standard – 2012-04-11. Union for Ethical BioTrade, Amsterdam

[USC] United States Congress (1994) Public Law 103-417: Dietary Supplement Health and Education Act of 1994. 103rd Cong US, Washington, DC. 108 STAT;4325–4335

[USDA] United States Department of Agriculture (2000) 7 CFR part 205 National Organic Program; final rule. Fed Regist 65(246):80548–80684. Available at: https://www.govinfo.gov/content/pkg/FR-2000-12-21/pdf/00-32257.pdf

[USDA] United States Department of Agriculture (2011) Guidance: wild crop harvesting. USDA Agricultural Marketing Service, Washington, DC. Available at: https://www.ams.usda.gov/sites/default/files/media/5022.pdf

[USDA] United States Department of Agriculture (2016a) Guidance: natural resources and biodiversity conservation. USDA Agricultural Marketing Service, Washington, DC

[USDA] United States Department of Agriculture (2016b) Response to comments: natural resources and biodiversity conservation. USDA Agricultural Marketing Service, Washington, DC

[USFWS] U.S. Fish and Wildlife Service (1996) Species being considered for amendments to the appendices to the convention on international trade in endangered species of wild Fauna and Flora; request for information. Fed Regist 61(168):44324–44342. Available at: https://www.govinfo.gov/content/pkg/FR-1996-08-28/pdf/96-21976.pdf

USP Botanical Dietary Supplements and Herbal Medicines Expert Committee (Brinckmann JA, Dentali S, Fletcher E, Gafner S, Marles RJ) and USP Staff (Okunji C, Sarma N, Giancaspro GI) (2016) stimuli to the revision process: need for clear regulation of pesticide residue limits for articles of botanical origin. Pharmacopeial Forum 42(2), 17 pages

[USPC] United States Pharmacopoeial Convention (1820) The pharmacopoeia of the United States of America 1820. Charles Ewer, Boston

[USPC] United States Pharmacopoeial Convention (2016) Guideline for assigning titles to USP dietary supplement monographs. United States Pharmacopoeial Convention, Rockville

[USPC] United States Pharmacopoeial Convention (2018a) General chapter <2030> supplemental information for articles of Botanical origin. In: The United States Pharmacopeia, Forty-First Revision, and the National Formulary, Thirty-Sixth Edition. United States Pharmacopoeial Convention, Rockville

[USPC] United States Pharmacopoeial Convention (2018b) General Chapter <561> Articles of Botanical Origin. In: The United States Pharmacopeia, Forty-First Revision, and the National Formulary, Thirty-Sixth Edition. United States Pharmacopoeial Convention, Rockville

[USPC] United States Pharmacopoeial Convention (2019) 2019 dietary supplements compendium (DSC). United States Pharmacopoeial Convention, Rockville

[WHO, IUCN & WWF] World Health Organization, International Union for Conservation of Nature & World Wide Fund for Nature (1993) Guidelines on the conservation of medicinal plants. IUCN, Gland, in partnership with the WHO, Geneva, and WWF, Gland

[WHO] World Health Organization (2003a) WHO guidelines on good agricultural and collection practices (GACP) for medicinal plants. World Health Organization, Geneva. Available at: http://whqlibdoc.who.int/publications/2003/9241546271.pdf

[WHO] World Health Organization (2003b) The fifty-sixth world health assembly, WHA56.31, traditional medicine. World Health Organization, Geneva. Available at: http://apps.who.int/gb/archive/pdf_files/WHA56/ea56r31.pdf

Wu C (2017) Recent developments of marketing authorization of botanical drugs in the USA. Presented at European Medicines Agency (EMA) Seminar, September 14

Chapter 13
Authentication of Medicinal Plant Components in North America's NHP Industry Using Molecular Diagnostic Tools

Steven Newmaster, Subramanyam Ragupathy, and W. John Kress

Abstract Plants have been used to support healthy lifestyles throughout the world for thousands of years. Traditionally, consumer experience has driven the utility of specific plants through trusted, localized supply chains that were accountable in small regionalized communities. Contemporary supply chains are global, and very complex as botanical ingredients are handled by many stakeholders before a product reaches the consumer. Complexity and removal from directly accountable sources of the supply increases the chances of adulteration. Ethnobotomics is a novel approach that is poised to lead botanical discoveries and innovations, whilst providing a tool to detect and eliminate product adulteration. The concept for this new approach draws on the assemblage of ancient and contemporary body of knowledge concerning the variation in the biological diversity of medicinal plants as experienced by consumers within different cultures. The molecular diagnostic approach combines with modern genomic and metabolomic tools such as genomic DNA sequencing and NMR chemical barcodes to provide a measure of the natural genetic and chemical variation found among plants. It is this expression of the plants natural chemistry that is the mechanism for positive health benefits experienced by consumers. Genomic mechanisms are responsible for the expression this chemical diversity; combing both provides a powerful approach for understanding how plants can heal people. The results from molecular diagnostic testing needs to be available for consumers so they can make informed purchases of authentic medicinal plant products. Product traceability block chains are providing solutions that make data available to consumers through smart labeling applications easily accessed on a cell phone. This will provide confidence in the feedback data we get from consumer

S. Newmaster (✉) · S. Ragupathy
NHP Research Alliance, College of Biological Sciences, University of Guelph, Guelph, ON, Canada
e-mail: snewmast@uoguelph.ca

W. J. Kress
National Museum of Natural History, Smithsonian Institution, Washington, DC, USA
e-mail: kressj@si.edu

© Springer Nature Switzerland AG 2020   325
Á. Máthé (ed.), *Medicinal and Aromatic Plants of North America*, Medicinal and Aromatic Plants of the World 6,
https://doi.org/10.1007/978-3-030-44930-8_13

experiences that can drive further research hypotheses concerning the efficacy and underpinning mechanisms of how plants heal and support healthy lifestyles. We are at a time where ancient localized experiences and cotemporary global social networks are serving to assemble knowledge on the value of medicinal plants. We predict that the next generation of medicinal plant research will provide considerable scientific evidence to support health claims. You the consumer will become part of the scientific process building knowledge just as it was assembled through human experiences with plants 5000 years ago.

Keywords Authentic · Ethnobotomics · Molecular diagnostics · Adulteration

13.1   Introduction

We have learned that some plants heal based on experience from a trusted supply chain. Humans have experienced and utilized medicinal plants to sustain nutrition and good health for thousands of years. This has resulted in the assemblage of considerable traditional knowledge (TK) of medicinal plants among many disparate cultures, which has been passed through oral narratives and in later times illustrative or written communications. The context of this knowledge is enriched within localized cultural norms, linguistics, and ethno-taxonomy. The foundation of this assemblage of TK is sincerity, trust and transparency. Medicinal plant use and efficacy was initiated and built on human experience with a particular plant in a localized, intimate context in which the consumer knew and respected the healer who had specific knowledge of the source of the plant material. Through iterative, positive and negative responses from consumers, healers could confidently assemble knowledge on the experiential efficacy of medicinal plants; this is very similar to hypothetical inductive/deductive reasoning in which scientific knowledge is assembled. Theoretical statistical confidence is built into this experiential/experimental model in which multiple consumers shared their independent responses with the healer much akin to a human health trial. In our ethnobotanical research with First Nations we have seen this first hand and have been able to gather data and calculate statistical confidence intervals using pile sorting and consensus analysis. For example, the knowledge of a specific ethnotaxa was recorded to have medicinal efficacy among many individuals and the knowledge concerning the actual source of the plant, where it grew, its taxonomic identity and preparation was all validated and documented. This is an example of traditional, localized medicinal plant ingredient supply chain in which there is absolute transparency. This system is inherent to humans and is susceptible to misinformation, bias and dishonesty which dismantle the assemblage of reliable knowledge. If our personal experience with a plant is founded on a contaminated or adulterated sample, then it will be a negative experience. Traditional knowledge is resilient to fraud because within localized communities it is easy to recognize the quality assurances that verify that plant ingredients are authentic; they know the healer who is accountable to the local community and the source of the medicinal plant is also known. Unfortunately, sincerity, trust and transparency are difficult to assess in our modern global economy and trade of medicinal plants.

Contemporary supply chains for medicinal plants are very complicated and vulnerable to contamination and adulteration. Many stakeholders are involved including producers, supply brokers, manufacturers, distributors and retailers before a product makes it to a consumer who has little context of the origin of the medicinal plant, nor has a trusting relationship with any of the stakeholders. The modern consumer of natural health products is likely anonymous with respect to retailer whom is less accountable because there is superficial relationship with little personal accountability. It is not surprising that several market surveys have uncovered considerable contamination, mislabeling and adulteration in medicinal plant products (Newmaster et al. 2013; de Boer et al. 2015), which have captured international attention within mainstream media. Unfortunately, this fraudulent activity lowers consumer confidence in natural health products. More importantly, contaminated or adulterated medicinal plant products may pose a health risk and at a minimum would not be efficacious. For example, if you went to the store to try Ginkgo to help memory loss and it was adulterated with another plant (e.g., *Sophora japonica*) or did not even have Ginkgo in the product, then you would not have a positive experience; there are several examples of this specific case in the literature (Raclariu et al. 2018). This is unfortunate as there are considerable benefits to humans from medicinal plants as an alternative to pharmaceuticals. The question remains, how do we bring sincerity, trust and transparency back into the supply chain of medicinal plants?

13.2   A Growing Population Is Seeking Good Health from Plants, Which Increase Demands on the Supply Chain that Encourage Adulteration

The burgeoning international trade in herbal medicinal products has resulted in increased concern relating to the safety and efficacy of these products (McCabe 2009; Techen et al. 2014). International trade of medicinal plant and herbal products has been increasing at a rapid rate of approximately 15% annually (Walker and Applequist 2012). Approximately 29,000 herbal products are used by more than 1000 companies that have annual revenues exceeding $60 billion USD (Barnes et al. 2008), with the bulk of the herbal products, or at least the raw material, being sourced from biodiversity-rich countries in Asia, Africa and South America. Medicinal plant product exports from China are estimated to be around 120,000 tonnes compared to about 32,000 tonnes from India. Both markets are showing sustained growth (Khan and Rauf 2014). The export of medicinal plants worldwide was in the range of US$15 million (Olsen and Helles 2009). India continues to export large numbers of medicinal plants with a staggering 8000 medicinal plant species either cultivated or collected from the wild, of which approximately 960 species are in active trade, with 178 being traded in quantities exceeding 100 metric tonnes (Kumar and Jeanson 2011). It is widely believed that there might be rampant adulteration may occur in the raw herbal trade due to an increase in the demand for these herbal products. Many of these plant species are in short supply due to the lack of cultivation of the species or the scarcity of these medicinal plants in the wild (Mishra et al. 2016). The increasing demand and lack of supply increases the frequency of adulteration.

13.3 Adulteration Is a Considerable Issue that Is Inherent to Many Industries!

Adulteration of food and natural health products is common across many sectors. Recent scientific articles and market surveys have detected considerable adulteration and fraudulent product substitution in many market sectors including fish (Naaum et al. 2016; Shehata et al. 2018), meat (Ballin et al. 2009), coffee (Song et al. 2018), tea (Stoeckle et al. 2011; Faller et al. 2019), and medicinal plants (Newmaster et al. 2013; Rohman et al. 2019). A product is authentic if it contains the species that are listed in the ingredients on the label; verification of authenticity must be completed using a DNA-based "species identity" test. A product is contaminated if the DNA test identified additional species other than what is labeled on the tested product. Product substitution occurs when a DNA test identifies a species other than what is labeled on the tested product, AND there is NOT a positive DNA test for the main ingredients listed on the product label. Note that this strict definition does not consider whether it is an accidental misidentification of a bulk product or a fraudulent market substitution for a cheaper product. A product contains undisclosed fillers if a DNA test identifies a species for known herbal product filler species including rice (Oryza sativa), soybean (Glycine max) and wheat (Triticum spp.) etc. Fillers may be found in place of (that is, substitution) or in addition to the main product ingredient (contamination). DNA-based testing of random products from the market place provide considerable evidence of adulteration. Early marketplace surveys using DNA metabarcoding, a combination of high-throughput sequencing (HTS) and polymerase chain reaction based DNA amplification, provides simultaneous taxonomic identification of taxa from samples containing DNA from different origins (Taberlet et al. 2012). Coghlan et al. (2012) evaluated the species composition of 15 highly processed traditional Chinese medicines (TCM) and found that these contained species and genera included rare species in CITES list. Newmaster et al. (2013) tested 44 herbal products sold in North America using DNA barcoding and found that 59% contained species not listed on the label, and only 2 out of 12 screened companies had products free of substitution, contamination or unreported fillers. Techen et al. (2014) reviewed the literature and found alarming levels of substitution within marketed herbal products, for instance, 6% in saw palmetto herbal dietary supplements (Little and Jeanson 2013), 16% in ginkgo products (Little 2014), 25% in black cohosh (Baker 2012), 33% in herbal teas (Stoeckle et al. 2011), 50% in ginseng (Wallace et al. 2012), 37% in Senna and 50% in Cassia market natural health products (Seethapathy et al. 2015). Cheng et al. (2014) also found a high level of contamination within 27 investigated herbal preparations and Ivanova et al. (2016) tested 15 herbal supplements that contained species not listed on the product label. Higher levels of adulteration can be found in specific products such as in the studies by Raclariu et al. (2017a, b) that found adulteration in 62–68% products tested. Adulteration of natural health products is a serious issue that undermines the efficacy of medicinal plants. The issue is not only an economic one, as it also includes safety and health concerns. We must also consider undeclared rare species and loss of biodiversity as we harvest rare species that are not commercially sustainable. These rare species were never declared on the labels of products but are used in products to boost efficacy or relevant phytochemistry needed for label claims. Sometimes it is simply misidentification of closely related, rare species by wild crafters and collectors of wild plants for commercial products; this still is quite a common practice throughout the world. Unfortunately, in some cases we are consuming our biodiversity, which is not unique to plants as it has also been documented in the fish and seafood market (Naaum et al. 2016)!

If not controlled, economically motivated adulteration could adversely impact the trade of medicinal plant products in the global market (Downey 2016). Presently there is a lack of rigorous global standard protocols and practices to identify and evaluate the different plant species or plant parts used in herbal products (Mishra et al. 2016). It is only in recent years that the adverse consequences of species adulteration on the health and safety of consumers have come to light and demanded a more comprehensive approach to identity testing (de Boer et al. 2015). Novel molecular diagnostic techniques that employ disruptive DNA-based technologies have been used to reveal considerable adulteration in the consumer marketplace. Molecular diagnostic biotechnology is underpinned by the 'omics' science, which is a compilation of diverse technologies such as genomics, proteomics and metabolomics of which these recent biotechnologies have been used to identify species adulteration (Pinu et al. 2019).

13.4 Ethnobotomics Provides a Solution Founded on Traditional and Scientific Knowledge

Ethnobotomics is a novel approach that is poised to lead botanical discoveries and innovations in a new era of exploratory research, whilst providing a tool to detect and eliminate product adulteration. The concept for this new approach is founded on the concept of 'assemblage' of biodiversity knowledge, which includes a coming together of different ways of knowing and valorizing species variation, while adding value to both traditional knowledge (TK) and scientific knowledge (SK). This provides context for the foundations of our knowledge of medicinal plants through generations of experiential use to the publication of science-based experimental, efficacy-based research. Ethnobotomics draws on an ancient or contemporary body of knowledge concerning the variation in the biological diversity as experienced within different cultures; combined with modern genomic and metabolomic tools such as DNA sequencing and NMR chemical barcodes provides a measure of the natural genetic and chemical variation found among plants. This genomic and metabolomic variation is explored along a gradient of variation in which any organism inhabits including population ecotypes that have been recognized by remote cultures for 1000s of years. Plants have evolved to inhabit different ecosites and express different chemical phenotypes of which provide insights to the mechanism of medicinal efficacy or the explanation to how plants heal us. The vision for this approach is to understand how the diversity of life that surrounds us can serve society-at-large as we seek a healthy life-style from plant-based nutrition and medicine.

The term "ethnobotany" implicitly embodies the concept of interdisciplinary research, which is a key concept in understanding the broader scope of this ethnobotomics approach. The term "ethnobotany" is derived from ethnology (study of culture) and botany (study of plants); it is the scientific study of the relationships that exist between people and plants. Historically, ethnobotanists documented, described and explained the complex relationships between cultures and their utility of plants. This often included how plants are used, managed and perceived across human societies as foods, medicines, cosmetics, dyes, textiles, building materials, tools, clothing or within cultural divination, rituals and religion (Newmaster et al. 2007). Much of this research assumes that TK can be imposed upon a SK classification of living things. We suggest that this is a biased approach and call for a more unified approach that includes concept of 'assemblage' as a coming together of different ways of knowing and valorizing variation among plants. This approach seeks to add value to both aboriginal knowledge and modern science such as biodiversity genomics and metabolomics as they work together to potentially create new knowledge; the ultimate goal is to have conclusive evidence concerning how plants heal us and support good health. Exploring the ways in which these different knowledge practices are worked together will show how such inquiries contribute to the common aim of the protection of cultural and biological diversity. An interdisciplinary approach such as this will respond to the increasing urgent global imperatives to conserve both cultural and biological diversity as urged by the Convention of Biological Diversity, UNESCO's Declaration on the Rights of Indigenous People. This approach does is not only rooted in TK, but is evolving to include the experiences of a modern consumer whom is seeking plants to combat modern-day ailments or in support of good health and well-being.

There is a global effort to expedite the documentation and understanding of the planet's natural biodiversity and the scientific underpinnings of different biological classification systems. This includes studies that have documented aboriginal classification systems for medicinal plants. Our understanding of ethnobiological classification is more complex than originally thought. TK often includes multiple mechanisms of classification that goes beyond morphology and includes sensory perception, ecology and utilitarian characters (Ragupathy et al. 2008). This presents an impediment to utilizing these ancient classification systems for interpreting biodiversity because they are very complicated, which requires a great deal of time to fully comprehend, reconstruct and utilize. Ethnobotomics engages modern tools that can overcome taxonomic impediments to exploring diversity of medicinal plants. Contemporary genomics includes sampling of populations at different taxonomic levels for a variety of genomic regions. This provides a link between variation in taxa, sequence evolution and genomic structure and function, providing a good estimate of the evolutionary process. The approach integrates "Genomic Thinking" (high-volume, high-throughput) with the natural variation encountered in ecosystems to explore biological diversity. The recent development and application of DNA-based approaches enables biodiversity genomics and the development of new areas of research such as ethnobotomics that includes other sources of variation such as metabolomics; variation in plant metabolites. We are engaged in and encourage other researchers to initiate further ethnobotomic studies in other cultures to develop theoretically sophisticated insights concerning the encounter between 'local' and 'scientific' approaches to biodiversity knowledge. These will further contribute to a body of research on the social, cultural and political underpinnings biodiversity science; our understanding of the natural variation that surrounds us. Furthermore, the research will add to a unifying global effort to speed up the documentation and understanding of variation among medicinal plants, while concurrently respecting cultural heterogeneity as a vital component of biological diversity. This is aligned with the Convention on Biological Diversity that was signed by over 150 nations and thus the world's complex array of human-natural technological developments that can serve as a tool for understanding how plants heal us or the authentication of natural health products.

13.5 Molecular Diagnostic Biotechnology for the Authentication of Medicinal Ingredients

There are considerable gaps in the commercial molecular biotechnology available to support pharmacovigilance in the natural health products industry. The quality control methods recorded in the pharmacopoeias from different countries consist of the methods for testing the identity of ingredient preparations of medicinal plant products. The industry standard uses two approaches including (1) traditional morphology, and (2) analytical chemistry. Traditional morphology requires whole plant parts that provide characteristics (e.g., leave shape, cell type, flower structure etc.). Unfortunately, the modern supply chain of medicinal plants consists of dried, powered plant materials and botanical extracts of which both forms do not have any morphological characteristics to support verification of plant species identity. Analytical chemistry identifies a specific chemical against a known standard, which is a useful test for supporting label claims with respect to a particular phytochemical

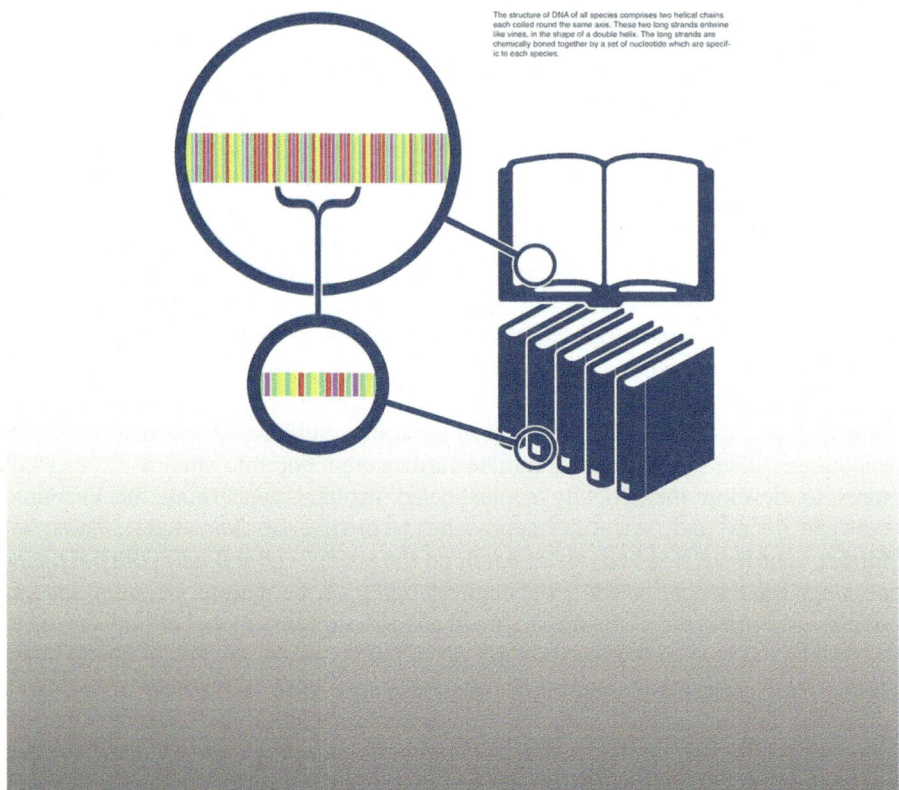

Fig. 13.1 Information contained by DNA sequences as classified in a hierarchical system

listed as an ingredient. The objective of an analytical chemistry test is to validate the presence of a specific phytochemical in a product. Although some phytochemicals are associated with specific plant species, other plant chemicals are common to other taxa. The chemical test identifies chemicals and is not fit for purpose as a plant species identity test because this technology is not a test of a species concept that encompasses inter/intraspecific variation and considers evolutionary mechanisms such as heredity, breeding mechanisms, hybridization, trait variation and natural selection.

Species identity is limited to either DNA-based molecular diagnostics or morphological techniques that are associated with species concepts. DNA sequences are likes words in a book that contain information and can be classified in a hierarchical system much like a book in a library (Fig. 13.1). The amino acids are like the alphabet and can be read as unique identifiers for individuals-species-families and higher order taxa. A DNA-based identity test is based on the evolutionary concept in which genetic traits are inherited and can be measured as variation within/among breeding populations. DNA is inherited, whereas plant chemicals are coded for by certain regions of DNA that could be inherited across different lineages. Therefore, the

Fig. 13.2 DNA-based methods across 2000 natural health product samples along a gradient of NHP manufacture processing botanicals for the common medicinal plant products distributed in North America

genome of a specific plant is unique and related to the ancestry of that plant, which is rooted in the species concept. Molecular diagnostic tools that have been used in the last few years for species ingredient testing include DNA barcoding Next Generation Sequencing (NGS), and the use of targeted PCR Probes (nucleotide signatures) (Kress 2017; Fazekas et al. 2012; Ragupathy et al. 2019) (Fig. 13.2). We provide here a table comparing different DNA-based methods across 2000 natural health product samples along a gradient of NHP manufacture processing botanicals for the common medicinal plant products distributed in North America. Our comprehensive analysis of variation in the success of each of the DNA methods clearly indicates that DNA barcoding is not a useful tool for processed materials of which are commonly distributed within the commoditized supply chain (Table 13.1). The low success of DNA barcoding and meta-barcoding NGS is associated with the effect that mechanical processing has on medicinal plant ingredients. Raw ingredients have an abundance of long sequences that are easily amplified and matched to

Table 13.1 Comparison of DNA biotechnology platforms on a gradient of manufacture processing from raw materials to botanical extracts

Based on 2000 sample tests	DNA technologies (percentage of success against reference material standards)			
Plant ingredient Processing	DNA Barcoding	NGS Meta-barcoding	Mini-Sequence	Nucleotide Signature (Probes)
Raw leaves	75%	85%	100%	100%
Ground powder	65%	80%	100%	100%
Sanitized (ground powder)	45%	75%	95%	100%
Extract	<10%	30%	70%	96%

classical barcode libraries. However, intensification of processing from grinding-sanitation-extraction has two impacts on DNA; it breaks it into shorter sequences and removes some of the DNA so there is less quantity as measured using a fluorometer (Faller et al. 2019). Contrary to some claims in the industry a specific DNA sequence itself not changed (i.e., base pair re-arrangement) by processing as evidenced by published research in which many highly processed extracts consistently had sequence reads that perfectly (100%) matched the reference material sequence libraries (Ragupathy et al. 2019). We developed mini-sequence (80–120 bp) libraries from vouchered reference materials archived in herbaria and present here a rachet phylogenetic tree for 1037 NA Medicinal plants listed in the Duke Handbook of Medicinal Plants (Duke 2002; Fig. 13.3). This is a unique library of short sequences that is differentiated from classical DNA barcode (rbcL, matK and ITS) libraries including metabarcoding, which use long sequence reads (400–600 bp) that are not present in highly processed medicinal plant products. In actual fact, there are enough short strands in highly processed plant ingredients that can be used to verify the identity of authentic medicinal plant species ingredients. The best success was documented using DNA probes to identify single nucleotide polymorphisms (SNPs) in the most intensive manufacturing of plant extracts.

There has been slow uptake of these molecular diagnostic tools in the natural health product industry. This is because industry quality assurance programs require validation of these new test methods, which need to be published in peer-reviewed scientific journals (Newmaster et al. 2019). It is not good manufacturing practice nor ethically appropriate for a manufacturer or commercial test lab to use only their proprietary method to internally validate their own methods (self-validation); they must adhere to external validation using a peer-reviewed method that is based on a scientific approach. To date there is no published validation method for DNA barcoding or NGS of medicinal plants due to a number of required criteria; the largest impediment is the lack of a positive control, PCR bias and absence correction factor bioinformatic algorithms for all medicinal plant species including the closely related adulterants.

We currently recommend the use of targeted PCR or DNA Probe-based tools for identity testing of medicinal plant ingredients. This should follow the

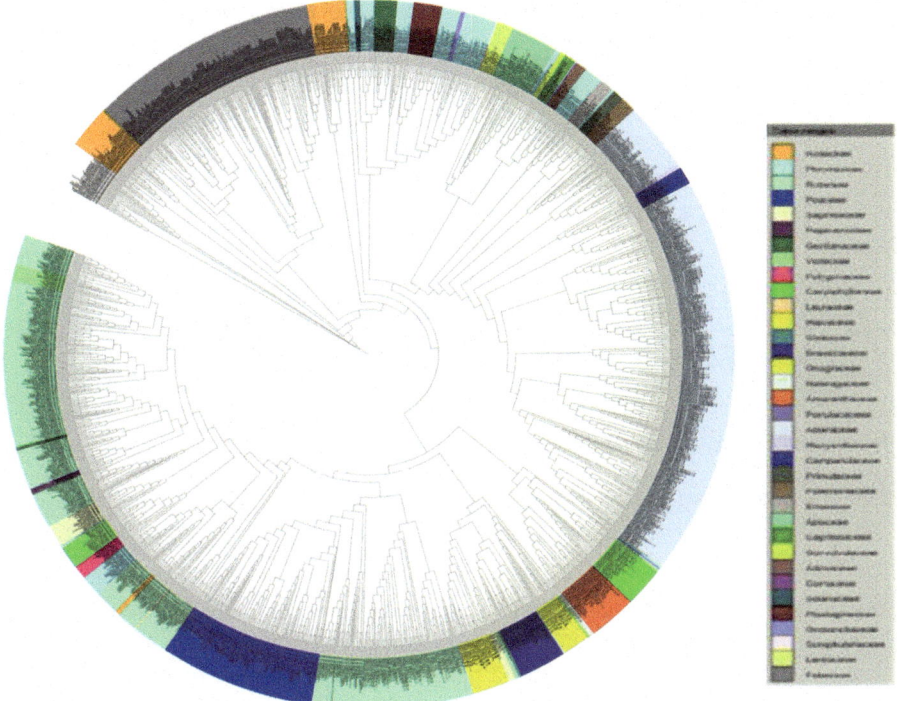

Fig. 13.3 Rachet phylogenetic tree for 1037 NA Medicinal plants. (Duke 2002, Handbook of Medicinal plants)

well-established methods that are already used to validate species ingredient identity throughout the supply chain in many sectors including foodborne pathogens (Malorny et al. 2004; Newmaster and Ragupathy 2014), probiotics (Newmaster et al. 2013), GMOs (29), meat (Claros et al. 2012), fish (Ripp et al. 2014), animal feed (Cavin et al. 2016) and allergens (Malorny et al. 2004). Recently published guidelines (Newmaster et al. 2019) have been developed for medicinal plants of which further research is needed to validate each botanical species ingredient (Shanmughanandhan et al. 2019). These validated identity tests have been developed for small portable PCR instruments that can provide quick results (<1 h) on-site at manufacturing facilities. The cost is affordable for industry at the level of testing every batch throughout the supply chain in a transparent traceability system (Fig. 13.4). An example includes the Hyris bCUBE mobile PCR platform that can be linked together in a swarm of machines along a secure block chain in which stakeholders can see real-time results and acquire Certificates of Analysis (COAs) in minutes. This type of transparent system provides an alternative species ingredient identity verification method in support of quality assurance programs. Quantitative PCR methods are in development with the goal of new methods that will provide an accurate estimate of the target ingredient relative to adulterants and

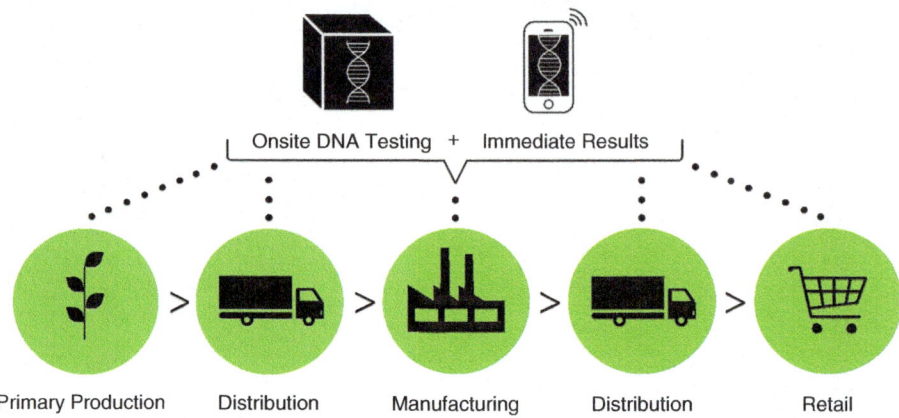

Fig. 13.4 DNA-Based transparent supply chain management

contaminants. Contaminant DNA is unavoidable as botanical ingredients come from farms and incidental DNA is almost certainly found in every product, at low levels; this incidental DNA is not a health hazard because it represents botanical ingredients in negligible amounts. Current research is determining how common incidental DNA is in food and NHP products and what is an acceptable limit for incidental DNA in products.

13.6   Conclusion – Transparency and Consumer Confidence

Products are now judged in the court of consumer confidence and brand damage from adulteration and contamination issues can be fatal for any corporation regardless of size. Smart consumers are demanding transparency and data based on real tests. They do not trust that the products they are buying are safe, healthy or of true value based on the constant media attention to food security issues and market surveys that identify considerable adulteration issues. The availability of data on the internet provides a source of knowledge that allows consumers to make educated purchase decisions. Molecular diagnostic methods are becoming important tools for quality assurance within the food and NHP industry and could be used to empower a transparent, consumer focused quality assurance system underpinned by real test data. We advocate that national authorities and industry should attempt to shift consumers' mind-sets on food safety concerns directly onto the products they purchase; consumers dictating the distribution of corporate market share based on evidence of quality assurance of botanical ingredients. Consumers could have access to transparent, quality assurance test data generated by validated molecular diagnostic tests of botanicals used in their desired natural health products or food supplements. Rather than expecting botanical ingredients and products to be authentic or safe, consumers should be empowered to verify the products they purchase have been adequately tested. Transparency can be enabled through real-time testing with on-site diagnostic tools such as PCR-based probes in which the results can be provided in a secure block chain like ledger. Ideally this could be accessed by a number of stakeholders including all members of the industrial supply chain from farm to fork of the consumer. Manufactures and retailers could easily check the ledger to see if producers and manufactures have tested the batches of ingredients they are purchasing for their brands. Consumers can check the ledger to see if manufactures and retailers verified the true identity of the ingredients in the products they seek to purchase. Regulatory agencies could peer into this transparent ledger of test-based evidence to ensure quality assurance for all consumers. Perhaps this is a large step for industry, but today's molecular biotechnology is poised to enable this vision for a transparent food chain quality system that crosses the void form industry to consumers. This would enable the age-old experiential system in which humans could experiment with safe, authentic medicinal plants in support of healthy lifestyles.

References

Baker DA (2012) DNA barcode identification of black cohosh herbal dietary supplements. J AOAC Int 95(4):2012–1034. https://doi.org/10.5740/jaoacint.11-261

Ballin NZ, Vogensen FK, Karlsson AH (2009) Species determination – can we detect and quantify meat adulteration? Meat Sci 83(2):165–174. https://doi.org/10.1016/j.meatsci.2009.06.003

Barnes PM, Bloom B, Nahin RL (2008) Complementary and alternative medicine use among adults and children: United States, 2007. Natl Health Stat Rep 12:1–23

Cavin C, Cottenet G, Blancpain C et al (2016) Food adulteration: from vulnerability assessment to new analytical solutions. Chimia 70(5):329–333

Cheng W, Su X, Chen X et al (2014) Chinese medicine preparation based on high-throughput sequencing: the story for Luiwei Dihuang Wan. Sci Rep 4:5147. https://doi.org/10.1038/srep05147

Claros MG, Bautista R, Guerrero-Fernandez D et al (2012) Why assembling plant genome sequences is so challenging. Biology 1(2):439–459

Coghlan ML, Haile J, Houston J et al (2012) Deep sequencing on plant and animal DNA contained within traditional Chinese medicines reveals legality issues and health safety concerns. PLoS Genet. https://doi.org/10.1371/journal.pgen.1002657

De Boer HJ, Ichim MC, Newmaster SG (2015) DNA barcoding and pharmacovigilance of herbal medicines. Drug Saf 38(7):611–620

Downey G (2016) Advances in food authenticity testing, 1st edn. Woodhead Publishing, London

Duke JA (2002) Handbook of medicinal herbs, 2nd edn. CPR Press, New York

Faller AC, Ragupathy S, Shanmughanandhan D et al (2019) DNA quality and quantity analysis of Camellia sinensis through processing from fresh leaves to a green tea extract. J AOAC Int. https://doi.org/10.5740/jaoacint.18-0318

Fazekas AJ, Kuzmina M, Newmaster SG et al (2012) DNA barcoding methods for land plants. In: Kress WJ, Erikson DL (eds) DNA barcodes: methods and protocols. Springer, New York

Ivanova VN, Kuzmina ML, Braukmann TWA et al (2016) Authentication of herbal supplements using next-generation sequencing. PLoS. https://doi.org/10.1371/journal.pone.0156426

Khan H, Rauf A (2014) Medicinal plants: economic perspective and recent developments. World Appl Sci J 31(11):1925–1929

Kress WJ (2017) Plant DNA barcodes: applications today and in the future. J Syst Evol 55(4):291–307

Kumar MR, Janagam D (2011) Export and import pattern of medicinal plants in India. Indian J Sco Technol 4(3):245–248

Little DP (2014) Authentication of Ginkgo biloba herbal dietary supplements using DNA barcoding. Genome 57(9):513–516

Little DP, Jeanson ML (2013) DNA barcode authentication of saw palmetto herbal dietary supplements. Sci Rep 3:3518

Malorny B, Paccassoni E, Fach P et al (2004) Diagnostic real-time PCR for detection of Salmonella in food. Appl Environ Microbiol 70:7046–7052

McCabe S (2009) Complementary herbal and alternative drugs in clinical practice. Perspect Psychiatr Care 38:98–107

Mishra P, Kumar A, Nagireddy A et al (2016) DNA barcoding: an efficient tool to overcome authentication challenges in the herbal market. Plant Biotechnol J 14(1):8–21

Naaum AM, Warner K, Mariani S et al (2016) Seafood mislabeling incidence and impacts. In: Naaum AM, Hanner RH (eds) Seafood authenticity and traceability. Academic, Amsterdam, pp 3–26

Newmaster SG, Ragupathy S (2014) Ethnobotany genomics. In: Vemulpad S, Jamie J (eds) Recent advances in plant-based, traditional and natural medicines. CPR Press, Waretown

Newmaster SG, Ragupathy S, Balasubramaniyam NC et al (2007) The multi-mechanistic taxonomy of the Irulas in Tamil Nadu, South India. J Enthnobiol 27(2):233–255

Newmaster SG, Grguric M, Shanmughanandhan D, Ramalingam S, Ragupathy S (2013) DNA barcoding detects contamination and substitution in North American herbal products. BMC Med 11:222–235

Newmaster SG, Shanmughanandhan D, Kesanakurti P et al (2019) Recommendations for validation of real-time PCR methods for molecular diagnostic identification of botanicals. J AOAC Int. https://doi.org/10.5740/jaoacint.18-0321

Olsen CS, Helles F (2009) Market efficiency and benefit distribution in medicinal plant markets: empirical evidence from South Asia. Int J Biodivers Sci Manag 5(2):53–62

Pinu FR, Beale DJ, Paten AM (2019) Systems biology and multi-omics integration: viewpoints from the metabolomics research community. Meta 9(4):76. https://doi.org/10.3390/metabo9040076

Raclairu AC, Heinrich M, Ichim MC et al (2018) Benefits and limitations of DNA barcoding and metabarcoding in herbal product authentication. Phytochem Anal 29:123–128. https://doi.org/10.1002/pca.2732

Raclariu AC, Mocan A, Popa MO et al (2017a) Veronica officinalis product authentication using DNA metabarcoding and HPLC-MS reveals widespread adulteration with Veronica chamaedrys. Front Pharmacol 8:378. https://doi.org/10.3389/fphar.2017.00378

Raclariu AC, Paltinean R, Vlase L et al (2017b) Comparative authentication of hypericum perforatum herbal products using DNA metabarcoding, TLC and HPLC-MS. Sci Rep 7(1):1291. https://doi.org/10.1038/s41598-017-10389-w

Ragupathy S, Newmaster SG, Murugesan M et al (2008) Consensus of the "Malasars" traditional aboriginal knowledge of medicinal plants in the Velliangiri Holy Hills. J Ethnobiol Ethnomed 4(1):81

Ragupathy S, Faller AC, Shanmughanandhan D et al (2019) Exploring DNA quantity and quality from raw materials to botanical extracts. Heliyon 5(6). https://doi.org/10.1016/j.hwliyon.2019.e01935

Ripp F, Krombholz CF, Liu Y et al (2014) All-Food-Seq (AFS): a quantifiable screen for species in biological samples by deep DNA sequencing. BMC Genomics 15:639

Rohman A, Windarsih A, Motalib Hossain MA et al (2019) Application of near- and mid-infrared spectroscopy combined with chemometrics for discrimination and authentication of herbal products: a review. J Appl Pharm Sci 9(3):137–147

Seethapathy GS, Ganesh D, Santhosh Kumar JU et al (2015) Assessing product adulteration in natural health products for laxative yielding plants, Cassia, Senna and Chamaecrista, in southern India using DNA barcoding. Int J Legal Med 129(4):693–700

Shanmughanandhan D, Ragupathy S, Kesanakurti P, Jeevitha S, Della Noce I and Steven G. Newmaster. (2019) Validated identity test method for Ginkgo biloba NHPs using DNA-based species-specific hydrolysis PCR probe. J AOAC Int. https://doi.org/10.5740/jaoacint.18-0319

Shehata HR, Naaum AM, Garduño RA et al (2018) DNA barcoding as a regulatory tool for seafood authentication in Canada. Food Control 92:147–153

Song HY, Jang HW, Debnath T et al (2018) Analytical method to detect adulteration of ground roasted coffee. J Food Sci Technol 54(1):256–262. https://doi.org/10.1111/ijfs.13942

Stoeckle MY, Gamble CC, Kirpekar R et al (2011) Commercial teas highlight plant DNA barcode identification successes and obstacles. Sci Rep 1:42. https://doi.org/10.1038/srep00042

Taberlet P, Coissac E, Pompanon F et al (2012) Towards next-generation biodiversity assessment using DNA metabarcoding. Mol Ecol 21(8):2045–2050. https://doi.org/10.1111/j.1365-294X.2012.05470.x

Techen N, Parveen I, Pan Z et al (2014) DNA barcoding of medicinal plant material for identification. Curr Opin Biotechnol 25:103–110

Walker KM, Applequist WL (2012) Adulteration of selected unprocessed botanicals in the US retail herbal trade. Econ Bot 66:321–327

Wallace LJ, Biolard SMAL, Eagle SHC et al (2012) DNA barcodes for everyday life: routine authentication of natural health products. Food Res Int 49:446–452

Index

Printed by Printforce, the Netherlands